Atlantis Thinking Machines

Volume 5

Series editor

Kai-Uwe Kühnberger, Osnabrück, Germany

For further volumes:
http://www.atlantis-press.com

Aims and Scope of the Series

This series publishes books resulting from theoretical research on and reproductions of general Artificial Intelligence (AI). The book series focuses on the establishment of new theories and paradigms in AI. At the same time, the series aims at exploring multiple scientific angles and methodologies, including results from research in cognitive science, neuroscience, theoretical and experimental AI, biology and from innovative interdisciplinary methodologies.

For more information on this series and our other book series, please visit our website at: www.atlantis-press.com/publications/books

AMSTERDAM—PARIS—BEIJING
ATLANTIS PRESS
Atlantis Press
29, avenue Laumière
75019 Paris, France

Ben Goertzel · Cassio Pennachin
Nil Geisweiller

Engineering General Intelligence, Part 1

A Path to Advanced AGI via Embodied Learning and Cognitive Synergy

With contributions by the OpenCog Team

ATLANTIS
PRESS

Ben Goertzel
G/F 51C Lung Mei Village
Tai Po
Hong Kong
People's Republic of China

Cassio Pennachin
Igenesis
Belo Horizonte, Minas Gerais
Brazil

Nil Geisweiller
Samokov
Bulgaria

ISSN 1877-3273
ISBN 978-94-6239-026-3   ISBN 978-94-6239-027-0   (eBook)
DOI 10.2991/978-94-6239-027-0

Library of Congress Control Number: 2013953280

Printed on acid-free paper

Preface

This is a large, two-part book with an even larger goal: To outline a practical approach to engineering software systems with general intelligence at the human level and ultimately beyond. Machines with flexible problem-solving ability, open-ended learning capability, creativity, and eventually their own kind of genius.

Part 1 of the book (Volume 5 in the Atlantis Thinking Machines book series), reviews various critical conceptual issues related to the nature of intelligence and mind. It then sketches the broad outlines of a novel, integrative architecture for Artificial General Intelligence (AGI) called CogPrime... and describes an approach for giving a young AGI system (CogPrime or otherwise) appropriate experience, so that it can develop its own smarts, creativity, and wisdom through its own experience. Along the way a formal theory of general intelligence is sketched, and a broad roadmap leading from here to human-level artificial intelligence. Hints are also given regarding how to eventually, potentially create machines advancing beyond human level—including some frankly futuristic speculations about strongly self-modifying AGI architectures with flexibility far exceeding that of the human brain.

Part 2 of the book (Volume 6 in the Atlantis Thinking Machines book series), then digs far deeper into the details of CogPrime's multiple structures, processes, and functions, culminating in a general argument as to why we believe CogPrime will be able to achieve general intelligence at the level of the smartest humans (and potentially greater), and a detailed discussion of how a CogPrime-powered virtual agent or robot would handle some simple practical tasks such as social play with blocks in a preschool context. It first describes the CogPrime software architecture and knowledge representation in detail; then reviews the cognitive cycle via which CogPrime perceives and acts in the world and reflects on itself; and next turns to various forms of learning: procedural, declarative (e.g., inference), simulative, and integrative. Methods of enabling natural language functionality in CogPrime are then discussed; and then the volume concludes with a chapter summarizing the argument that CogPrime can lead to human-level (and eventually perhaps greater) AGI, and a chapter giving a thought experiment describing the internal dynamics via which a completed CogPrime system might solve the problem of obeying the request "Build me something with blocks that I haven't seen before."

The chapters here are written to be read in linear order—and if consumed thus, they tell a coherent story about how to get from here to advanced AGI.

However, we suggest the impatient reader may wish to take a quick look at the final chapter of Part 2, after reading Chaps. 1–3 of Part 1. This final chapter gives a broad overview of why we think the CogPrime design will work, in a way that depends on the technical details of the previous chapters, but (we believe) not so sensitively as to be incomprehensible without them.

This is admittedly an unusual sort of book, mixing demonstrated conclusions with unproved conjectures in a complex way, all oriented toward an extraordinarily ambitious goal. Further, the chapters are somewhat variant in their levels of detail—some very nitty-gritty, some more high level, with much of the variation due to how much concrete work has been done on the topic of the chapter at time of writing. However, it is important to understand that the ideas presented here are not mere armchair speculation—they are currently being used as the basis for an open-source software project called OpenCog, which is being worked on by software developers around the world. Right now OpenCog embodies only a percentage of the overall CogPrime design as described here. But if OpenCog continues to attract sufficient funding or volunteer interest, then the ideas presented in these volumes will be validated or refuted via practice (As a related note: here and there in this book, we will refer to the "current" CogPrime implementation (in the OpenCog framework); in all cases this refers to OpenCog as of late 2013).

To state one believes one knows a workable path to creating a human-level (and potentially greater) general intelligence is to make a dramatic statement, given the conventional way of thinking about the topic in the contemporary scientific community. However, we feel that once a little more time has passed, the topic will lose its drama (if not its interest and importance), and it will be widely accepted that there are *many* ways to create intelligent machines—some simpler and some more complicated; some more brain-like or human-like and some less so; some more efficient and some more wasteful of resources; etc. We have little doubt that, from the perspective of AGI science 50 or 100 years hence (and probably even 10–20 years hence), the specific designs presented here will seem awkward, messy, inefficient, and circuitous in various respects. But that is how science and engineering progress. Given the current state of knowledge and understanding, having any concrete, comprehensive design, and plan for creating AGI is a significant step forward; and it is in this spirit that we present here our thinking about the CogPrime architecture and the nature of general intelligence.

In the words of Sir Edmund Hillary, the first to scale Everest: "Nothing Venture, Nothing Win."

Prehistory of the Book

The writing of this book began in earnest in 2001, at which point it was informally referred to as "The Novamente Book." The original "Novamente Book" manuscript ultimately got too big for its own britches, and subdivided into a number of different works—*The Hidden Pattern* (Goertzel 2006), a philosophy of mind book published in 2006; *Probabilistic Logic Networks* (Goertzel et al. 2008), a more technical work published in 2008; *Real World Reasoning* (Goertzel et al. 2011), a sequel to *Probabilistic Logic Networks* published in 2011; and the two parts of this book.

The ideas described in this book have been the collaborative creation of multiple overlapping communities of people over a long period of time. The vast bulk of the writing here was done by Ben Goertzel; but Cassio Pennachin and Nil Geisweiller made sufficient writing, thinking, and editing contributions over the years to more than merit their inclusion as coauthors. Further, many of the chapters here have coauthors beyond the three main coauthors of the book; and the set of chapter coauthors does not exhaust the set of significant contributors to the ideas presented.

The core concepts of the CogPrime design and the underlying theory were conceived by Ben Goertzel in the period 1995–1996 when he was a Research Fellow at the University of Western Australia; but those early ideas have been elaborated and improved by many more people than can be listed here (as well as by Ben's ongoing thinking and research). The collaborative design process ultimately resulting in CogPrime started in 1997 when Intelligenesis Corp. was formed—the Webmind AI Engine created in Intelligenesis's research group during 1997–2001 was the predecessor to the Novamente Cognition Engine created at Novamente LLC during 2001–2008, which was the predecessor to CogPrime.

Online Appendices

Just one more thing before getting started! This book originally had even more chapters than the ones currently presented in Parts 1 and 2. In order to decrease length and increase focus, however, a number of chapters dealing with peripheral—yet still relevant and interesting—matters were moved to online appendices. These may be downloaded in a single PDF file at http://goertzel.org/engineering_general_Intelligence_appendices_B-H.pdf. The titles of these appendices are:

- Appendix A: Possible Worlds Semantics and Experiential Semantics
- Appendix B: Steps Toward a Formal Theory of Cognitive Structure and Dynamics
- Appendix C: Emergent Reflexive Mental Structures
- Appendix D: GOLEM: Toward an AGI Meta-Architecture Enabling Both Goal Preservation and Radical Self-Improvement
- Appendix E: Lojban++: A Novel Linguistic Mechanism for Teaching AGI Systems
- Appendix F: Possible Worlds Semantics and Experiential Semantics
- Appendix G: PLN and the Brain
- Appendix H: Propositions About Environments in Which CogPrime Components Are Useful

None of these are critical to understanding the key ideas in the book, which is why they were relegated to online appendices. However, reading them will deepen your understanding of the conceptual and formal perspectives underlying the CogPrime design. These appendices are referred to here and there in the text of the main book.

References

B. Goertzel, *The Hidden Pattern* (Brown Walker, Boca Raton, 2006)

B. Goertzel, M. Ikle, I. Goertzel, A. Heljakka, *Probabilistic Logic Networks* (Springer, Heidelberg, 2008)

B. Goertzel, N. Geisweiller, L. Coelho, P. Janicic, C. Pennachin, *Real World Reasoning* (Atlantis, Oxford, 2011)

Acknowledgments

For sake of simplicity, this acknowledgments section is presented from the perspective of the primary author, Ben Goertzel. Ben will thus begin by expressing his thanks to his primary coauthors, Cassio Pennachin (Collaborator since 1998) and Nil Geisweiller (Collaborator since 2005). Without outstandingly insightful, deep-thinking colleagues like you, the ideas presented here—let alone the book itself—would not have developed nearly as effectively as what has happened. Similar thanks also go to the other OpenCog Collaborators who have coauthored various chapters of the book.

Beyond the coauthors, huge gratitude must also be extended to everyone who has been involved with the OpenCog project, and/or was involved in Novamente LLC and Webmind Inc. before that. We are grateful to all of you for your collaboration and intellectual companionship!

Building a thinking machine is a huge project, too big for any one human; it will take a team and I'm happy to be part of a great one. It is through the genius of human collectives, going beyond any individual human mind, that genius machines are going to be created.

A tiny, incomplete sample from the long list of those others deserving thanks is:

- Ken Silverman and Gwendalin Qi Aranya (formerly Gwen Goertzel), both of whom listened to me talk at inordinate length about many of the ideas presented here a long, long time before anyone else was interested in listening. Ken and I schemed some AGI designs at Simon's Rock College in 1983, years before we worked together on the Webmind AI Engine.
- Allan Combs, who got me thinking about consciousness in various different ways, at a very early point in my career. I'm very pleased to still count Allan as a friend and sometime Collaborator! Fred Abraham as well, for introducing me to the intersection of chaos theory and cognition, with a wonderful flair. George Christos, a deep AI/math/physics thinker from Perth, for reawakening my interest in attractor neural nets and their cognitive implications, in the mid-1990s.
- All of the 130 staff of Webmind Inc. during 1998–2001 while that remarkable, ambitious, peculiar AGI-oriented firm existed. Special shout-outs to the "Voice of Reason" Pei Wang and the "Siberian Madmind" Anton Kolonin, Mike Ross, Cate Hartley, Karin Verspoor, and the tragically prematurely deceased Jeff

Pressing (compared to whom we are all mental midgets), who all made serious conceptual contributions to my thinking about AGI. Lisa Pazer and Andy Siciliano who made Webmind happen on the business side. And of course Cassio Pennachin, a coauthor of this book; and Ken Silverman, who coarchitected the whole Webmind system and vision with me from the start.

- The Webmind Diehards, who helped begin the Novamente project that succeeded Webmind beginning in 2001: Cassio Pennachin, Stephan Vladimir Bugaj, Takuo Henmi, Matthew Ikle', Thiago Maia, Andre Senna, Guilherme Lamacie, and Saulo Pinto.
- Those who helped get the Novamente project off the ground and keep it progressing over the years, including some of the Webmind Diehards and also Moshe Looks, Bruce Klein, Izabela Lyon Freire, Chris Poulin, Murilo Queiroz, Predrag Janicic, David Hart, Ari Heljakka, Hugo Pinto, Deborah Duong, Paul Prueitt, Glenn Tarbox, Nil Geisweiller and Cassio Pennachin (the coauthors of this book), Sibley Verbeck, Jeff Reed, Pejman Makhfi, Welter Silva, Lukasz Kaiser, and more.
- Izabela Lyon Freire, for endless wide-ranging discussions on AI, philosophy of mind and related topics, which helped concretize many of the ideas here.
- David Hart, for his various, inspired IT and editing help over the years, including lots of editing and formatting help on an earlier version of this book which took the form of a Wikibook.
- A group of volunteers who helped proofread a draft of this manuscript, including Brett Woodward, Terry Ward, Dario Garcia Gasulla, Tom Heiman as well as others. Tim Josling's efforts in this regard were especially thorough and helpful.
- All those who have helped with the OpenCog system, including Linas Vepstas, Joel Pitt, Jared Wigmore/Jade O'Neill, Zhenhua Cai, Deheng Huang, Shujing Ke, Lake Watkins, Alex van der Peet, Samir Araujo, Fabricio Silva, Yang Ye, Shuo Chen, Michel Drenthe, Ted Sanders, Gustavo Gama and of course Nil, and Cassio again. Tyler Emerson and Eliezer Yudkowsky, for choosing to have the Singularity Institute for AI (now MIRI) provide seed funding for OpenCog.
- The numerous members of the AGI community who have tossed around AGI ideas with me since the first AGI conference in 2006, including but definitely not limited to: Stan Franklin, Juergen Schmidhuber, Marcus Hutter, Kai-Uwe Küehnberger, Stephen Reed, Blerim Enruli, Kristinn Thorisson, Joscha Bach, Abram Demski, Itamar Arel, Mark Waser, Randal Koene, Paul Rosenbloom, Zhongzhi Shi, Steve Omohundro, Bill Hibbard, Eray Ozkural, Brandon Rohrer, Ben Johnston, John Laird, Shane Legg, Selmer Bringsjord, Anders Sandberg, Alexei Samsonovich, Wlodek Duch, and more.
- The inimitable "Artilect Warrior" Hugo de Garis, who (when he was working at Xiamen University) got me started working on AGI in the Orient (and introduced me to my wife Ruiting in the process). And Changle Zhou, who brought Hugo to Xiamen and generously shared his brilliant research students with Hugo and me. And Min Jiang, Collaborator of Hugo and Changle, a deep AGI thinker who is helping with OpenCog theory and practice at time of writing.

- Gino Yu, who got me started working on AGI here in Hong Kong, where I am living at time of writing. As of 2013 the bulk of OpenCog work is occurring in Hong Kong via a research grant that Gino and I obtained together.
- Dan Stoicescu, whose funding helped Novamente through some tough times.
- Jeffrey Epstein, whose visionary funding of my AGI research has helped me through a number of tight spots over the years. At time of writing, Jeffrey is helping support the OpenCog Hong Kong project.
- Zeger Karssen, Founder of Atlantis Press, who conceived the *Thinking Machines* book series in which this book appears, and who has been a strong supporter of the AGI conference series from the beginning.
- My wonderful wife Ruiting Lian, a source of fantastic amounts of positive energy for me since we became involved 4 years ago. Ruiting has listened to me discuss the ideas contained here time and time again, often with judicious and insightful feedback (as she is an excellent AI Researcher in her own right); and has been wonderfully tolerant of me diverting numerous evenings and weekends to getting this book finished (as well as to other AGI-related pursuits). And my parents Ted and Carol and kids Zar, Zeb, and Zade, who have also indulged me in discussions on many of the themes discussed here on countless occasions! And my dear, departed grandfather Leo Zwell, for getting me started in science.
- Crunchkin and Pumpkin, for regularly getting me away from the desk to stroll around the village where we live; many of my best ideas about AGI and other topics have emerged while walking with my furry four-legged family members.

September 2013   Ben Goertzel

Contents

Acronyms

AA	Attention Allocation
ADF	Automatically Defined Function (in the context of Genetic Programming)
AF	Attentional Focus
AGI	Artificial General Intelligence
AV	Attention Value
BD	Behavior Description
C-space	Configuration Space
CBV	Coherent Blended Volition
CEV	Coherent Extrapolated Volition
CGGP	Contextually Guided Greedy Parsing
CSDLN	Compositional Spatiotemporal Deep Learning Network
CT	Combo Tree
ECAN	Economic Attention Network
ECP	Embodied Communication Prior
EPW	Experiential Possible Worlds (semantics)
FCA	Formal Concept Analysis
FI	Fisher Information
FIM	Frequent Itemset Mining
FOI	First Order Inference
FOPL	First Order Predicate Logic
FOPLN	First Order PLN
FS-MOSES	Feature Selection MOSES (i.e. MOSES with feature selection integrated a la LIFES)
GA	Genetic Algorithms
GB	Global Brain
GEOP	Goal Evaluator Operating Procedure (in a GOLEM context)
GIS	Geospatial Information System
GOLEM	Goal-Oriented LEarning Meta-architecture
GP	Genetic Programming
HOI	Higher-Order Inference
HOPLN	Higher-Order PLN

HR	Historical Repository (in a GOLEM context)
HTM	Hierarchical Temporal Memory
IA	(Allen) Interval Algebra (an algebra of temporal intervals)
IRC	Imitation/Reinforcement/Correction (Learning)
LIFES	Learning-Integrated Feature Selection
LTI	Long Term Importance
MA	Mind Agent
MOSES	Meta-Optimizing Semantic Evolutionary Search
MSH	Mirror System Hypothesis
NARS	Non-Axiomatic Reasoning System
NLGen	A specific software component within OpenCog, which provides one way of dealing with Natural Language Generation
OCP	OpenCogPrime
OP	Operating Program (in a GOLEM context)
PEPL	Probabilistic Evolutionary Procedure Learning (e.g. MOSES)
PLN	Probabilistic Logic Networks
RCC	Region Connection Calculus
RelEx	A specific software component within OpenCog, which provides one way of dealing with natural language Relationship Extraction
SAT	Boolean SATisfaction, as a mathematical/computational problem
SMEPH	Self-Modifying Evolving Probabilistic Hypergraph
SRAM	Simple Realistic Agents Model
STI	Short Term Importance
STV	Simple Truth Value
TV	Truth Value
VLTI	Very Long Term Importances
WSPS	Whole-Sentence Purely-Syntactic Parsing

Chapter 1
Introduction

1.1 AI Returns to Its Roots

Our goal in this book is straightforward, albeit ambitious: to present a conceptual and technical design for a thinking machine, a software program capable of the same qualitative sort of general intelligence as human beings. It's not certain exactly how far the design outlined here will be able to take us, but it seems plausible that once fully implemented, tuned and tested, it will be able to achieve general intelligence at the human level and in some respects beyond.

Our ultimate aim is Artificial General Intelligence construed in the broadest sense, including artificial creativity and artificial genius. We feel it is important to emphasize the extremely broad potential of Artificial General Intelligence systems. The human brain is not built to be modified, except via the slow process of evolution. Engineered AGI systems, built according to designs like the one outlined here, will be much more susceptible to rapid improvement from their initial state. It seems reasonable to us to expect that, relatively shortly after achieving the first roughly human-level AGI system, AGI systems with various sorts of beyond-human-level capabilities will be achieved.

Though these long-term goals are core to our motivations, we will spend much of our time here explaining how we think we can make AGI systems do relatively simple things, like the things human children do in preschool. Chapter 3 describes a thought-experiment involving a robot playing with blocks, responding to the request "Build me something I haven't seen before". We believe that preschool creativity contains the seeds of, and the core structures and dynamics underlying, adult human level genius... and new, as yet unforeseen forms of artificial innovation.

Much of the book focuses on a specific AGI architecture, which we call CogPrime, and which is currently in the midst of implementation using the OpenCog software framework. CogPrime is large and complex and embodies a host of specific decisions regarding the various aspects of intelligence. We don't view CogPrime as the unique path to advanced AGI, nor as the ultimate end-all of AGI research. We feel confident there are multiple possible paths to advanced AGI, and that in following any of

B. Goertzel et al., *Engineering General Intelligence, Part 1*,
Atlantis Thinking Machines 5, DOI: 10.2991/978-94-6239-027-0_1,
© Atlantis Press and the authors 2014

these paths, multiple theoretical and practical lessons will be learned, leading to modifications of the ideas possessed while along the early stages of the path. But our goal here is to articulate **one** path that we believe makes sense to follow, one overall design that we believe can work.

1.2 AGI Versus Narrow AI

An outsider to the AI field might think this sort of book commonplace in the research literature, but insiders know that's far from the truth. The field of Artificial Intelligence (AI) was founded in the mid 1950s with the aim of constructing "thinking machines"—that is, computer systems with human-like general intelligence, including humanoid robots that not only look but act and think with intelligence equal to and ultimately greater than human beings. But in the intervening years, the field has drifted far from its ambitious roots, and this book represents part of a movement aimed at restoring the initial goals of the AI field, but in a manner powered by new tools and new ideas far beyond those available half a century ago.

After the first generation of AI researchers found the task of creating human-level AGI very difficult given the technology of their time, the AI field shifted focus toward what Ray Kurzweil has called "narrow AI"—the understanding of particular specialized aspects of intelligence; and the creation of AI systems displaying intelligence regarding specific tasks in relatively narrow domains. In recent years, however, the situation has been changing. More and more researchers have recognized the necessity—and feasibility—of returning to the original goals of the field.

In the decades since the 1950s, cognitive science and neuroscience have taught us a lot about what a cognitive architecture needs to look like to support roughly human-like general intelligence. Computer hardware has advanced to the point where we can build distributed systems containing large amounts of RAM and large numbers of processors, carrying out complex tasks in real time. The AI field has spawned a host of ingenious algorithms and data structures, which have been successfully deployed for a huge variety of purposes.

Due to all this progress, increasingly, there has been a call for a transition from the current focus on highly specialized "narrow AI" problem solving systems, back to confronting the more difficult issues of "human level intelligence" and more broadly "artificial general intelligence (AGI)". Recent years have seen a growing number of special sessions, workshops and conferences devoted specifically to AGI, including the annual BICA (Biologically Inspired Cognitive Architectures) AAAI Symposium, and the international AGI conference series (one in 2006, and annual since 2008). And, even more exciting, as reviewed in Chap. 6, there are a number of contemporary projects focused directly and explicitly on AGI (sometimes under the name "AGI", sometimes using related terms such as "Human Level Intelligence").

In spite of all this progress, however, we feel that no one has yet clearly articulated a detailed, systematic design for an AGI, with potential to yield general intelligence at the human level and ultimately beyond. In this spirit, our main goal in this lengthy two-part book is to outline a novel *design for a thinking machine*—an AGI design which we believe has the capability to produce software systems with intelligence at the human adult level and ultimately beyond. Many of the technical details of this design have been previously presented online in a wikibook [Goewiki]; and the basic ideas of the design have been presented briefly in a series of conference papers [GPSL03, GPPG06, Goe09c]. But the overall design has not been presented in a coherent and systematic way before this book. In order to frame this design properly, we also present a considerable number of broader theoretical and conceptual ideas here, some more and some less technical in nature.

1.3 CogPrime

The AGI design presented here has not previously been granted a name independently of its particular software implementations, but for the purposes of this book it needs one, so we've christened it **CogPrime**. This fits with the name "OpenCogPrime" that has already been used to describe the software implementation of CogPrime within the open-source OpenCog AGI software framework. The OpenCogPrime software, right now, implements only a small fraction of the CogPrime design as described here. However, OpenCog was designed specifically to enable efficient, scalable implementation of the full CogPrime design (as well as to serve as a more general framework for AGI R&D); and work currently proceeds in this direction, though there is a lot of work still to be done and many challenges remain.[1]

The CogPrime design is more comprehensive and thorough than anything that has been presented in the literature previously, including the work of others reviewed in Chap 6. It covers all the key aspects of human intelligence, and explains how they interoperate and how they can be implemented in digital computer software. Volume 5 of this work outlines CogPrime at a high level, and makes a number of more general points about artificial general intelligence and the path thereto; then Part 2 digs deeply into the technical particulars of CogPrime. Even Part 2, however, doesn't explain all the details of CogPrime that have been worked out so far, and it definitely doesn't explain all the implementation details that have gone into designing and building OpenCogPrime. Creating a thinking machine is a large task, and even the intermediate level of detail takes up a lot of pages.

---

[1] This brings up a terminological note: At several places in this Volume and the next we will refer to the current CogPrime or OpenCog implementation; in all cases this refers to OpenCog as of late 2013. We realize the risk of mentioning the state of our software system at time of writing: for future readers this may give the wrong impression, because if our project goes well, more and more of CogPrime will get implemented and tested as time goes on (e.g. within the OpenCog framework, under active development at time of writing). However, not mentioning the current implementation at all seems an even worse course to us, since we feel readers will be interested to know which of our ideas—at time of writing—have been honed via practice and which have not. Online resources such as http://www.opencog.org may be consulted by readers curious about the current state of the main OpenCog implementation; though in future forks of the code may be created, or other systems may be built using some or all of the ideas in this book, etc.

1.4 The Secret Sauce

There is no consensus on why all the related technological and scientific progress mentioned above has not yet yielded AI software systems with human-like general intelligence (or even greater levels of brilliance!). However, we hypothesize that the core reason boils down to the following three points:

- Intelligence depends on the emergence of certain high-level structures and dynamics across a system's whole knowledge base;
- We have not discovered any one algorithm or approach capable of yielding the emergence of these structures;
- Achieving the emergence of these structures within a system formed by integrating a number of different AI algorithms and structures requires careful attention to the manner in which these algorithms and structures are integrated; and so far the integration has not been done in the correct way.

The human brain appears to be an integration of an assemblage of diverse structures and dynamics, built using common components and arranged according to a sensible cognitive architecture. However, its algorithms and structures have been honed by evolution to work closely together—they are very tightly inter-adapted, in the same way that the different organs of the body are adapted to work together. Due to their close interoperation they give rise to the overall systemic behaviors that characterize human-like general intelligence. We believe that the main missing ingredient in AI so far is **cognitive synergy**: the fitting-together of different intelligent components into an appropriate cognitive architecture, in such a way that the components richly and dynamically support and assist each other, interrelating very closely in a similar manner to the components of the brain or body and thus giving rise to appropriate emergent structures and dynamics. This leads us to one of the central hypotheses underlying the CogPrime approach to AGI: that **the cognitive synergy ensuing from integrating multiple symbolic and subsymbolic learning and memory components in an appropriate cognitive architecture and environment, can yield robust intelligence at the human level and ultimately beyond**.

The reason this sort of intimate integration has not yet been explored much is that it's difficult on multiple levels, requiring the design of an architecture and its component algorithms with a view toward the structures and dynamics that will arise in the system once it is coupled with an appropriate environment. Typically, the AI algorithms and structures corresponding to different cognitive functions have been developed based on divergent theoretical principles, by disparate communities of researchers, and have been tuned for effective performance on different tasks in different environments. Making such diverse components work together in a truly synergetic and cooperative way is a tall order, yet we believe that this—rather than some particular algorithm, structure or architectural principle—is the "secret sauce" needed to create human-level AGI based on technologies available today.

1.5 Extraordinary Proof?

There is a saying that "extraordinary claims require extraordinary proof" and by that standard, if one believes that having a design for an advanced AGI is an extraordinary claim, this book must be rated a failure. We don't offer extraordinary proof that CogPrime, once fully implemented and educated, will be capable of human-level general intelligence and more.

It would be nice if we could offer mathematical proof that CogPrime has the potential we think it does, but at the current time mathematics is simply not up to the job. We'll pursue this direction briefly in Chap. 8 and other chapters, where we'll clarify exactly what kind of mathematical claim "CogPrime has the potential for human-level intelligence" turns out to be. Once this has been clarified, it will be clear that current mathematical knowledge does not yet let us evaluate, or even fully formalize, this kind of claim. Perhaps one day rigorous and detailed analyses of practical AGI designs will be feasible—and we look forward to that day—but it's not here yet.

Also, it would of course be profoundly exciting if we could offer dramatic practical demonstrations of CogPrime's capabilities. We do have a partial software implementation, in the OpenCogPrime system, but currently the things OpenCogPrime does are too simple to really serve as proofs of CogPrime's power for advanced AGI. We have used some CogPrime ideas in the OpenCog framework to do things like natural language understanding and data mining, and to control virtual dogs in online virtual worlds; and this has been very useful work in multiple senses. It has taught us more about the CogPrime design; it has produced some useful software systems; and it constitutes fractional work building toward a full OpenCog based implementation of CogPrime. However, to date, the things OpenCogPrime has done are all things that could have been done in different ways without the CogPrime architecture (though perhaps not as elegantly nor with as much room for interesting expansion).

The bottom line is that building an AGI is a big job. Software companies like Microsoft spend dozens to hundreds of man-years building software products like word processors and operating systems, so it should be no surprise that creating a digital intelligence is also a relatively large-scale software engineering project. As time advances and software tools improve, the number of man-hours required to develop advanced AGI gradually decreases—but right now, as we write these words, it's still a rather big job. In the OpenCogPrime project we are making a serious attempt to create a CogPrime based AGI using an open-source development methodology, with the open-source Linux operating system as one of our inspirations. But the open-source methodology doesn't work magic either, and it remains a large project, currently at an early stage. I emphasize this point so that readers lacking software engineering expertise don't take the currently fairly limited capabilities of OpenCogPrime as somehow a damning indictment of the potential of the CogPrime design. The design is one thing, the implementation another—and the OpenCogPrime implementation currently encompasses perhaps one third to one half of the key ideas in this book.

So we don't have extraordinary proof to offer. What we aim to offer instead are clearly-constructed conceptual and technical arguments as to why we think the CogPrime design has dramatic AGI potential.

It is also possible to push back a bit on the common intuition that having a design for human-level AGI is such an "extraordinary claim". It may be extraordinary relative to contemporary science and culture, but we have a strong feeling that the AGI problem is not difficult in the same ways that most people (including most AI researchers) think it is. We suspect that in hindsight, after human-level AGI has been achieved, people will look back in shock that it took humanity so long to come up with a workable AGI design. As you'll understand once you've finished Part 1 of the book, we don't think general intelligence is nearly as "extraordinary" and mysterious as it's commonly made out to be. Yes, building a thinking machine is hard—but humanity has done a lot of other hard things before. It may seem difficult to believe that human-level general intelligence could be achieved by something as simple as a collection of algorithms linked together in an appropriate way and used to control an agent. But we suggest that, once the first powerful AGI systems are produced, it will become apparent that engineering human-level minds is not so profoundly different from engineering other complex systems.

All in all, we'll consider the book successful if a significant percentage of open-minded, appropriately-educated readers come away from it scratching their chins and pondering: *"Hmm. You know, that just might work"*. and a small percentage come away thinking *"Now that's an initiative I'd really like to help with!"*.

1.6 Potential Approaches to AGI

In principle, there is a large number of approaches one might take to building an AGI, starting from the knowledge, software and machinery now available. This is not the place to review them in detail, but a brief list seems apropos, including commentary on why these are not the approaches we have chosen for our own research. Our intent here is not to insult or dismiss these other potential approaches, but merely to indicate why, as researchers with limited time and resources, we have made a different choice regarding where to focus our own energies.

1.6.1 Build AGI from Narrow AI

Most of the AI programs around today are "narrow AI" programs—they carry out one particular kind of task intelligently. One could try to make an advanced AGI by combining a bunch of enhanced narrow AI programs inside some kind of overall framework.

However, we're rather skeptical of this approach because none of these narrow AI programs have the ability to generalize across domains—and we don't see how combining them or extending them is going to cause this to magically emerge.

1.6.2 Enhancing Chatbots

One could seek to make an advanced AGI by taking a chatbot, and trying to improve its code to make it actually understand what it's talking about. We have some direct experience with this route, as in 2010 our AI consulting firm was contracted to improve Ray Kurzweil's online chatbot "Ramona". Our new Ramona understands a lot more than the previous Ramona version or a typical chatbot, due to using Wikipedia and other online resources, but still it's far from an AGI.

A more ambitious attempt in this direction was Jason Hutchens' a-i.com project, which sought to create a human child level AGI via development and teaching of a statistical learning based chatbot (rather than the typical rule-based kind). The difficulty with this approach, however, is that the architecture of a chatbot is fundamentally different from the architecture of a generally intelligent mind. Much of what's important about the human mind is not directly observable in conversations, so if you start from conversation and try to work toward an AGI architecture from there, you're likely to miss many critical aspects.

1.6.3 Emulating the Brain

One can approach AGI by trying to figure out how the brain works, using brain imaging and other tools from neuroscience, and then emulating the brain in hardware or software.

One rather substantial problem with this approach is that we don't really understand how the brain works yet, because our software for measuring the brain is still relatively crude. There is no brain scanning method that combines high spatial and temporal accuracy, and none is likely to come about for a decade or two. So to do brain-emulation AGI seriously, one needs to wait a while until brain scanning technology improves.

Current AI methods like neural nets that are loosely based on the brain, are really not brain-like enough to make a serious claim at emulating the brain's approach to general intelligence. We don't yet have any real understanding of how the brain represents abstract knowledge, for example, or how it does reasoning (though the authors, like many others, have made some speculations in this regard [GMIH08]).

Another problem with this approach is that once you're done, what you get is something with a very humanlike mind, and we already have enough of those! However, this is perhaps not such a serious objection, because a digital-computer-based version of a human mind could be studied much more thoroughly than a biology-based human mind. We could observe its dynamics in real-time in perfect precision, and could then learn things that would allow us to build other sorts of digital minds.

1.6.4 Evolve an AGI

Another approach is to try to run an evolutionary process inside the computer, and wait for advanced AGI to evolve.

One problem with this is that we don't know how evolution works all that well. There's a field of artificial life, but so far its results have been fairly disappointing. It's not yet clear how much one can vary on the chemical structures that underly real biology, and still get powerful evolution like we see in real biology. If we need good artificial chemistry to get good artificial biology, then do we need good artificial physics to get good artificial chemistry?

Another problem with this approach, of course, is that it might take a really long time. Evolution took billions of years on Earth, using a massive amount of computational power. To make the evolutionary approach to AGI effective, one would need some radical innovations to the evolutionary process (such as, perhaps, using probabilistic methods like BOA [Pel05] or MOSES [Loo06] in place of traditional evolution).

1.6.5 Derive an AGI Design Mathematically

One can try to use the mathematical theory of intelligence to figure out how to make advanced AGI.

This interests us greatly, but there's a huge gap between the rigorous math of intelligence as it exists today and anything of practical value. As we'll discuss in Chap. 8, most of the rigorous math of intelligence right now is about how to make AI on computers with dramatically unrealistic amounts of memory or processing power. When one tries to create a theoretical understanding of real-world general intelligence, one arrives at quite different sorts of considerations, as we will roughly outline in Chap. 11. Ideally we would like to be able to study the CogPrime design using a rigorous mathematical theory of real-world general intelligence, but at the moment that's not realistic. The best we can do is to conceptually analyze CogPrime and its various components in terms of relevant mathematical and theoretical ideas; and perform analysis of CogPrime's individual structures and components at varying levels of rigor.

1.6.6 Use Heuristic Computer Science Methods

The computer science field contains a number of abstract formalisms, algorithms and structures that have relevance beyond specific narrow AI applications, yet aren't necessarily understood as thoroughly as would be required to integrate them into the rigorous mathematical theory of intelligence. Based on these formalisms, algorithms and structures, a number of "single formalism/algorithm focused" AGI approaches have been outlined, some of which will be reviewed in Chap. 6. For example Pei Wang's NARS ("Non-Axiomatic Reasoning System") approach is based on a specific logic which he argues to be the "logic of general intelligence"—so, while his system contains many other aspects than this logic, he considers this logic to be the crux of the system and the source of its potential power as an AGI system.

The basic intuition on the part of these "single formalism/algorithm focused" researchers seems to be that there is one key formalism or algorithm underlying intelligence, and if you achieve this key aspect in your AGI program, you're going to get something that fundamentally thinks like a person, even if it has some differences due to its different implementation and embodiment. On the other hand, it's also possible that this idea is philosophically incorrect: that there is no one key formalism, algorithm, structure or idea underlying general intelligence. The CogPrime approach is based on the intuition that to achieve human-level, roughly human-like general intelligence based on feasible computational resources, one needs an appropriate heterogeneous combination of algorithms and structures, each coping with different types of knowledge and different aspects of the problem of achieving goals in complex environments.

1.6.7 Integrative Cognitive Architecture

Finally, to create advanced AGI one can try to build some sort of integrative cognitive architecture: a software system with multiple components that each carry out some cognitive function, and that connect together in a specific way to try to yield overall intelligence.

Cognitive science gives us some guidance about the overall architecture, and computer science and neuroscience give us a lot of ideas about what to put in the different components. But still this approach is very complex and there is a lot of need for creative invention.

This is the approach we consider most "serious" at present (at least until neuroscience advances further). And, as will be discussed in depth in these pages, this is the approach we've chosen: CogPrime is an integrative AGI architecture.

1.6.8 Can Digital Computers Really Be Intelligent?

All the AGI approaches we've just mentioned assume that it's possible to make AGI on digital computers. While we suspect this is correct, we must note that it isn't proven.

It might be that—as Penrose [Pen96], Hameroff [Ham87] and others have argued—we need quantum computers or quantum gravity computers to make AGI. However, there is no evidence of this at this stage. Of course the brain like all matter is described by quantum mechanics, but this doesn't imply that the brain is a "macroscopic quantum system" in a strong sense (like, say, a Bose-Einstein condensate). And even if the brain does use quantum phenomena in a dramatic way to carry out some of its cognitive processes (a hypothesis for which there is no current evidence), this doesn't imply that these quantum phenomena are *necessary* in order to carry out the given cognitive processes. For example there is evidence that birds use quantum nonlocal phenomena to carry out navigation based on the Earth's magnetic fields [Gau11]; yet scientists have built instruments that carry out the same functions without using any special quantum effects. The importance of quantum phenomena in biology (except via their obvious role in giving rise to biological phenomena describable via classical physics) remains a subject of debate [Abb08].

Quantum "magic" aside, it is also conceivable that building AGI is fundamentally impossible for some *other* reason we don't understand. Without getting religious about it, it is rationally quite possible that some aspects of the universe are beyond the scope of scientific methods. Science is fundamentally about recognizing patterns in finite sets of bits (e.g. finite sets of finite-precision observations), whereas mathematics recognizes many sets much larger than this. Selmer Bringsjord [Bri03], and other advocates of "hypercomputing" approaches to intelligence, argue that the human mind depends on massively large infinite sets and therefore can never be simulated on digital computers nor understood via finite sets of finite-precision measurements such as science deals with.

But again, while this sort of possibility is interesting to speculate about, there's no real reason to believe it at this time. Brain science and AI are both very young sciences and the "working hypothesis" that digital computers can manifest advanced AGI has hardly been explored at all yet, relative to what will be possible in the next decades as computers get more and more powerful and our understanding of neuroscience and cognitive science gets more and more complete. The CogPrime AGI design presented here is based on this working hypothesis.

Many of the ideas in the book are actually independent of the "mind can be implemented digitally" working hypothesis, and could apply to AGI systems built on analog, quantum or other non-digital frameworks—but we will not pursue these possibilities here. For the moment, outlining an AGI design for digital computers is hard enough! Regardless of speculations about quantum computing in the brain, it seems clear that AGI on quantum computers is part of our future and will be a powerful thing; but the description of a CogPrime analogue for quantum computers will be left for a later work.

1.7 Five Key Words

As noted, the CogPrime approach lies squarely in the integrative cognitive architecture camp. But it is not a haphazard or opportunistic combination of algorithms and data structures. At bottom it is motivated by the *patternist* philosophy of mind laid out in Ben Goertzel's book *The Hidden Pattern* [Goe06a], which was in large part a summary and reformulation of ideas presented in a series of books published earlier by the same author [Goe94], [Goe93], [Goe93], [Goe97], [Goe01]. A few of the core ideas of this philosophy are laid out in Chap. 5, though that chapter is by no means a thorough summary.

One way to summarize some of the most important yet commonsensical parts of the patternist philosophy of mind, in an AGI context, is to list five words: **perception, memory, prediction, action, goals**.

In a phrase: **"A mind uses perception and memory to make predictions about which actions will help it achieve its goals"**.

This ties in with the ideas of many other thinkers, including Jeff Hawkins' "memory/prediction" theory [HB06], and it also speaks directly to the formal characterization of intelligence presented in Chap. 8: general intelligence as "the ability to achieve complex goals in complex environments".

Naturally the goals involved in the above phrase may be explicit or implicit to the intelligent agent, and they may shift over time as the agent develops.

Perception is taken to mean pattern recognition: the recognition of (novel or familiar) patterns in the environment or in the system itself. Memory is the storage of already-recognized patterns, enabling recollection or regeneration of these patterns as needed. Action is the formation of patterns in the body and world. Prediction is the utilization of temporal patterns to guess what perceptions will be seen in the future, and what actions will achieve what effects in the future—in essence, prediction consists of temporal pattern recognition, plus the (implicit or explicit) assumption that the universe possesses a "habitual tendency" according to which previously observed patterns continue to apply.

1.7.1 Memory and Cognition in CogPrime

Each of these five concepts has a lot of depth to it, and we won't say too much about them in this brief introductory overview; but we will take a little time to say something about memory in particular.

As we'll see in Chap. 8, one of the things that the mathematical theory of general intelligence makes clear is that, if you assume your AI system has a huge amount of computational resources, then creating general intelligence is not a big trick. Given enough computing power, a very brief and simple program can achieve any computable goal in any computable environment, quite effectively. Marcus Hutter's AIXI^{tl} design [Hut05] gives one way of doing this, backed up by rigorous mathematics. Put informally, what this means is: the problem of AGI is really a problem of coping with inadequate compute resources, just as the problem of natural intelligence is really a problem of coping with inadequate energetic resources.

One of the key ideas underlying CogPrime is a principle called *cognitive synergy*, which explains how real-world minds achieve general intelligence using limited resources, by appropriately organizing and utilizing their memories.

This principle says that there are many different kinds of memory in the mind: sensory, episodic, procedural, declarative, attentional, intentional. Each of them has certain learning processes associated with it; for example, reasoning is associated with declarative memory. Synergy arises here in the way the learning processes associated with each kind of memory have got to help each other out when they get stuck, rather than working at cross-purposes.

Cognitive synergy is a fundamental principle of *general intelligence*—it doesn't tend to play a central role when you're building narrow-AI systems.

In the CogPrime approach all the different kinds of memory are linked together in a single meta-representation, a sort of combined semantic/neural network called the AtomSpace. It represents everything from perceptions and actions to abstract relationships and concepts and even a system's model of itself and others. When specialized representations are used for other types of knowledge (e.g. program trees for procedural knowledge, spatiotemporal hierarchies for perceptual knowledge) then the knowledge stored outside the AtomSpace is represented via tokens (Atoms) in the AtomSpace, allowing it to be located by various cognitive processes, and associated with other memory items of any type.

So for instance an OpenCog AI system has an AtomSpace, plus some specialized knowledge stores linked into the AtomSpace; and it also has specific algorithms acting on the AtomSpace and appropriate specialized stores corresponding to each type of memory. Each of these algorithms is complex and has its own story; for instance (an incomplete list, for more detail see the following section of this Introduction):

- Declarative knowledge is handled using Probabilistic Logic Networks, described in Chap. 16 (Part 2) and others;
- Procedural knowledge is handled using MOSES, a probabilistic evolutionary learning algorithm described in Chap. 3 (Part 2) and others;
- Attentional knowledge is handled by ECAN (economic attention allocation), described in Chap. 5 (Part 2) and others;
- OpenCog contains a language comprehension system called RelEx that takes English sentences and turns them into nodes and links in the AtomSpace. It's currently being extended to handle Chinese. RelEx handles mostly declarative knowledge but also involves some procedural knowledge for linguistic phenomena like reference resolution and semantic disambiguation.

But the crux of the CogPrime cognitive architecture is not any particular cognitive process, but rather the way they all work together using cognitive synergy.

1.8 Virtually and Robotically Embodied AI

Another issue that will arise frequently in these pages is embodiment. There's a lot of debate in the AI community over whether embodiment is necessary for advanced AGI or not. Personally, we doubt it's necessary but we think it's extremely convenient, and are thus considerably interested in both virtual world and robotic embodiment. The CogPrime architecture itself is neutral on the issue of embodiment, and it could be used to build a mathematical theorem prover or an intelligent chat bot just as easily as an embodied AGI system. However, most of our attention has gone into figuring out how to use CogPrime to control embodied agents in virtual worlds, or else (to a lesser extent) physical robots. For instance, during 2011–2012 we are involved in a Hong Kong government funded project using OpenCog to control video game agents in a simple game world modeled on the game Minecraft [Goe11x].

Current virtual world technology has significant limitations that make them far less than ideal from an AGI perspective, and in Chap. 17 we will discuss how they can be remedied. However, for the medium-term future virtual worlds are not going to match the natural world in terms of richness and complexity—and so there's also something to be said for physical robots that interact with all the messiness of the real world.

With this in mind, in the Artificial Brain Lab at Xiamen University in 2009–2010, we conducted some experiments using OpenCog to control the Nao humanoid robot [Goe09c]. The goal of that work was to take the same code that controls the virtual dog and use it to control the physical robot. But it's harder because in this context we need to do real vision processing and real motor control. A similar project is being undertaken in Hong Kong at time of writing, involving a collaboration between OpenCog AI developers and David Hanson's robotics group. One of the key ideas involved in this project is explicit integration of subsymbolic and more symbolic subsystems. For instance, one can use a purely subsymbolic, hierarchical pattern recognition network for vision processing, and then link its internal structures into the nodes and links in the AtomSpace that represent concepts. So the subsymbolic and symbolic systems can work harmoniously and productively together, a notion we will review in more detail in Chap. 8 (Part 2).

1.9 Language Learning

One of the subtler aspects of our current approach to teaching CogPrime is language learning. Three relatively crisp and simple approaches to language learning would be:

- Build a language processing system using hand-coded grammatical rules, based on linguistic theory;
- Train a language processing system using supervised, unsupervised or semisupervised learning, based on computational linguistics;

- Have an AI system learn language via experience, based on imitation and reinforcement and experimentation, without any built-in distinction between linguistic
  behaviors and other behaviors.

While the third approach is conceptually appealing, our current approach in Cog-
Prime (described in a series of chapters in Part 2) is none of the above, but rather a
combination of the above. OpenCog contains a natural language processing system
built using a combination of the rule-based and statistical approaches, which has
reasonably adequate functionality; and our plan is to use it as an initial condition for
ongoing adaptive improvement based on embodied communicative experience.

1.10 AGI Ethics

When discussing AGI work with the general public, ethical concerns often arise.
Science fiction films like the *Terminator* series have raised public awareness of the
possible dangers of advanced AGI systems without correspondingly advanced ethics.
Non-profit organizations like the Singularity Institute for AI (http://www.singinst.
org) have arisen specifically to raise attention about, and foster research on, these
potential dangers.

Our main focus here is on how to create AGI, not how to teach an AGI human
ethical principles. However, we will address the latter issue explicitly in Chap. 13,
and we do think it's important to emphasize that AGI ethics has been at the center
of the design process throughout the conception and development of CogPrime and
OpenCog.

Broadly speaking there are (at least) two major threats related to advanced AGI.
One is that people might use AGIs for bad ends; and the other is that, even if an AGI
is made with the best intentions, it might reprogram itself in a way that causes it to
do something terrible. If it's smarter than us, we might be watching it carefully while
it does this, and have no idea what's going on.

The best way to deal with this second "bad AGI" problem is to build ethics into
your AGI architecture—and we have done this with CogPrime, via creating a goal
structure that explicitly supports ethics-directed behavior, and via creating an overall
architecture that supports "ethical synergy" along with cognitive synergy. In short,
the notion of ethical synergy is that there are different kinds of ethical thinking
associated with the different kinds of memory and you want to be sure your AGI has
all of them, and that it uses them together effectively.

In order to create AGI that is not only intelligent but beneficial to other sentient
beings, ethics has got to be part of the design and the roadmap. As we teach our AGI
systems, we need to lead them through a series of instructional and evaluative tasks
that move from a primitive level to the mature human level—in intelligence, but also
in ethical judgment.

1.11 Structure of the Book

The book is divided into two parts. The technical particulars of CogPrime are discussed in Part 2; what we deal with in Part 1 are important preliminary and related matters such as:

- The nature of real-world general intelligence, both conceptually and from the perspective of formal modeling ("Artificial and Natural General Intelligence").
- The nature of cognitive and ethical development for humans and AGIs ("Cognitive and Ethical Development").
- The high-level properties of CogPrime, including the overall architecture and the various sorts of memory involved ("Networks for Explicit and Implicit Knowledge Representation").
- What kind of path may viably lead us from here to AGI, with focus laid on preschool-type environments that easily foster humanlike cognitive development. Various advanced aspects of AGI systems, such as the network and algebraic structures that may emerge from them, the ways in which they may self-modify, and the degree to which their initial design may constrain or guide their future state even after long periods of radical self-improvement ("A Path to Human-Level AGI").

One point made repeatedly throughout Part 1, which is worth emphasizing here, is the current lack of a really rigorous and thorough general technical theory of general intelligence. Such a theory, if complete, would be incredibly helpful for understanding complex AGI architectures like CogPrime. Lacking such a theory, we must work on CogPrime and other such systems using a combination of theory, experiment and intuition. This is not a bad thing, but it will be very helpful if the theory and practice of AGI are able to grow collaboratively together.

1.12 Key Claims of the Book

We will wrap up this Introduction with a systematic list of some of the key claims to be argued for in these pages. Not all the terms and ideas in these claims have been mentioned in the preceding portions of this Introduction, but we hope they will be reasonably clear to the reader anyway, at least in a general sense. This list of claims will be revisited in Chap. 31 (Part 2) near the end of Part 2, where we will look back at the ideas and arguments that have been put forth in favor of them, in the intervening chapters.

In essence this is a list of claims such that, if the reader accepts these claims, they should probably accept that the CogPrime approach to AGI is a viable one. On the other hand if the reader rejects one or more of these claims, they may find one or more aspects of CogPrime unacceptable for some reason.

Without further ado, now, the claims:

1. General intelligence (at the human level and ultimately beyond) can be achieved via creating a computational system that seeks to achieve its goals, via using perception and memory to predict which actions will achieve its goals in the contexts in which it finds itself.
2. To achieve general intelligence in the context of human-intelligence-friendly environments and goals using feasible computational resources, it's important that an AGI system can handle different kinds of memory (declarative, procedural, episodic, sensory, intentional, attentional) in customized but interoperable ways.
3. Cognitive synergy: It's important that the cognitive processes associated with different kinds of memory can appeal to each other for assistance in overcoming bottlenecks in a manner that enables each cognitive process to act in a manner that is sensitive to the particularities of each others' internal representations, and that doesn't impose unreasonable delays on the overall cognitive dynamics.
4. As a general principle, neither purely localized nor purely global memory is sufficient for general intelligence under feasible computational resources; "glocal" memory will be required.
5. To achieve human-like general intelligence, it's important for an intelligent agent to have sensory data and motoric affordances that roughly emulate those available to humans. We don't know exactly how close this emulation needs to be, which means that our AGI systems and platforms need to support fairly flexible experimentation with virtual-world and/or robotic infrastructures.
6. To work toward adult human-level, roughly human-like general intelligence, one fairly easily comprehensible path is to use environments and goals reminiscent of human childhood, and seek to advance one's AGI system along a path roughly comparable to that followed by human children.
7. It is most effective to teach an AGI system aimed at roughly human-like general intelligence via a mix of spontaneous learning and explicit instruction, and to instruct it via a combination of imitation, reinforcement and correction, and a combination of linguistic and nonlinguistic instruction.
8. One effective approach to teaching an AGI system human language is to supply it with some in-built linguistic facility, in the form of rule-based and statistical-linguistics-based NLP systems, and then allow it to improve and revise this facility based on experience.
9. An AGI system with adequate mechanisms for handling the key types of knowledge mentioned above, and the capability to explicitly recognize large-scale patterns in itself, should, **upon sustained interaction with an appropriate environment in pursuit of appropriate goals**, emerge a variety of complex structures in its internal knowledge network, including, but not limited to:

   - a hierarchical network, representing both a spatiotemporal hierarchy and an approximate "default inheritance" hierarchy, cross-linked
   - a heterarchical network of associativity, roughly aligned with the hierarchical network
   - a self network which is an approximate micro image of the whole network

- inter-reflecting networks modeling self and others, reflecting a "mirrorhouse" design pattern

10. Given the strengths and weaknesses of current and near-future digital computers,

   a. A (loosely) neural-symbolic network is a good representation for directly storing many kinds of memory, and interfacing between those that it doesn't store directly;
   b. Uncertain logic is a good way to handle declarative knowledge. To deal with the problems facing a human-level AGI, an uncertain logic must integrate imprecise probability and fuzziness with a broad scope of logical constructs. PLN is one good realization.
   c. Programs are a good way to represent procedures (both cognitive and physical-action, but perhaps not including low-level motor-control procedures).
   d. Evolutionary program learning is a good way to handle difficult program learning problems. Probabilistic learning on normalized programs is one effective approach to evolutionary program learning. MOSES is one good realization of this approach.
   e. Multistart hill-climbing, with a strong Occam prior, is a good way to handle relatively straightforward program learning problems.
   f. Activation spreading and Hebbian learning comprise a reasonable way to handle attentional knowledge (though other approaches, with greater overhead cost, may provide better accuracy and may be appropriate in some situations).

      - Artificial economics is an effective approach to activation spreading and Hebbian learning in the context of neural-symbolic networks;
      - ECAN is one good realization of artificial economics;
      - A good trade-off between comprehensiveness and efficiency is to focus on two kinds of attention: processor attention (represented in CogPrime by ShortTermImportance) and memory attention (represented in CogPrime by LongTermImportance).

   g. Simulation is a good way to handle episodic knowledge (remembered and imagined). Running an internal world simulation engine is an effective way to handle simulation.
   h. Hybridization of one's integrative neural-symbolic system with a spatiotemporally hierarchical deep learning system is an effective way to handle representation and learning of low-level sensorimotor knowledge. DeSTIN is one example of a deep learning system of this nature that can be effective in this context.
   i. One effective way to handle goals is to represent them declaratively, and allocate attention among them economically. CogPrime's PLN/ECAN based framework for handling intentional knowledge is one good realization.

11. It is important for an intelligent system to have some way of recognizing large-scale patterns in itself, and then embodying these patterns as new, localized knowledge items in its memory. Given the use of a neural-symbolic network for knowledge representation, a graph-mining based "map formation" heuristic is one good way to do this.

12. Occam's Razor: Intelligence is closely tied to the creation of procedures that achieve goals in environments *in the simplest possible way*. Each of an AGI system's cognitive algorithms should embody a simplicity bias in some explicit or implicit form.

13. An AGI system, if supplied with a commonsensically ethical goal system and an intentional component based on rigorous uncertain inference, should be able to reliably achieve a much higher level of commonsensically ethical behavior than any human being.

14. Once sufficiently advanced, an AGI system with a logic-based declarative knowledge approach and a program-learning-based procedural knowledge approach should be able to radically self-improve via a variety of methods, including supercompilation and automated theorem-proving.

Part I
Overview of the CogPrime Architecture

Chapter 2
A Brief Overview of CogPrime

2.1 Introduction

Just as there are many different approaches to human flight—airplanes, helicopters, balloons, spacecraft, and doubtless many methods no person has thought of yet—similarly, there are likely many different approaches to advanced artificial general intelligence. All the different approaches to flight exploit the same core principles of aerodynamics in different ways; and similarly, the various different approaches to AGI will exploit the same core principles of general intelligence in different ways.

The thorough presentation of the CogPrime design is the job of Part 2 of this book—where, not only are the algorithms and structures involved in CogPrime reviewed in more detail, but their relationship to the theoretical ideas underlying CogPrime is pursued more deeply. The job of this chapter is a smaller one: to give a high-level overview of some key aspects the CogPrime architecture at a mostly nontechnical level, so as to enable you to approach Part 2 with a little more idea of what to expect. The remainder of Part 1, following this chapter, will present various theoretical notions enabling the particulars, intent and consequences of the CogPrime design to be more thoroughly understood.

2.2 High-Level Architecture of CogPrime

Figures 2.1, 2.2, 2.4 and 2.5 depict the high-level architecture of CogPrime, which involves the use of multiple cognitive processes associated with multiple types of memory to enable an intelligent agent to execute the procedures that it believes have the best probability of working toward its goals in its current context. In a robot preschool context, for example, the top-level goals will be simple things such as pleasing the teacher, learning new information and skills, and protecting the robot's body. Figure 2.3 shows part of the architecture via which cognitive processes interact with each other, via commonly acting on the AtomSpace knowledge repository.

B. Goertzel et al., *Engineering General Intelligence, Part 1*, Atlantis Thinking Machines 5, DOI: 10.2991/978-94-6239-027-0_2, © Atlantis Press and the authors 2014

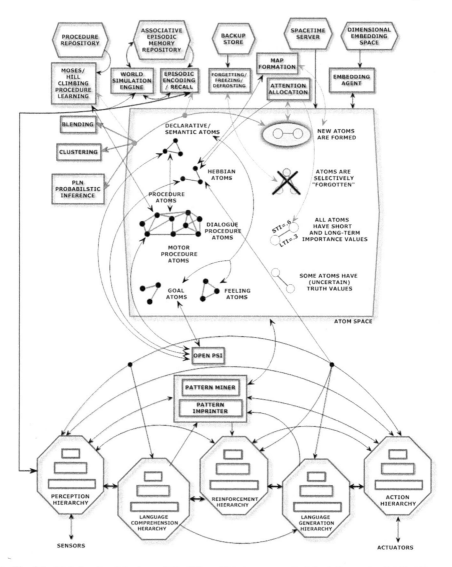

Fig. 2.1 High-level architecture of CogPrime. This is a conceptual depiction, not a detailed flow-chart (which would be too complex for a single image). Figures 2.2, 2.4 and 2.5 highlight specific aspects of this diagram

Comparing these diagrams to the integrative human cognitive architecture diagrams to be given in Chap. 7, one sees the main difference is that the CogPrime diagrams commit to specific structures (e.g. knowledge representations) and processes, whereas the generic integrative architecture diagram refers merely to types of structures and processes. For instance, the integrative diagram refers generally to declarative knowledge and learning, whereas the CogPrime diagram refers to PLN, as a specific system for reasoning and learning about declarative knowledge. Table 2.1 articulates the key connections between the components of the CogPrime diagram and those of the integrative diagram, thus indicating the general cognitive functions instantiated by each of the CogPrime components.

2.3 Current and Prior Applications of OpenCog

Before digging deeper into the theory, and elaborating some of the dynamics under-lying the above diagrams, we pause to briefly discuss some of the practicalities of work done with the OpenCog system currently implementing parts of the CogPrime architecture.

OpenCog, the open-source software framework underlying the "OpenCogPrime" (currently partial) implementation of the CogPrime architecture, has been used for commercial applications in the area of natural language processing and data min-ing; for instance, see [GPPG06] where OpenCogPrime's PLN reasoning and RelEx language processing are combined to do automated biological hypothesis generation based on information gathered from PubMed abstracts. Most relevantly to the present work, it has also been used to control virtual agents in virtual worlds [GEA08].

Prototype work done during 2007–2008 involved using an OpenCog variant called the OpenPetBrain to control virtual dogs in a virtual world (see Fig. 2.6 for a screenshot of an OpenPetBrain-controlled virtual dog). While these OpenCog virtual dogs did not display intelligence closely comparable to that of real dogs (or human children), they did demonstrate a variety of interesting and relevant functionalities including:

- learning new behaviors based on imitation and reinforcement
- responding to natural language commands and questions, with appropriate actions and natural language replies
- spontaneous exploration of their world, remembering their experiences and using them to bias future learning and linguistic interaction.

One current OpenCog initiative involves extending the virtual dog work via using OpenCog to control virtual agents in a game world inspired by the game Minecraft. These agents are initially specifically concerned with achieving goals in a game world via constructing structures with blocks and carrying out simple English com-munications. Representative example tasks would be:

- Learning to build steps or ladders to get desired objects that are high up
- Learning to build a shelter to protect itself from aggressors
- Learning to build structures resembling structures that it's shown (even if the available materials are a bit different)
- Learning how to build bridges to cross chasms.

Of course, the AI significance of learning tasks like this all depends on what kind of feedback the system is given, and how complex its environment is. It would be

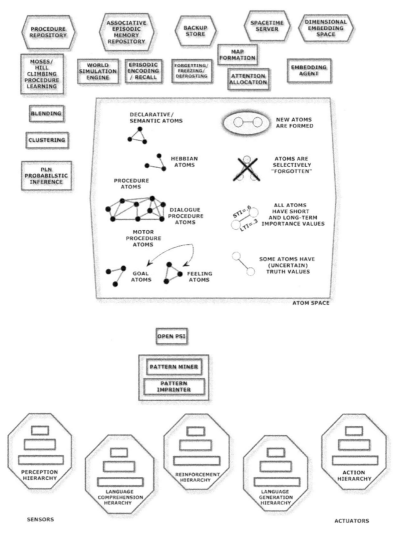

Fig. 2.2 Key explicitly implemented processes of CogPrime. The large box at the center is the Atomspace, the system's central store of various forms of (long-term and working) memory, which contains a weighted labeled hypergraph whose nodes and links are "Atoms" of various sorts. The hexagonal boxes at the bottom denote various hierarchies devoted to recognition and generation of patterns: perception, action and linguistic. Intervening between these recognition/generation hierarchies and the Atomspace, we have a pattern mining/imprinting component (that recognizes patterns in the hierarchies and passes them to the Atomspace; and imprints patterns from the Atomspace on the hierarchies); and also OpenPsi, a special dynamical framework for choosing actions based on motivations. Above the Atomspace we have a host of cognitive processes, which act on the Atomspace, some continually and some only as context dictates, carrying out various sorts of learning and reasoning (pertinent to various sorts of memory) that help the system fulfill its goals and motivations

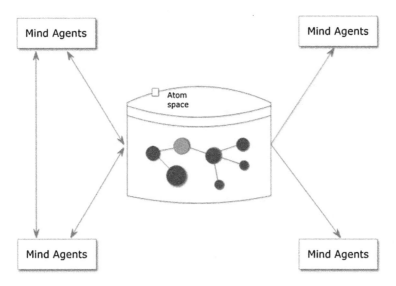

Fig. 2.3 MindAgents and AtomSpace in OpenCog. This is a conceptual depiction of one way cognitive processes may interact in OpenCog—they may be wrapped in MindAgent objects, which interact via cooperatively acting on the AtomSpace

relatively simple to make an AI system do things like this in a trivial and highly specialized way, but that is not the intent of the project the goal is to have the system learn to carry out tasks like this using general learning mechanisms and a general cognitive architecture, based on embodied experience and only scant feedback from human teachers. If successful, this will provide an outstanding platform for ongoing AGI development, as well as a visually appealing and immediately meaningful demo for OpenCog.

Specific, particularly simple tasks that are the focus of this project team's current work at time of writing include:

- Watch another character build steps to reach a high-up object
- Figure out via imitation of this that, in a different context, building steps to reach a high up object may be a good idea
- Also figure out that, if it wants a certain high-up object but there are no materials for building steps available, finding some other way to get elevated will be a good idea that may help it get the object.

2.3.1 Transitioning from Virtual Agents to a Physical Robot

Preliminary experiments have also been conducted using OpenCog to control a Nao robot as well as a virtual dog [GdG08]. This involves hybridizing OpenCog with a

Table 2.1 Connections between the CogPrime architecture diagram and the integrative architecture diagram

CogPrime component	Int. diag. sub-diagram	Int. diag. component
Procedure repository	Long-term memory	Procedural
Procedure repository	Working memory	Active procedural
Associative episodic memory	Long-term memory	Episodic
Associative episodic memory	Working memory	Transient episodic
Backup store	Long-term memory	No correlate: a function not necessarily possessed by the human mind
Spacetime server	Long-term memory	Declarative and sensorimotor
Dimensional embedding space	No clear correlate: a tool for helping multiple types of LTM	
Dimensional embedding agent	No clear correlate	
Blending	Long-term and working memory	Concept formation
Clustering	Long-term and working memory	Concept formation
PLN probabilistic inference	Long-term and working memory	Reasoning and plan learning/optimization
MOSES/Hillclimbing	Long-term and working memory	Procedure learning
World simulation	Long-term and working memory	Simulation
Episodic encoding/recall	Long-term g memory	Story-telling
Episodic encoding/recall	Working memory	Consolidation
Forgetting/freezing/defrosting	Long-term and working memory	No correlate: a function not necessarily possessed by the human mind
Map formation	Long-term memory	Concept formation and pattern mining
Attention allocation	Long-term and working memory	Hebbian/attentional learning
Attention allocation	High-level mind architecture	Reinforcement
Attention allocation	Working memory	Perceptual associative memory and local association
AtomSpace	High-level mind architecture	No clear correlate: a general tool for representing memory including long-term and working, plus some of perception and action
AtomSpace	Working memory	Global workspace (the high-STI portion of AtomSpace) and other workspaces
Declarative atoms	Long-term and working memory	Declarative and sensorimotor

(continued)

Table 2.1 (continued)

CogPrime component	Int. diag. sub-diagram	Int. diag. component
Procedure atoms	Long-term and working memory	Procedural
Hebbian atoms	Long-term and working memory	Attentional
Goal atoms	Long-term and working memory	Intentional
Feeling atoms	Long-term and working memory	Spanning declarative, intentional and sensorimotor
OpenPsi	High-level mind architecture	Motivation/action selection
OpenPsi	Working memory	Action selection
Pattern miner	High-level mind architecture	Arrows between perception and working and long-term memory
Pattern miner	Working memory	Arrows between sensory memory and perceptual associative and transient episodic memory
Pattern imprinter	Working memory	Arrows between action selection and sensorimotor memory, and between the latter and perception/action subsystems
Pattern imprinter	High-level mind architecture	Arrows pointing to action subsystem from working and long-term memories
Perception hierarchy	High-level mind architecture	Perception subsystems
Perception hierarchy	Working memory	Perception/action subsystems and sensory and sensorimotor memory
Language comprehension hierarchy	Language	Comprehension hierarchy
Language generation hierarchy	Language	Generation hierarchy
Reinforcement hierarchy	High-level mind architecture	Reinforcement
Reinforcement hierarchy	Action	Reinforcement hierarchy
Action hierarchy	Action	Collection of specialized action hierarchies

There is a row for each component in the CogPrime architecture diagram, which tells the corresponding sub-diagrams and components of the integrative architecture diagram given in Chap. 7. Note that the description "Long Term and Working Memory" indicates occurrence in two separate sub diagrams of the integrative diagram, "Long Term Memory" and "Working Memory"

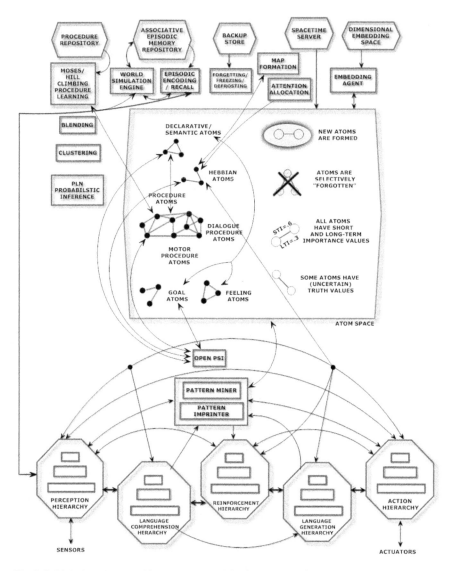

Fig. 2.4 Links between cognitive processes and the Atomspace. The cognitive processes depicted all act on the Atomspace, in the sense that they operate by observing certain Atoms in the Atomspace and then modifying (or in rare cases deleting) them, and potentially adding new Atoms as well. Atoms represent all forms of knowledge, but some forms of knowledge are additionally represented by external data stores connected to the Atomspace, such as the Procedure Repository; these are also shown as linked to the Atomspace

separate (but interlinked) subsystem handling low-level perception and action. In the experiments done so far, this has been accomplished in an extremely simplistic way. How to do this right is a topic treated in detail in Chap. 8 of Part 2.

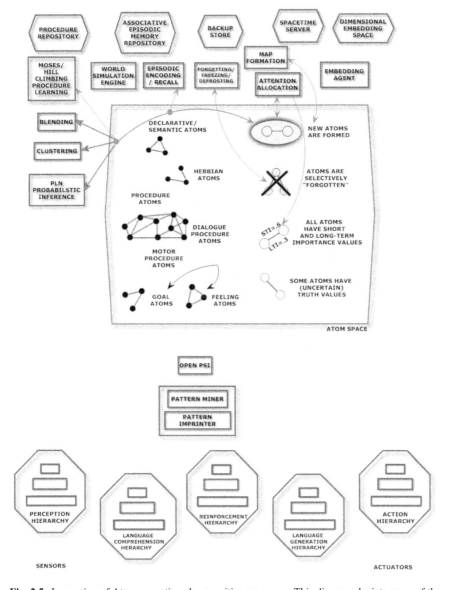

Fig. 2.5 Invocation of Atom operations by cognitive processes. This diagram depicts some of the Atom modification, creation and deletion operations carried out by the abstract cognitive processes in the CogPrime architecture

We suspect that reasonable level of capability will be achievable by simply interposing DeSTIN, as described in Chaps. 8–10 of Part 2 (or some other similar system), as a perception/action "black box" between OpenCog and a robot. Some preliminary experiments in this direction have already been carried out, connecting the

Fig. 2.6 Screenshot of OpenCog-controlled virtual dog

OpenPetBrain to a Nao robot using simpler, less capable software than DeSTIN in
the intermediary role (off-the-shelf speech-to-text, text-to-speech and visual object
recognition software).

However, we also suspect that to achieve robustly intelligent robotics we must
go beyond this approach, and connect robot perception and actuation software with
OpenCogPrime in a "white box" manner that allows intimate dynamic feedback
between perceptual, motoric, cognitive and linguistic functions. We will achieve this
via the creation and real-time utilization of links between the nodes in CogPrime's
and DeSTIN's internal networks (a topic to be explored in more depth later in this
chapter).

2.4 Memory Types and Associated Cognitive Processes
   in CogPrime

Now we return to the basic description of the CogPrime approach, turning to aspects
of the relationship between structure and dynamics. Architecture diagrams are all
very well, but, ultimately it is dynamics that makes an architecture come alive. Intelligence is all about learning, which is by definition about change, about dynamical
response to the environment and internal self-organizing dynamics.

CogPrime relies on multiple memory types and, as discussed above, is founded on the premise that the right course in architecting a pragmatic, roughly human-like AGI system is to handle different types of memory differently in terms of both structure and dynamics.

CogPrime's memory types are the declarative, procedural, sensory, and episodic memory types that are widely discussed in cognitive neuroscience [TC05], plus attentional memory for allocating system resources generically, and intentional memory for allocating system resources in a goal-directed way. Table 2.2 overviews these memory types, giving key references and indicating the corresponding cognitive processes, and also indicating which of the generic patternist cognitive dynamics each cognitive process corresponds to (pattern creation, association, etc.). Figure 2.7 illustrates the relationships between several of the key memory types in the context of a simple situation involving an OpenCogPrime-controlled agent in a virtual world.

In terms of patternist cognitive theory, the multiple types of memory in CogPrime should be considered as specialized ways of storing particular types of patterns, optimized for spacetime efficiency. The cognitive processes associated with a certain type of memory deal with creating and recognizing patterns of the type for which the memory is specialized. While in principle all the different sorts of pattern could be handled in a unified memory and processing architecture, the sort of specialization used in CogPrime is necessary in order to achieve acceptable efficient general intelligence using currently available computational resources. And as we have argued in detail in Chap. 8, efficiency is not a side-issue but rather the essence of real-world AGI (since as Hutter (2005) has shown, if one casts efficiency aside, arbitrary levels of general intelligence can be achieved via a trivially simple program).

The essence of the CogPrime design lies in the way the structures and processes associated with each type of memory are designed to work together in a closely coupled way, yielding cooperative intelligence going beyond what could be achieved by an architecture merely containing the same structures and processes in separate "black boxes".

The inter-cognitive-process interactions in OpenCog are designed so that

- conversion between different types of memory is possible, though sometimes computationally costly (e.g. an item of declarative knowledge may with some effort be interpreted procedurally or episodically, etc.)
- when a learning process concerned centrally with one type of memory encounters a situation where it learns very slowly, it can often resolve the issue by converting some of the relevant knowledge into a different type of memory: i.e. **cognitive synergy**.

Table 2.2 Memory types and cognitive processes in CogPrime

Memory type	Specific cognitive processes	General cognitive functions
Declarative	Probabilistic logic networks (PLN) [GMIH08]; conceptual blending [FT02]	Pattern creation
Procedural	MOSES (a novel probabilistic evolutionary program learning algorithm) [Loo06]	Pattern creation
Episodic	Internal simulation engine [GEA08]	Association, pattern creation
Attentional	Economic attention networks (ECAN) [GPI+10]	Association, credit assignment
Intentional	Probabilistic goal hierarchy refined by PLN and ECAN, structured according to MicroPsi [Bac09]	Credit assignment, pattern creation
Sensory	DeSTIN, and corresponding structures in the atomspace	Association, attention allocation, pattern creation, credit assignment

The third column indicates the general cognitive function that each specific cognitive process carries out, according to the patternist theory of cognition

2.4.1 Cognitive Synergy in PLN

To put a little meat on the bones of the "cognitive synergy" idea, discussed briefly above and more extensively in latter chapters, we now elaborate a little on the role it plays in the interaction between procedural and declarative learning.

While the probabilistic evolutionary learning algorithm MOSES handles much of CogPrime's procedural learning, and CogPrime's internal simulation engine handles most episodic knowledge, CogPrime's primary tool for handling declarative knowledge is an uncertain inference framework called Probabilistic Logic Networks (PLN). The complexities of PLN are the topic of a lengthy technical monograph [GMIH08], and are summarized in Chap. 16 (Part 2); here we will eschew most details and focus mainly on pointing out how PLN seeks to achieve efficient inference control via integration with other cognitive processes.

As a logic, PLN is broadly integrative: it combines certain term logic rules with more standard predicate logic rules, and utilizes both fuzzy truth values and a variant of imprecise probabilities called *indefinite probabilities*. PLN mathematics tells how these uncertain truth values propagate through its logic rules, so that uncertain premises give rise to conclusions with reasonably accurately estimated uncertainty values. This careful management of uncertainty is critical for the application of

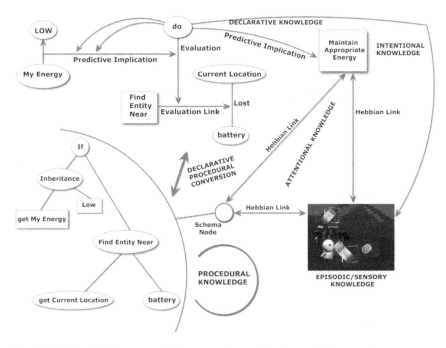

Fig. 2.7 Relationship between multiple memory types. The bottom left corner shows a program tree, constituting procedural knowledge. The upper left shows declarative nodes and links in the Atomspace. The upper right corner shows a relevant system goal. The lower right corner contains an image symbolizing relevant episodic and sensory knowledge. All the various types of knowledge link to each other and can be approximatively converted to each other

logical inference in the robotics context, where most knowledge is abstracted from experience and is hence highly uncertain.

PLN can be used in either forward or backward chaining mode; and in the language introduced above, it can be used for either analysis or synthesis. As an example, we will consider backward chaining analysis, exemplified by the problem of a robot preschool-student trying to determine whether a new playmate "Bob" is likely to be a regular visitor to is preschool or not (evaluating the truth value of the implication *Bob → regular_visitor*). The basic backward chaining process for PLN analysis looks like:

1. Given an implication $L \equiv A \rightarrow B$ whose truth value must be estimated (for instance $L \equiv Concept \wedge Procedure \rightarrow Goal$ as discussed above), create a list $(A_1, ..., A_n)$ of *(inference rule, stored knowledge)* pairs that might be used to produce L
2. Using analogical reasoning to prior inferences, assign each A_i a probability of success

- If some of the A_i are estimated to have reasonable probability of success at generating reasonably confident estimates of L's truth value, then invoke Step 1 with A_i in place of L (at this point the inference process becomes recursive)
- If none of the A_i looks sufficiently likely to succeed, then inference has "gotten stuck" and another cognitive process should be invoked, e.g.
  - **Concept creation** may be used to infer new concepts related to A and B, and then Step 1 may be revisited, in the hope of finding a new, more promising A_i involving one of the new concepts
  - **MOSES** may be invoked with one of several special goals, e.g. the goal of finding a procedure P so that $P(X)$ predicts whether $X \rightarrow B$. If MOSES finds such a procedure P then this can be converted to declarative knowledge understandable by PLN and Step 1 may be revisited....
  - **Simulations** may be run in CogPrime's internal simulation engine, so as to observe the truth value of $A \rightarrow B$ in the simulations; and then Step 1 may be revisited....

The combinatorial explosion of inference control is combatted by the capability to defer to other cognitive processes when the inference control procedure is unable to make a sufficiently confident choice of which inference steps to take next. Note that just as MOSES may rely on PLN to model its evolving populations of procedures, PLN may rely on MOSES to create complex knowledge about the terms in its logical implications. This is just one example of the multiple ways in which the different cognitive processes in CogPrime interact synergetically; a more thorough treatment of these interactions is given in [Goe09a].

In the "new playmate" example, the interesting case is where the robot initially seems not to know enough about Bob to make a solid inferential judgment (so that none of the A_i seem particularly promising). For instance, it might carry out a number of possible inferences and not come to any reasonably confident conclusion, so that the reason none of the A_i seem promising is that all the decent-looking ones have been tried already. So it might then recourse to MOSES, simulation or concept creation.

For instance, the PLN controller could make a list of everyone who has been a regular visitor, and everyone who has not been, and pose MOSES the task of figuring out a procedure for distinguishing these two categories. This procedure could then be used directly to make the needed assessment, or else be translated into logical rules to be used within PLN inference. For example, perhaps MOSES would discover that older males wearing ties tend not to become regular visitors. If the new playmate is an older male wearing a tie, this is directly applicable. But if the current playmate is wearing a tuxedo, then PLN may be helpful via reasoning that even though a tuxedo is not a tie, it's a similar form of fancy dress—so PLN may extend the MOSES-learned rule to the present case and infer that the new playmate is not likely to be a regular visitor.

2.5 Goal-Oriented Dynamics in CogPrime

CogPrime's dynamics has both goal-oriented and "spontaneous" aspects; here for simplicity's sake we will focus on the goal-oriented ones. The basic goal-oriented dynamic of the CogPrime system, within which the various types of memory are utilized, is driven by implications known as "cognitive schematics", which take the form

$$Context \land Procedure \to Goal < p >$$

(summarized $C \land P \to G$). Semi-formally, this implication may be interpreted to mean: "If the context C appears to hold currently, then if I enact the procedure P, I can expect to achieve the goal G with certainty p". Cognitive synergy means that the learning processes corresponding to the different types of memory actively cooperate in figuring out what procedures will achieve the system's goals in the relevant contexts within its environment.

CogPrime's cognitive schematic is significantly similar to production rules in classical architectures like SOAR and ACT-R (as reviewed in Chap. 6; however, there are significant differences which are important to CogPrime's functionality. Unlike with classical production rules systems, uncertainty is core to CogPrime's knowledge representation, and each CogPrime cognitive schematic is labeled with an uncertain truth value, which is critical to its utilization by CogPrime's cognitive processes. Also, in CogPrime, cognitive schematics may be incomplete, missing one or two of the terms, which may then be filled in by various cognitive processes (generally in an uncertain way). A stronger similarity is to MicroPsi's triplets; the differences in this case are more low-level and technical and have already been mentioned in Chap. 6.

Finally, the biggest difference between CogPrime's cognitive schematics and production rules or other similar constructs, is that in CogPrime this level of knowledge representation is not the only important one. CLARION [SZ04], as will be reviewed in Chap. 6 below, is an example of a cognitive architecture that uses production rules for explicit knowledge representation and then uses a totally separate subsymbolic knowledge store for implicit knowledge. In CogPrime both explicit and implicit knowledge are stored in the same graph of nodes and links, with

- explicit knowledge stored in probabilistic logic based nodes and links such as cognitive schematics (see Fig. 2.8 for a depiction of some explicit linguistic knowledge.)
- implicit knowledge stored in patterns of activity among these same nodes and links, defined via the activity of the "importance" values (see Fig. 2.9 for an illustrative example thereof) associated with nodes and links and propagated by the ECAN attention allocation process.

The meaning of a cognitive schematic in CogPrime is hence not entirely encapsulated in its explicit logical form, but resides largely in the activity patterns that ECAN causes its activation or exploration to give rise to. And this fact is important because the synergetic interactions of system components are in large part modulated by

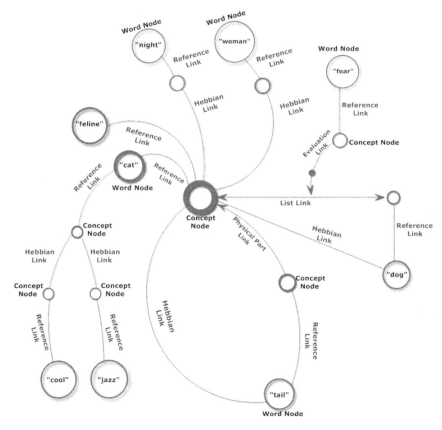

Fig. 2.8 Example of explicit knowledge in the Atomspace. One simple example of explicitly represented knowledge in the Atomspace is linguistic knowledge, such as words and the concepts directly linked to them. Not all of a CogPrime system's concepts correlate to words, but some do

ECAN activity. Without the real-time combination of explicit and implicit knowledge in the system's knowledge graph, the synergetic interaction of different cognitive processes would not work so smoothly, and the emergence of effective high-level hierarchical, heterarchical and self structures would be less likely.

2.6 Analysis and Synthesis Processes in CogPrime

We now return to CogPrime's fundamental cognitive dynamics, using examples from the "virtual dog" application to motivate the discussion.

The cognitive schematic *Context ∧ Procedure → Goal* leads to a conceptualization of the internal action of an intelligent system as involving two key categories of learning:

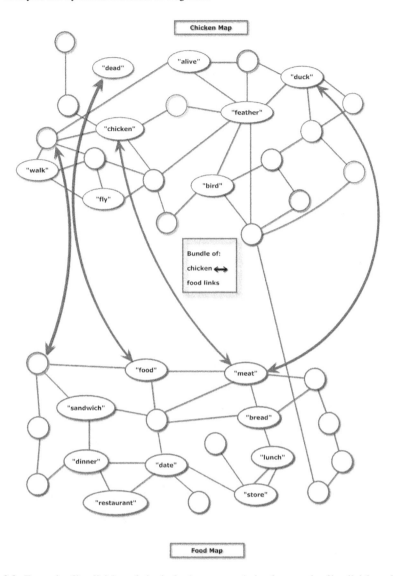

Fig. 2.9 Example of implicit knowledge in the Atomspace. A simple example of implicit knowledge in the Atomspace. The "chicken" and "food" concepts are represented by "maps" of ConceptNodes interconnected by HebbianLinks, where the latter tend to form between ConceptNodes that are often simultaneously important. The bundle of links between nodes in the chicken map and nodes in the food map, represents an "implicit, emergent link" between the two concept maps. This diagram also illustrates "glocal" knowledge representation, in that the chicken and food concepts are each represented by individual nodes, but also by distributed maps. The "chicken" ConceptNode, when important, will tend to make the rest of the map important—and vice versa. Part of the overall chicken concept possessed by the system is expressed by the explicit links coming out of the chicken ConceptNode, and part is represented only by the distributed chicken map as a whole

- **Analysis**: Estimating the probability p of a posited $C \wedge P \to G$ relationship
- **Synthesis**: Filling in one or two of the variables in the cognitive schematic, given assumptions regarding the remaining variables, and directed by the goal of maximizing the probability of the cognitive schematic.

More specifically, where synthesis is concerned,

- The MOSES probabilistic evolutionary program learning algorithm is applied to find P, given fixed C and G. Internal simulation is also used, for the purpose of creating a simulation embodying C and seeing which P lead to the simulated achievement of G.

  – *Example: A virtual dog learns a procedure P to please its owner (the goal G) in the context C where there is a ball or stick present and the owner is saying "fetch".*

- PLN inference, acting on declarative knowledge, is used for choosing C, given fixed P and G (also incorporating sensory and episodic knowledge as appropriate). Simulation may also be used for this purpose.

  – *Example: A virtual dog wants to achieve the goal G of getting food, and it knows that the procedure P of begging has been successful at this before, so it seeks a context C where begging can be expected to get it food. Probably this will be a context involving a friendly person.*

- PLN-based goal refinement is used to create new subgoals G to sit on the right hand side of instances of the cognitive schematic.

  – *Example: Given that a virtual dog has a goal of finding food, it may learn a subgoal of following other dogs, due to observing that other dogs are often heading toward their food.*

- Concept formation heuristics are used for choosing G and for fueling goal refinement, but especially for choosing C (via providing new candidates for C). They are also used for choosing P, via a process called "predicate schematization" that turns logical predicates (declarative knowledge) into procedures.

  – *Example: At first a virtual dog may have a hard time predicting which other dogs are going to be mean to it. But it may eventually observe common features among a number of mean dogs, and thus form its own concept of "pit bull," without anyone ever teaching it this concept explicitly.*

Where analysis is concerned:

- PLN inference, acting on declarative knowledge, is used for estimating the probability of the implication in the cognitive schematic, given fixed C, P and G. Episodic knowledge is also used in this regard, via enabling estimation of the probability via simple similarity matching against past experience. Simulation is also used: multiple simulations may be run, and statistics may be captured therefrom.

- *Example: To estimate the degree to which asking Bob for food (the procedure P is "asking for food", the context C is "being with Bob") will achieve the goal G of getting food, the virtual dog may study its memory to see what happened on previous occasions where it or other dogs asked Bob for food or other things, and then integrate the evidence from these occasions.*

- Procedural knowledge, mapped into declarative knowledge and then acted on by PLN inference, can be useful for estimating the probability of the implication $C \wedge P \rightarrow G$, in cases where the probability of $C \wedge P_1 \rightarrow G$ is known for some P_1 related to P.

  - *Example: knowledge of the internal similarity between the procedure of asking for food and the procedure of asking for toys, allows the virtual dog to reason that if asking Bob for toys has been successful, maybe asking Bob for food will be successful too.*

- Inference, acting on declarative or sensory knowledge, can be useful for estimating the probability of the implication $C \wedge P \rightarrow G$, in cases where the probability of $C_1 \wedge P \rightarrow G$ is known for some C_1 related to C.

  - *Example: if Bob and Jim have a lot of features in common, and Bob often responds positively when asked for food, then maybe Jim will too.*

- Inference can be used similarly for estimating the probability of the implication $C \wedge P \rightarrow G$, in cases where the probability of $C \wedge P \rightarrow G_1$ is known for some G_1 related to G. Concept creation can be useful indirectly in calculating these probability estimates, via providing new concepts that can be used to make useful inference trails more compact and hence easier to construct.

  - *Example: The dog may reason that because Jack likes to play, and Jack and Jill are both children, maybe Jill likes to play too. It can carry out this reasoning only if its concept creation process has invented the concept of "child" via analysis of observed data.*

In these examples we have focused on cases where two terms in the cognitive schematic are fixed and the third must be filled in; but just as often, the situation is that only one of the terms is fixed. For instance, if we fix G, sometimes the best approach will be to collectively learn C and P. This requires either a procedure learning method that works interactively with a declarative-knowledge-focused concept learning or reasoning method; or a declarative learning method that works interactively with a procedure learning method. That is, it requires the sort of cognitive synergy built into the CogPrime design.

2.7 Conclusion

To thoroughly describe a comprehensive, integrative AGI architecture in a brief chapter would be an impossible task; all we have attempted here is a brief overview, to be elaborated on in Part 2 of this book. We do not expect this brief summary to be enough to convince the skeptical reader that the approach described here has a reasonable odds of success at achieving its stated goals. However, we hope to have given the reader at least a rough idea of *what sort of AGI design* we are advocating, and *why and in what sense we believe it can lead to advanced artificial general intelligence.* For more details on the structure, dynamics and underlying concepts of CogPrime, the reader is encouraged to proceed to Part 2—after completing Part 1, of course. Please be patient—building a thinking machine is a big topic, and we have a lot to say about it!

Chapter 3
Build Me Something I Haven't Seen:
A CogPrime Thought Experiment

3.1 Introduction

AGI design necessarily leads one into some rather abstract spaces—but being a human-like intelligence in the everyday world is a pretty concrete thing. If the Cog-Prime research program is successful, it will result not just in abstract ideas and equations, but rather in real AGI robots carrying out tasks in the world, and AGI agents in virtual worlds and online digital spaces conducting important business, do-ing science, entertaining and being entertained by us, and so forth. With this in mind, in this chapter we will ground the discussion firmly in the concrete and everyday, and pursue a thought experiment of the form "How would a completed CogPrime system carry out this specific task?"

The task we will use for this thought-experiment is one we will a running example now and then in the following chapters. We consider the case of a robotically or virtually embodied CogPrime system, operating in a preschool type environment, interacting with a human whom it already knows and given the task of "Build me something with blocks that I haven't seen before."

This target task is fairly simple, but it is complex enough to involve essentially every one of CogPrime's processes, interacting in a unified way. It involves simple, grounded creativity of the sort that normal human children display every day—and which, we conjecture, is structurally and dynamically basically the same as the creativity underlying the genius of adult human creators like Einstein, Dali, Dostoevsky, Hendrix, and so forth ... and as the creativity that will power massively capable genius machines in future.

We will consider the case of a simple interaction based on the above task where:

1. The human teacher tells the CogPrime agent "Build me something with blocks that I haven't seen before."
2. After a few false starts, the agent builds something it thinks is appropriate and says "Do you like it?"
3. The human teacher says "It's beautiful. What is it?"

B. Goertzel et al., *Engineering General Intelligence, Part 1*,
Atlantis Thinking Machines 5, DOI: 10.2991/978-94-6239-027-0_3,
© Atlantis Press and the authors 2014

4. The agent says "It's a car man" [and indeed, the construct has 4 wheels and a chassis vaguely like a car, but also a torso, arms and head vaguely like a person].

Of course, a complex system like CogPrime could carry out an interaction like this internally in many different ways, and what is roughly described here is just one among many possibilities.

First we will enumerate a number of CogPrime processes and explain some ways that each one may help CogPrime carry out the target task. Then we will give a more evocative narrative, conveying the dynamics that would occur in CogPrime while carrying out the target task, and mentioning how each of the enumerated cognitive processes as it arises in the narrative.

Coming as it does at the beginning of the book, yet referring to numerous cognitive processes and structures that will only be defined later in the book, the chapter will necessarily be somewhat opaque to the reader with no prior exposure to CogPrime ideas. However, it has been placed here at the start for a reason—so serve as motivation and conceptual guidance for the complex and voluminous material to follow. You may wish to skim this chapter over relatively lightly the first time around, getting a general idea of how the different processes and structures in CogPrime fit together in a practical context—and then return to the chapter again once you've finished with Part Two (Part 2) and have a fuller understanding of what all the different parts of the design are supposed to do and how they're supposed to work.

3.2 Roles of Selected Cognitive Processes

Now we review a number of the more interesting CogPrime cognitive processes to be reviewed in the following chapters of the book, for each one indicating one or more of the roles it might play in helping a CogPrime system carry out the target task. Note that this list is incomplete in many senses, e.g. it doesn't list all the cognitive processes, nor all the roles played by the ones listed. The purpose is to give an evocative sense of the roles played by the different parts of the design in carrying out the task.

- Chapter 1 Part 2 (OpenCog Framework)

  - **Freezing/Defrosting**.
    When the agent builds a structure from blocks and decides it's not good enough to show off to the teacher, what does it do with the detailed ideas and thought process underlying the structure it built? If it doesn't like the structure so much, it may just leave this to the generic forgetting process. But if it likes the structure a lot, it may want to increase the VLTI (Very Long Term Importance) of the Atoms related to the structure in question, to be sure that these are stored on disk or other long-term storage, even after they're deemed sufficiently irrelevant to be pushed out of RAM by the forgetting mechanism.

When given the target task, the agent may decide to revive from disk the mind-states it went through when building crowd-pleasing structures from blocks before, so as to provide it with guidance.

- Chapter 4 of Part 2 (Emotion, Motivation, Attention and Control)

  - **Cognitive Cycle**.
    While building with blocks, the agent's cognitive cycle will be dominated by perceiving, acting on, and thinking about the blocks it is building with.
    When interacting with the teacher, then interaction-relevant linguistic, perceptual and gestural processes will also enter into the cognitive cycle.

  - **Emotion**. The agent's emotions will fluctuate naturally as it carries out the task.
    If it has a goal of pleasing the teacher, then it will experience happiness as its expectation of pleasing the teacher increases.
    If it has a goal of experiencing novelty, then it will experience happiness as it creates structures that are novel in its experience.
    If it has a goal of learning, then it will experience happiness as it learns new things about blocks construction.
    On the other hand, it will experience unhappiness as its experienced or predicted satisfaction of these goals decreases.

  - **Action Selection**
    In dialoguing with the teacher, action selection will select one or more DialogueController schema to control the conversational interaction (based on which DC schema have proved most effective in prior similar situations.
    · When the agent wants to know the teacher's opinion of its construct, what is happening internally is that the "please teacher" Goal Atom gets a link of the conceptual form (Implication "find out teacher's opinion of my current construct" "please teacher"). This link may be created by PLN inference, probably largely by analogy to previously encountered similar situations. Then, GoalImportance is spread from the "please teacher" Goal Atom to the "find out teacher's opinion of my current construct" Atom (via the mechanism of sending an RFS package to the latter Atom). More inference causes a link (Implication "ask the teacher for their opinion of my current construct" "find out teacher's opinion of my current construct") to be formed, and the "ask the teacher for their opinion of my current construct" Atom to get GoalImportance also. Then Predicate Schematization causes the predicate "ask the teacher for their opinion of my current construct" to get turned into an actionable schema, which gets GoalImportance, and which gets pushed into the ActiveSchemaPool via Goal-driven action selection. Once the schema version of "ask the teacher for their opinion of my current construct" is in the ActiveSchemaPool, it then invokes natural language generation Tasks, which lead to the formulation of an English sentence such as "Do you like it?"

· When the teacher asks "It's beautiful. What is it?", then the NL comprehension MindAgent identifies this as a question, and the "please teacher" Goal Atom gets a link of the conceptual form (Implication "answer the question the teacher just asked" "please teacher"). This follows simply from the knowledge (Implication ("teacher has just asked a question" AND "I answer the teacher's question") ("please teacher")), or else from more complex knowledge refining this Implication. From this point, things proceed much as in the case "Do you like it?" described just above. Consider a schema such as "pick up a red cube and place it on top of the long red block currently at the top of the structure" (let's call this P). Once P is placed in the ActiveSchemaPool, then it runs and generates more specific procedures, such as the ones needed to find a red cube, to move the agent's arm toward the red cube and grasp it, etc. But the execution of these specific low-level procedures is done via the ExecutionManager, analogously to the execution of the specifics of generating a natural language sentence from a collection of semantic relationships. Loosely speaking, reaching for the red cube and turning simple relationships into a simple sentences, are considered as "automated processes" not requiring holistic engagement of the agent's mind. What the generic, more holistic Action Selection mechanism does in the present context is to figure out to put P in the ActiveSchemaPool in the first place. This occurs because of a chain such as: P predictively implies (with a certain probabilistic weight) "completion of the car-man structure", which in turn predictively implies "completion of a structure that is novel to the teacher", which in turn predictively implies "please the teacher", which in turn implies "please others", which is assumed an Ubergoal (a top-level system goal).

– **Goal Atoms**. As the above items make clear, the scenario in question requires the initial Goal Atoms to be specialized, via the creation of more and more particular subgoals suiting the situation at hand.

– **Context Atoms**.

Knowledge of the context the agent is in can help it disambiguate language it hears, e.g. knowing the context is blocks-building helps it understand which sense of the word "blocks" is meant.

On the other hand, if the context is that the teacher is in a bad mood, then the agent might know via experience that in this context, the strength of (Implication "ask the teacher for their opinion of my current construct" "find out teacher's opinion of my current construct") is lower than in other contexts.

– **Context Formation**.

A context like blocks-building or "teacher in a bad mood" may be formed by clustering over multiple experience-sets, i.e. forming Atoms that refer to spatiotemporally grouped sets of percepts/concepts/actions, and grouping together similar Atoms of this nature into clusters.

The Atom referring to the cluster of experience-sets involving blocks-building will then survive as an Atom if it gets involved in relationships that are important or have surprising truth values. If many relationships have significantly different truth-value inside the blocks-building context than outside it, this means it's likely that the blocks-building ConceptNode will remain as an Atom with reasonably high LTI, so it can be used as a context in future.

- **Time-Dependence of Goals**. Many of the agent's goals in this scenario have different importances over different time scales. For instance "please the teacher" is important on multiple time-scales: the agent wants to please the teacher in the near term but also in the longer term. But a goal like "answer the question the teacher just asked" has an intrinsic time-scale to it; if it's not fulfilled fairly rapidly then its importance goes away.

- Chapter 5 of Part 2 (Attention allocation)

  - **ShortTermImportance versus LongTermImportance**. While conversing, the concepts and immediately involved in the conversation (including the Atoms describing the agents in the conversation) have very high STI. While building, Atoms representing to the blocks and related ideas about the structures being built (e.g. images of cars and people perceived or imagined in the past) have very high STI. But the reason these Atoms are in RAM prior to having their STI boosted due to their involvement in the agent's activities, is because they had their LTI boosted at some point in the past. And after these Atoms leave the AttentionalFocus and their STI reduces, they will have boosted LTI and hence likely remain in RAM for a long while, to be involved in "background thought", and in case they're useful in the AttentionalFocus again.

  - **HebbianLink Formation**. As a single example, the car-man has both wheels and arms, so now a Hebbian association between wheels and arms will exist in the agent's memory, to potentially pop up again and guide future thinking. The very idea of a car-man likely emerged partly due to previously formed HebbianLinks—because people were often seen sitting in cars, the association between person and car existed, which made the car concept and the human concept natural candidates for blending.

  - **Data Mining the System Activity Table**. The HebbianLinks mentioned above may have been formed via mining the SystemActivityTable

  - **ECAN Based Associative Memory**. When the agent thinks about making a car, this spreads importance to various Atoms related to the car concept, and one thing this does is lead to the emergence of the car attractor into the AttentionalFocus. The different aspects of a car are represented by heavily interlinked Atoms, so that when some of them become important, there's a strong tendency for the others to also become important—and for "car" to then emerge as an attractor of importance dynamics.

  - **Schema Credit Assignment**.

    Suppose the agent has a subgoal of placing a certain blue block on top of a certain red block. It may use a particular motor schema for carrying out this action—involving, for instance, holding the blue block above the red block and then gradually lowering it. If this schema results in success (rather than in, say, knocking down the red block), then it should get rewarded via having its STI and LTI boosted and also having the strength of the link between it and the subgoal increased.

Next, suppose that a certain cognitive schema (say, the schema of running multiple related simulations and averaging the results, to estimate the success probability of a motor procedure) was used to arrive at the motor schema in question. Then this cognitive schema may get passed some importance from the motor schema, and it will get the strength of its link to the goal increased. In this way credit passes backwards from the goal to the various schema directly or indirectly involved in fulfilling it.

– **Forgetting**. If the agent builds many structures from blocks during its lifespan, it will accumulate a large amount of perceptual memory.

- Chapter 6 of Part 2 (Goal and Action Selection). Much of the use of the material in this chapter was covered above in the bullet point for Chap. 6 of Part 2, but a few more notes are:

  – **Transfer of RFS Between Goals**. Above it was noted that the link (Implication "ask the teacher for their opinion of my current construct" "find out teacher's opinion of my current construct") might be formed and used as a channel for GoalImportance spreading.

  – **Schema Activation**. Supposing the agent is building a man-car, it may have car-building schema and man-building schema in its ActiveSchemaPool at the same time, and it may enact both of them in an interleaved manner. But if each tend to require two hands for their real-time enaction, then schema activation will have to pass back and forth between the two of them, so that at any one time, one is active whereas the other one is sitting in the ActiveSchemaPool waiting to get activated.

  – **Goal Based Schema Learning**. To take a fairly low-level example, suppose the agent has the (sub)goal of making an arm for a blocks-based person (or man-car), given the presence of a blocks-based torso. Suppose it finds a long block that seems suitable to be an arm. It then has the problem of figuring out how to attach the arm to the body. It may try out several procedures in its internal simulation world, until it finds one that works: *hold the arm in the right position while one end of it rests on top of some block that is part of the torso, then place some other block on top of that end, then slightly release the arm and see if it falls. If it doesn't fall, leave it. If it seems about to fall, then place something heavier atop it, or shove it further in toward the center of the torso.* The procedure learning process could be MOSES here, or it could be PLN.

- Chapter 7 of Part 2 (Procedure Evaluation)

  – **Inference Based Procedure Evaluation**. A procedure for man-building such as "first put up feet, then put up legs, then put up torso, then put up arms and head" may be synthesized from logical knowledge (via predicate schematization) but without filling in the details of how to carry out the individual steps, such as "put up legs." If a procedure with abstract (ungrounded) schema like PutUpTorso is chosen for execution and placed into the ActiveSchemaPool, then in the course of execution, inferential procedure evaluation must be used to figure out how to make the abstract schema actionable. The GoalDrivenActionSelection MindAgent must make the choice whether to put a not-fully-grounded schema into the ActiveSchemaPool, rather than grounding it first and then making it active; this is the sort of choice that may be made effectively via learned cognitive schema.

- Chapter 8 of Part 2 (Perception and Action)

  - **ExperienceDB**. No person remembers every blocks structure they ever saw or built, except maybe some autists. But a CogPrime can store all this information fairly easily, in its ExperienceDB, even if it doesn't keep it all in RAM in its AtomSpace. It can also store everything anyone ever said about blocks structures in its vicinity.
  - **Perceptual Pattern Mining**.
  - **Object Recognition**. Recognizing structures made of blocks as cars, people, houses, etc. requires fairly abstract object recognition, involving identifying the key shapes and features involved in an object-type, rather than just going by simple visual similarity.
  - **Hierarchical Perception Networks**. If the room is well-lit, it's easy to visually identify individual blocks within a blocks structure. If the room is darker, then more top-down processing may be needed—identifying the overall shape of the blocks structure may guide one in making out the individual blocks.
  - **Hierarchical Action Networks**. Top-down action processing tells the agent that, if it wants to pick up a block, it should move its arm in such a way as to get its hand near the block, and then move its hand. But if it's still learning how to do that sort of motion, more likely it will do this, but then start moving its hand and find that it's hard to get a grip on the block—and then have to go back and move its arm a little differently. Iterating between broader arm/hand movements and more fine-grained hand/finger movements is an instance of information iteratively passing up and down a hierarchical action network.
  - **Coupling of Perception and Action Networks**. Picking up a block in the dark is a perfect example of rich coupling of perception and action networks. Feeling the block with the fingers helps with identifying blocks that can't be clearly seen.

- Chapter 12 of Part 2 (Procedure Learning)

  - **Specification Based Procedure Learning**.
    Suppose the agent has never seen a horse, but the teacher builds a number of blocks structures and calls them horses, and draws a number of pictures and calls them horses. This may cause a procedure learning problem to be spawned, where the fitness function is accuracy at distinguishing horses from non-horses.

Learning to pick up a block is specification-based procedure learning, where the specification is to pick up the block and grip it and move it without knocking down the other stuff near the block.

- **Representation Building**.

    In the midst of building a procedure to recognize horses, MOSES would experimentally vary program nodes recognizing visual features into other program nodes recognizing other visual features

    In the midst of building a procedure to pick up blocks, MOSES would experimentally vary program nodes representing physical movements into other nodes representing physical movements

    In both of these cases, MOSES would also carry out the standard experimental variations of mathematical and control operators according to its standard representation-building framework

- Chapter 13 Part 2 (Imitative, Reinforcement and Corrective Learning)

    - **Reinforcement Learning**.

        Motor procedures for placing blocks (in simulations or reality) will get rewarded if they don't result in the blocks structure falling down, punished otherwise.

        Procedures leading to the teacher being pleased, in internal simulations (or in repeated trials of scenarios like the one under consideration), will get rewarded; procedures leading to the teacher being displeased will get punished.

    - **Imitation Learning**. If the agent has seen others build with blocks before, it may summon these memories and then imitate the actions it has seen others take.

    - **Corrective Learning**. This would occur if the teacher intervened in the agent's block-building and guided him physically—e.g. steadying his shaky arm to prevent him from knocking the blocks structure over.

- Chapter 14 of Part 2 (Hillclimbing)

    - **Complexity Penalty**. In learning procedures for manipulating blocks, the complexity penalty will militate against procedures that contain extraneous steps.

- Chapter 15 of Part 2 (Probabilistic Evolutionary Procedure Learning)

    - **Supplying Evolutionary Learning with Long-Term Memory**. Suppose the agent has previously built people from clay, but never from blocks. It may then have learned a "classification model" predicting which clay people will look appealing to humans, and which won't. It may then transfer this knowledge, using PLN, to form a classification model predicting which blocks-people will look appealing to humans, and which won't.

    - **Fitness Function Estimation via Integrative Intelligence**. To estimate the fitness of a procedure for, say, putting an arm on a blocks-built human, the agent may try out the procedure in the internal simulation world; or it may use PLN inference to reason by analogy to prior physical situations it's observed. These allow fitness to be estimated without actually trying out the procedure in the environment.

- Chapter 16 of Part 2 (Probabilistic Logic Networks)

  – **Deduction**. This is a tall skinny structure; tall skinny structures fall down easily; thus this structure may fall down easily.
  – **Induction**. This teacher is talkative; this teacher is friendly; therefore the talkative are generally friendly.
  – **Abduction**. This structure has a head and arms and torso; a person has a head and arms and torso; therefore this structure is a person.
  – **PLN Forward Chaining**. What properties might a car-man have, based on inference from the properties of cars and the properties of men?
  – **PLN Backward Chaining**.
      An inference target might be: *Find X so that X looks something like a wheel and can be attached to this blocks-chassis, and I can find four fairly similar copies.*
      Or: *Find the truth value of the proposition that this structure looks like a car.*
  – **Indefinite Truth Values**. Consider the deductive inference "This is a tall skinny structure; tall skinny structures fall down easily; thus this structure may fall down easily." In this case, the confidence of the second premise may be greater than the confidence of the first premise, which may result in an intermediate confidence for the conclusion, according to the propagation of indefinite probabilities through the PLN deduction rule.
  – **Intensional Inference**. Is the blocks-structure a person? According to the definition of intensional inheritance, it shares many informative properties with people (e.g. having arms, torso and head), so to a significant extent, it is a person.
  – **Confidence Decay**. The agent's confidence in propositions regarding building things with blocks should remain nearly constant. The agent's confidence in propositions regarding the teacher's taste should decay more rapidly. This should occur because the agent should observe that, in general, propositions regarding physical object manipulation tend to retain fairly constant truth value, whereas propositions regarding human tastes tend to have more rapidly decaying truth value.

- Chapter 17 Part 2 (Spatiotemporal Inference)

  – **Temporal Reasoning**. Suppose, after the teacher asks "What is it?", the agent needs to think a while to figure out a good answer. But maybe the agent knows that it's rude to pause too long before answering something to a direct question. Temporal reasoning helps figure out "how long is too long" to wait before answering.
  – **Spatial Reasoning**. Suppose the agent puts shoes on the wheels of the car. This is a joke relying on the understanding that wheels hold a car up, whereas feet hold a person up, and the structure is a car-man. But it also relies on the spatial inferences that: the car's wheels are in the right position for the man's feet (below

the torso); and, the wheels are below the car's chassis just like a person's feet are below its torso.

- Chapter 18 of Part 2 (Inference Control)

  - **Evaluator Choice as a Bandit Problem**. In doing inference regarding how to make a suitably humanlike arm for the blocks-man, there may be a choice between multiple inference pathways, perhaps one that relies on analogy to other situations building arms, versus one that relies on more general reasoning about lengths and weights of blocks. The choice between these two pathways will be made randomly with a certain probabilistic bias assigned to each one, via prior experience.
  - **Inference Pattern Mining**. The probabilities used in choosing which inference path to take, are determined in part by prior experience—e.g. maybe it's the case that in prior situations of building complex blocks structures, analogy has proved a better guide than naive physics, thus the prior probability of the analogy inference pathway will be nudged up.
  - **PLN and Bayes Nets**. What's the probability that the blocks-man's hat will fall off if the man-car is pushed a little bit to simulate driving? This question could be resolved in many ways (e.g. by internal simulation), but one possibility is inference. If this is resolved by inference, it's the sort of conditional probability calculation that could potentially be done faster if a lot of the probabilistic knowledge from the AtomSpace were summarized in a Bayes Net. Updating the Bayes net structure can be slow, so this is probably not appropriate for knowledge that is rapidly shifting; but knowledge about properties of blocks structures may be fairly persistent after the agent has gained a fair bit of knowledge by playing with blocks a lot.

- Chapter 19 of Part 2 (Pattern Mining)

  - **Greedy Pattern Mining**.
    "Push a tall structure of blocks and it tends to fall down" is the sort of repetitive pattern that could easily be extracted from a historical record of perceptions and (the agent's and others') actions via simple greedy pattern mining algorithm.
    If there is a block that is shaped like a baby's rattle, with a long slender handle and then a circular shape at the end, then greedy pattern mining may be helpful due to having recognized the pattern that structures like this are sometimes rattles—and also that structures like this are often stuck together, with the handle part connected sturdily to the circular part.
  - **Evolutionary Pattern Mining**. "Push a tall structure of blocks with a wide base and a gradual narrowing toward the top and it may not fall too badly" is a more complex pattern that may not be found via greedy mining, unless the agent has dealt with a lot of pyramids.

- Chapter 20 of Part 2 (Concept Formation)

  - **Formal Concept Analysis**. Suppose there are many long, slender blocks of different colors and different shapes (some cylindrical, some purely rectangular for example). Learning this sort of concept based on common features is exactly what FCA is good at (and when the features are defined fuzzily or probabilistically, it's exactly what uncertain FCA is good at). Learning the property of "slender" itself is another example of something uncertain FCA is good at—it would learn this if there were many concepts that preferentially involved slender things (even though formed on the basis of concepts other than slenderness)
  - **Conceptual Blending**. The concept of a "car-man" or "man-car" is an obvious instance of conceptual blending. The agents know that building a man won't surprise the teacher, and nor will building a car ... but both "man" and "car" may pop to the forefront of its mind (i.e. get a briefly high STI) when it thinks about what to build. But since it knows it has to do something new or surprising, there may be a cognitive schema that boosts the amount of funds to the ConceptBlending MindAgent, causing it to be extra-active. In any event, the ConceptBlending agent seeks to find ways to combine important concepts; and then PLN explores these to see which ones may be able to achieve the given goal of surprising the teacher (which includes subgoals such as actually being buildable).

- Chapter 21 of Part 2 (Dimensional Embedding)

  - **Dimensional Embedding**. When the agent needs to search its memory for a previously seen blocks structure similar to the currently observed one—or for a previously articulated thought similar to the one it's currently trying to articulate—then it needs to a search through its large memory for "an entity similar to X" (where X is a structure or a thought). This kind of search can be quite computationally difficult—but if the entities in question have been projected into an embedding space, then it's quite rapid. (The cost is shifted to the continual maintenance of the embedding space, and its periodic updating; and there is some error incurred in the projection, but in many cases this error is not a show-stopper).
  - **Embedding Based Inference Control**. Rapid search for answers to similarity or inheritance queries can be key for guiding inference in appropriate directions; for instance reasoning about how to build a structure with certain properties, can benefit greatly from rapid search for previously-encountered substructures currently structurally or functionally similar to the substructures one desires to build.

- Chapter 22 of Part 2 (Simulation and Episodic Memory)

  - **Fitness Estimation via Simulation**. One way to estimate whether a certain blocks structure is likely to fall down or not, is to build it in one's "mind's eye" and see if the physics engine in one's mind's-eye causes it to fall down. This is something that in many cases will work better for CogPrime than for humans, because CogPrime has a more mathematically accurate physics engine than the human mind does; however, in cases that rely heavily on naive physics rather than, say, direct applications of Newton's Laws, then CogPrime's simulation engine may underperform the typical human mind.

– **Concept Formation via Simulation**. Objects may be joined into categories using uncertain FCA, based on features that they are identified to have via "simulation experiments" rather than physical world observations. For instance, it may be observed that pyramid-shaped structures fall less easily than pencil-shaped tower structures—and the concepts corresponding to these two categories may be formed—from experiments run in the internal simulation world, perhaps inspired by isolated observations in the physical world.

– **Episodic Memory**. Previous situations in which the agent has seen similar structures built, or been given similar problems to solve, may be brought to mind as "episodic movies" playing in the agent's memory. By watching what happens in these replayed episodic movies, the agent may learn new declarative or procedural knowledge about what to do. For example, maybe there was some situation in the agent's past where it saw someone asked to do something surprising, and that someone created something funny. This might (via a simple PLN step) bias the agent to create something now, which it has reason to suspect will cause others to laugh.

- Chapter 23 of Part 2 (Integrative Procedure Learning)

  – **Concept-Driven Procedure Learning**. Learning the concept of "horse", as discussed above in the context of Chap. 12 of Part 2, is an example of this.
  – **Predicate Schematization**. The synthesis of a schema for man-building, as discussed above in the context of Chap. 7 of Part 2, is an example of this.

- Chapter 24 of Part 2 (Map Formation)

  – **Map Formation**. The notion of a car involves many aspects: the physical appearance of cars, the way people get in and out of cars, the ways cars drive, the noises they make, etc. All these aspects are represented by Atoms that are part of the car map, and are richly interconnected via HebbianLinks as well as other links.
  – **Map Encapsulation**. The car map forms implicitly via the interaction of multiple cognitive dynamics, especially ECAN. But then the Anticoagulation MindAgent may do its pattern mining and recognize this map explicitly, and form a PredicateNode encapsulating it. This PredicateNode may then be used in PLN inference, conceptual blending, and so forth (e.g. helping with the formation of a concept like car-man via blending).

- Chapter 26 of Part 2 (Natural Language Comprehension)

  – **Experience Based Disambiguation**. The particular dialogue involved in the present example doesn't require any nontrivial word sense disambiguation. But it does require parse selection, and semantic interpretation selection:

In "Build me something with blocks," the agent has no trouble understanding that "blocks" means "toy building blocks" rather than, say, "city blocks", based on many possible mechanisms, but most simply importance spreading. "Build me something with blocks" has at least three interpretations: the building could be carried out using blocks with a tool; or the thing built could be presented alongside blocks; or the thing built could be composed of blocks. The latter is the most commonsensical interpretation for most humans, but that is because we have heard the phrase "building with blocks" used in a similarly grounded way before (as well as other similar phrases such as "playing with Legos", etc., whose meaning helps militate toward the right interpretation via PLN inference and importance spreading). So here we have a simple example of experience-based disambiguation, where experiences at various distances of association from the current one are used to help select the correct parse.

A subtler form of semantic disambiguation is involved in interpreting the clause "that I haven't seen before." A literal-minded interpretation would say that this requirement is fulfilled by any blocks construction that's not precisely identical to one the teacher has seen before. But of course, any sensible human knows this is an idiomatic clause that means "significantly different from anything I've seen before." This could be determined by the CogPrime agent if it has heard the idiomatic clause before, or if it's heard a similar idiomatic phrase such as "something I've never done before." Or, even if the agent has never heard such an idiom before, it could potentially figure out the intended meaning simply because the literal-minded interpretation would be a pointless thing for the teacher to say. So if it knows the teacher usually doesn't add useless modificatory clauses onto their statements, then potentially the agent could guess the correct meaning of the phrase.

- Chapter 28 of Part 2 (Language Generation)

  - **Experience-Based Knowledge Selection for Language Generation**. When the teacher asks "What is it?", the agent must decide what sort of answer to give. Within the confines of the QuestionAnswering DialogueController, the agent could answer "A structure of blocks", or "A part of the physical world", or "A thing", or "Mine." (Or, if it were running another DC, it could answer more broadly, e.g. "None of your business," etc.). However, the QA DC tells it that, in the present context, the most likely desired answer is one that the teacher doesn't already know; and the most important property of the structure that the teacher doesn't obviously already know is the fact that it depicts a "car man." Also, memory of prior conversations may bring up statements like "It's a horse" in reference to a horse built of blocks, or a drawing of a horse, etc.

  - **Experience-Based Guidance of Word and Syntax Choice**. The choice of phrase "car man" requires some choices to be made. The agent could just as well say "It's a man with a car for feet" or "It's a car with a human upper body and head" or "It's a car centaur", etc. A bias toward simple expressions would lead to "car man." If the teacher were known to prefer complex expressions, then the agent might be biased toward expressing the idea in a different way.

- Chapter 30 of Part 2 (Natural Language Dialogue)

  - **Adaptation of Dialogue Controllers**. The QuestionAsking and QuestionAnswering DialogueControllers both get reinforcement from this interaction, for the specific internal rules that led to the given statements being made.

3.3 A Semi-Narrative Treatment

Now we describe how a CogPrime system might carry out the specified task in a semi-narrative form, weaving in the material from the previous section as we go along, and making some more basic points as well. The semi-narrative covers most but not all of the bullet points from the previous section, but with some of the technical details removed; and it introduces a handful of new examples not given in the bullet points.

The reason this is called a semi-narrative rather than a narrative is that there is no particular linear order to the processes occurring in each phase of the situation described here. CogPrime's internal cognitive processes do not occur in a linear narrative; rather, what we have is a complex network of interlocking events. But still, describing some of these events concretely in a manner correlated with the different stages of a simple interaction, may have some expository value.

The human teacher tells the CogPrime agent "Build me something with blocks that I haven't seen before."

Upon hearing this, the agent's cognitive cycles are dominated by language processing and retrieval from episodic and sensory memory.

The agent may decide to revive from disk the mind-states it went through when building human-pleasing structures from blocks before, so as to provide it with guidance

It will likely experience the emotion of happiness, because it anticipates the pleasure of getting rewarded for the task in future.

The ubergoal of pleasing the teacher gets active (gets funded significantly with STI currency), as it becomes apparent there are fairly clear ways of fulfilling that goal (via the subgoal S of building blocks structures that will get positive response from the teacher). Other ubergoals like gaining knowledge are not funded as much with STI currency just now, as they are not immediately relevant.

Action selection, based on ImplicationLinks derived via PLN (between various possible activities and the subgoal S) causes it to start experimentally building some blocks structures. Past experience with building (turned into ImplicationLinks via mining the SystemActivityTable) tells it that it may want to build a little bit in its internal simulation world before building in the external world, causing STI currently to flow to the simulation MindAgent.

The Atom corresponding to the context blocks-building gets high STI and is pushed into the AttentionalFocus, making it likely that many future inferences will occur in this context. Other Atoms related to this one also get high STI (the ones in the blocks-building map, and others that are especially related to blocks-building in this particular context).

After a few false starts, the agent builds something it thinks is appropriate and says "Do you like it?"

Now that the agent has decided what to do to fulfill its well-funded goal, its cognitive cycles are dominated by action, perception and related memory access and concept creation.

An obvious subgoal is spawned: build a new structure now, and make this particular structure under construction appealing and novel to the teacher. This subgoal has a shorter time scale than the high level goal. The subgoal gets some currency from its supergoal using the mechanism of RFS spreading.

Action selection must tell it when to continue building the same structure and when to try a new one, as well as more micro level choices.

Atoms related to the currently pursued blocks structure get high STI.

After a failed structure (a "false start") is disassembled, the corresponding Atoms lose STI dramatically (leaving AF) but may still have significant LTI, so they can be recalled later as appropriate. They may also have VLTI so they will be saved to disk later on if other things push them out of RAM due to getting higher LTI.

Meanwhile everything that's experienced from the external world goes into the ExperienceDB.

Atoms representing different parts of aspects of the same blocks structure will get Hebbian links between them, which will guide future reasoning and importance spreading.

Importance spreading helps the system go from an idea for something to build (say, a rock or a car) to the specific plans and ideas about how to build it, via increasing the STI of the Atoms that will be involved in these plans and ideas.

If something apparently good is done in building a blocks structure, then other processes and actions that helped lead to or support that good thing, get passed some STI from the Atoms representing the good thing, and also may get linked to the Goal Atom representing "good" in this context. This leads to reinforcement learning.

The agent may play with building structures and then seeing what they most look like, thus exercising abstract object recognition (that uses procedures learned by MOSES or hillclimbing, or uncertain relations learned by inference, to guess what object category a given observed collection of percepts most likely falls into).

Since the agent has been asked to come up with something surprising, it knows it should probably try to formulate some new concepts—because it has learned in the past, via SystemActivityTable mining, that often newly formed concepts are surprising to others. So, more STI currency is given to concept formation MindAgents, such as the ConceptualBlending Mind Agent (which, along with a lot of stuff that gets thrown out or stored for later use, comes up with "car-man").

When the notion of "car" is brought to mind, the distributed map of nodes corresponding to "car" get high STI. When car-man is formed, it is reasoned about (producing new Atoms), but it also serves as a nexus of importance-spreading, causing the creation of a distributed car-man map.

If the goal of making an arm for a man-car occurs, then goal-driven schema learning may be done to learn a procedure for arm-making (where the actual learning is done by MOSES or hill-climbing).

If the agent is building a man-car, it may have man-building and car-building schema in its ActiveSchemaPool at the same time, and SchemaActivation may spread back and forth between the different modules of these two schema.

If the agent wants to build a horse, but has never seen a horse made of blocks (only various pictures and movies of horses), it may uses MOSES or hillclimbing internally to solve the problem of creating a horse-recognizer or a horse-generator which embodies appropriate abstract properties of horses. Here as in all cases of procedure learning, a complexity penalty rewards simpler programs, from among all programs that approximately fulfill the goals of the learning process.

If a procedure being executed has some abstract parts, then these may be executed by inferential procedure evaluation (which makes the abstract parts concrete on the fly in the course of execution).

To guess the fitness of a procedure for doing something (say, building an arm or recognizing a horse), inference or simulation may be used, as well as direct evaluation in the world.

Deductive, inductive and abductive PLN inference may be used in figuring out what a blocks structure will look or act like before building it (it's tall and thin so it may fall down; it won't be bilaterally symmetric so it won't look much like a person; etc.)

Backward-chaining inference control will help figure out how to assemble something matching a certain specification e.g. how to build a chassis based on knowledge of what a chassis looks like. Forward chaining inference (critically including intensional relationships) will be used to estimate the properties that the teacher will perceive a given specific structure to have. Spatial and temporal algebra will be used extensively in this reasoning, within the PLN framework.

Coordinating different parts of the body—say an arm and a hand—will involve importance spreading (both up and down) within the hierarchical action network, and from this network to the hierarchical perception network and the heterarchical cognitive network.

In looking up Atoms in the AtomSpace, some have truth values whose confidences have decayed significantly (e.g. those regarding the teacher's tastes), whereas others have confidences that have hardly decayed at all (e.g. those regarding general physical properties of blocks).

Finding previous blocks structures similar to the current one (useful for guiding building by analogy to past experience) may be done rapidly by searching the system's internal dimensional-embedding space.

As the building process occurs, patterns mined via past experience (tall things often fall down) are used within various cognitive processes (reasoning, procedure learning, concept creation, etc.); and new pattern mining also occurs based on the new observations made as different structures are build and experimented with and destroyed.

Simulation of teacher reactions, based on inference from prior examples, helps with the evaluation of possible structures, and also of procedures for creating structures.

As the agent does all this, it experiences the emotion of curiosity (likely among other emotions), because as it builds each new structure it has questions about what it will look like and how the teacher would react to it.

The human teacher says "It's beautiful. What is it?" The agent says "It's a car man"

Now that the building is done and the teacher says something, the agent's cognitive cycles are dominated by language understanding and generation. The Atom representing the context of talking to the teacher gets high STI, and is used as the context for many ensuing inferences.

Comprehension of "it" uses anaphor resolution based on a combination of ECAN and PLN inference based on a combination of previously interpreted language and observation of the external world situation.

The agent experiences the emotion of happiness because the teacher has called its creation beautiful, which is recognizes as a positive evaluation—so the agent knows one of its ubergoals ("please the teacher") has been significantly fulfilled.

The goal of pleasing the teacher causes the system to want to answer the question. So the QuestionAnswering DialogueController schema gets paid a lot and gets put into the ActiveSchemaPool. In reaction to the question asked, this DC chooses a semantic graph to speak, then invokes NL generation to say it.

NL generation chooses the most compact expression that seems to adequately convey the intended meaning, so it decides on "car man" as the best simple verbalization to match the newly created conceptual blend that it thinks effectively describes the newly created blocks structure.

The positive feedback from the user leads to reinforcement of the Atoms and processes that led to the construction of the blocks structure that has been judged beautiful (via importance spreading and SystemActivityTable mining).

3.4 Conclusion

The simple situation considered in this chapter is complex enough to involve nearly all the different cognitive processes in the CogPrime system—and many interactions between these processes. This fact illustrates one of the main difficulties of designing, building and testing an artificial mind like CogPrime—until nearly all of the system is build and made to operate in an integrated way, it's hard to do any meaningful test of the system. Testing PLN or MOSES or conceptual blending in isolation may be interesting computer science, but it doesn't tell you much about CogPrime as a design for a thinking machine.

According to the CogPrime approach, getting a simple child-like interaction like "build me something with blocks that I haven't seen before" to work properly requires a holistic, integrated cognitive system. Once one has built a system capable of this sort of simple interaction then, according to the theory underlying CogPrime, one is not that far from a system with adult human-level intelligence. And once one has an adult human-level AGI built according to a highly flexible design like CogPrime, given the potential of such systems to self-analyze and self-modify, one is not far off from a dramatically powerful Genius Machine. Of course there will be a lot of work to do to get from a child-level system to an adult-level system—it won't necessarily unfold as "automatically" as seems to happen with a human child, because CogPrime lacks the suite of developmental processes and mechanisms that the young human brain has. But still, a child CogPrime mind capable of doing the things outlined in this chapter will have all the basic components and interactions in place, all the ones that are needed for a much more advanced artificial mind.

Of course, one could concoct a narrow-AI system carrying out the specific activities described in this chapter, much more simply than one could build a CogPrime system capable of doing these activities. But that's not the point—the point of this chapter is not to explain how to achieve some particular narrow set of activities "by any means necessary", but rather to explain how these activities might be achieved within the CogPrime framework, which has been designed with much more generality in mind.

It would be worthwhile to elaborate a number of other situations similar to the one described in this chapter, and to work through the various cognitive processes and structures in CogPrime carefully in the context of each of these situations. In fact this sort of exercise has frequently been carried out informally in the context of developing CogPrime. But the burden of this book is already large enough, so we will leave this for future works—emphasizing that it is via intimate interplay between concrete considerations like the ones presented in this chapter, and general algorithmic and conceptual considerations as presented in most of the chapters of this book, that we have the greatest hope of creating advanced AGI. The value of this sort of interplay actually follows from the theory of real-world general intelligence presented several of the following chapters in Part 1. Thoroughly general intelligence is only possible given unrealistic computational resources, so real-world general intelligence is about achieving high generality given limited resources relative to the specific classes of environments relevant to a given agent. Specific situations like building surprising things with blocks are particularly important insofar as they embody broader information about the classes of environments relevant to broadly *human-like* general intelligence.

No doubt, once a CogPrime system is completed, the specifics of its handling of the situation described here will differ somewhat from the treatment presented in this chapter. Furthermore, the final CogPrime system may differ algorithmically and structurally in some respects from the specifics given in this book—it would be surprising if the process of building, testing and interacting with CogPrime didn't teach us some new things about various of the topics covered. But our conjecture is that, if sufficient effort is deployed appropriately, then a system much like the CogPrime system described in this book will be able to handle the situation described in this chapter in a roughly similar manner to the one described in this chapter—and that this will serve as a natural precursor to much more dramatic AGI achievements.

Part II
Artificial and Natural General Intelligence

Chapter 4
What is Human-Like General Intelligence?

4.1 Introduction

CogPrime, the AGI architecture on which the bulk of this book focuses, is aimed at the creation of artificial general intelligence that is vaguely human-like in nature, and possesses capabilities at the human level and ultimately beyond.

Obviously this description begs some foundational questions, such as, for starters: What is "general intelligence"? What is "human-like general intelligence"? What is "intelligence" at all?

Perhaps in the future there will exist a rigorous theory of general intelligence which applies usefully to real-world biological and digital intelligences. In later chapters we will give some ideas in this direction. But such a theory is currently nascent at best. So, given the present state of science, these two questions about intelligence must be handled via a combination of formal and informal methods. This brief, informal chapter attempts to explain our view on the nature of intelligence in sufficient detail to place the discussion of CogPrime in appropriate context, without trying to resolve all the subtleties.

Psychologists sometimes define human general intelligence using IQ tests and related instruments—so one might wonder: why not just go with that? But these sorts of intelligence testing approaches have difficulty even extending to humans from diverse cultures [HHPO12] [Fis01]. So it's clear that to ground AGI approaches that are not based on precise modeling of human cognition, one requires a more fundamental understanding of the nature of general intelligence. On the other hand, if one conceives intelligence too broadly and mathematically, there's a risk of leaving the real human world too far behind. In this chapter (followed up in Chaps. 10 and 8 with more rigor), we present a highly abstract understanding of intelligence-in-general, and then portray human-like general intelligence as a (particularly relevant) special case.

B. Goertzel et al., *Engineering General Intelligence, Part 1*,
Atlantis Thinking Machines 5, DOI: 10.2991/978-94-6239-027-0_4,
© Atlantis Press and the authors 2014

4.1.1 What is General Intelligence?

Many attempts to characterize general intelligence have been made; Legg and Hutter [LH07a] review over 70! Our preferred abstract characterization of intelligence is: **the capability of a system to choose actions maximizing its goal-achievement, based on its perceptions and memories, and making reasonably efficient use of its computational resources** [Goe10b]. A general intelligence is then understood as one that can do this for a variety of complex goals in a variety of complex environments.

However, apart from positing definitions, it is difficult to say anything nontrivial about general intelligence *in general*. Marcus Hutter [Hut05a] has demonstrated, using a characterization of general intelligence similar to the one above, that a very simple algorithm called AIXI^{tl} can demonstrate arbitrarily high levels of general intelligence, if given sufficiently immense computational resources. This is interesting because it shows that (if we assume the universe can effectively be modeled as a computational system) general intelligence is basically a problem of computational efficiency. The particular structures and dynamics that characterize real-world general intelligences like humans arise because of the need to achieve reasonable levels of intelligence using modest space and time resources.

The "patternist" theory of mind presented in [Goe06a] and briefly summarized in Chap. 5 presents a number of *emergent structures and dynamics* that are hypothesized to characterize pragmatic general intelligence, including such things as system-wide hierarchical and heterarchical knowledge networks, and a dynamic and self-maintaining self-model. Much of the thinking underlying CogPrime has centered on how to make multiple learning components combine to give rise to these emergent structures and dynamics.

4.1.2 What is Human-Like General Intelligence?

General principles like "complex goals in complex environments" and patternism are not sufficient to specify the nature of *human-like* general intelligence. Due to the harsh reality of computational resource restrictions, real-world general intelligences are necessarily biased to particular classes of environments. Human intelligence is biased toward the physical, social and linguistic environments in which humanity evolved, and if AI systems are to possess humanlike general intelligence they must to some extent share these biases.

But what are these biases, specifically? This is a large and complex question, which we seek to answer in a theoretically grounded way in Chap. 10. However, before turning to abstract theory, one may also approach the question in a pragmatic way, by looking at the categories of things that humans do to manifest their particular variety of general intelligence. This is the task of the following section.

4.2 Commonly Recognized Aspects of Human-Like Intelligence

It would be nice if we could give some sort of "standard model of human intelligence" in this chapter, to set the context for our approach to artificial general intelligence—but the truth is that there isn't any. What the cognitive science field has produced so far is better described as: a broad set of principles and platitudes, plus a long, loosely-organized list of ideas and results. Chapter 7 constitutes an attempt to present an integrative architecture diagram for human-like general intelligence, synthesizing the ideas of a number of different AGI and cognitive theorists. However, though the diagram given there attempts to be inclusive, it nonetheless contains many features that are accepted by only a plurality of the research community.

The following list of key aspects of human-like intelligence has a better claim at truly being generic and representing the consensus understanding of contemporary science. It was produced by a very simple method: starting with the Wikipedia page for cognitive psychology, and then adding a few items onto it based on scrutinizing the tables of contents of some top-ranked cognitive psychology textbooks. There is some redundancy among list items, and perhaps also some minor omissions (depending on how broadly one construes some of the items), but the point is to give a broad indication of human mental functions as standardly identified in the psychology field:

- Perception

  - General perception
  - Psychophysics
  - Pattern recognition (the ability to correctly interpret ambiguous sensory information)
  - Object and event recognition
  - Time sensation (awareness and estimation of the passage of time)

- Motor Control

  - Motor planning
  - Motor execution
  - Sensorimotor integration

- Categorization

  - Category induction and acquisition
  - Categorical judgement and classification
  - Category representation and structure
  - Similarity

- Memory

  - Aging and memory
  - Autobiographical memory
  - Constructive memory
  - Emotion and memory

- False memories
- Memory biases
- Long-term memory
- Episodic memory
- Semantic memory
- Procedural memory
- Short-term memory
- Sensory memory
- Working memory

- Knowledge representation

  - Mental imagery
  - Propositional encoding
  - Imagery versus propositions as representational mechanisms
  - Dual-coding theories
  - Mental models

- Language

  - Grammar and linguistics
  - Phonetics and phonology
  - Language acquisition

- Thinking

  - Choice
  - Concept formation
  - Judgment and decision making
  - Logic, formal and natural reasoning
  - Problem solving
  - Planning
  - Numerical cognition
  - Creativity

- Consciousness

  - Attention and Filtering (the ability to focus mental effort on specific stimuli whilst excluding other stimuli from consideration)
  - Access consciousness
  - Phenomenal consciousness

- Social Intelligence

  - Distributed Cognition
  - Empathy

If there's nothing surprising to you in the above list, I'm not surprised! If you've read a bit in the modern cognitive science literature, the list may even seem trivial. But it's worth reflecting that 50 years ago, no such list could have been produced with the same level of broad acceptance. And less than 100 years ago, the Western world's scientific understanding of the mind was dominated by Freudian thinking; and not too long after that, by behaviorist thinking, which argued that theorizing about what went on inside the mind made no sense, and science should focus entirely on analyzing external behavior. The progress of cognitive science hasn't made as many headlines as contemporaneous progress in neuroscience or computing hardware and software, but it's certainly been dramatic. One of the reasons that AGI is more achievable now than in the 1950s and 1960s when the AI field began, is that now we understand the structures and processes characterizing human thinking a lot better.

In spite of all the theoretical and empirical progress in the cognitive science field, however, there is still no consensus among experts on how the various aspects of intelligence in the above "human intelligence feature list" are achieved and interrelated. In these pages, however, for the purpose of motivating CogPrime, we assume a broad integrative understanding roughly as follows:

- **Perception**: There is significant evidence that human visual perception occurs using a spatiotemporal hierarchy of pattern recognition modules, in which higher-level modules deal with broader spacetime regions, roughly as in the DeSTIN AGI architecture discussed in Chap. 6. Further, there is evidence that each module carries out temporal predictive pattern recognition as well as static pattern recognition. Audition likely utilizes a similar hierarchy. Olfaction may use something more like a Hopfield attractor neural network, as described in Chap. 14. The networks corresponding to different sense modalities have multiple cross-linkages, more at the upper levels than the lower, and also link richly into the parts of the mind dealing with other functions.
- **Motor Control**: This appears to be handled by a spatiotemporal hierarchy as well, in which each level of the hierarchy corresponds to higher-level (in space and time) movements. The hierarchy is very tightly linked in with the perceptual hierarchies, allowing sensorimotor learning and coordination.
- **Memory**: There appear to be multiple distinct but tightly cross-linked memory systems, corresponding to different sorts of knowledge such as declarative (facts and beliefs), procedural, episodic, sensorimotor, attentional and intentional (goals).
- **Knowledge Representation**: There appear to be multiple base-level representational systems; at least one corresponding to each memory system, but perhaps more than that. Additionally there must be the capability to dynamically create new context-specific representational systems founded on the base representational system.
- **Language**: While there is surely some innate biasing in the human mind toward learning certain types of linguistic structure, it's also notable that language shares a great deal of structure with other aspects of intelligence like social roles [CB00] and the physical world [Cas07]. Language appears to be learned based on biases toward learning certain types of relational role systems; and language processing seems a complex mix of generic reasoning and pattern recognition processes with specialized acoustic and syntactic processing routines.

- **Consciousness** is pragmatically well-understood using Baars' "global workspace" theory, in which a small subset of the mind's content is summoned at each time into a "working memory" aka "workspace" aka "attentional focus" where it is heavily processed and used to guide action selection.
- **Thinking** is a diverse combination of processes encompassing things like categorization, (crisp and uncertain) reasoning, concept creation, pattern recognition, and others; these processes must work well with all the different types of memory and must effectively integrate knowledge in the global workspace with knowledge in long-term memory.
- **Social Intelligence** seems closely tied with language and also with self-modeling; we model ourselves in large part using the same specialized biases we use to help us model others.

None of the points in the above bullet list is particularly controversial, but neither are any of them universally agreed-upon by experts. However, in order to make any progress on AGI design one must make some commitments to particular cognition-theoretic understandings, at this level and ultimately at more precise levels as well. Further, general philosophical analyses like the patternist philosophy to be reviewed in the following chapter only provide limited guidance here. Patternism provides a filter for theories about specific cognitive functions—it rules out assemblages of cognitive-function-specific theories that don't fit together to yield a mind that could act effectively as a pattern-recognizing, goal-achieving system with the right internal emergent structures. But it's not a precise enough filter to serve as a sole guide for cognitive theory even at the high level.

The above list of points leads naturally into the integrative architecture diagram presented in Chap. 7. But that generic architecture diagram is fairly involved, and before presenting it, we will go through some more background regarding human-like intelligence (in the rest of this chapter), philosophy of mind (in Chap. 5) and contemporary AGI architectures (in Chap. 6).

4.3 Further Characterizations of Human-Like Intelligence

We now present a few complementary approaches to characterizing the key aspects of humanlike intelligence, drawn from different perspectives in the psychology and AI literature. These different approaches all overlap substantially, which is good, yet each gives a slightly different slant.

4.3.1 Competencies Characterizing Human-Like Intelligence

First we give a list of key competencies characterizing human level intelligence resulting from the AGI Roadmap Workshop held at the University of Knoxville in

October 2008,[1] which was organized by Ben Goertzel and Itamar Arel. In this list, each broad competency area is listed together with a number of specific competencies sub-areas within its scope:

1. **Perception**: vision, hearing, touch, proprioception, crossmodal
2. **Actuation**: physical skills, navigation, tool use
3. **Memory**: episodic, declarative, behavioral
4. **Learning**: imitation, reinforcement, interactive verbal instruction, written media, experimentation
5. **Reasoning**: deductive, abductive, inductive, causal, physical, associational, categorization
6. **Planning**: strategic, tactical, physical, social
7. **Attention**: visual, social, behavioral
8. **Motivation**: subgoal creation, affect-based motivation, control of emotions
9. **Emotion**: expressing emotion, understanding emotion
10. **Self**: self-awareness, self-control, other-awareness
11. **Social**: empathy, appropriate social behavior, social communication, social inference, group play, theory of mind
12. **Communication**: gestural, pictorial, verbal, language acquisition, cross-modal
13. **Quantitative**: counting, grounded arithmetic, comparison, measurement
14. **Building/Creation**: concept formation, verbal invention, physical construction, social group formation.

Clearly this list is getting at the same things as the textbook headings given in Sect. 4.2, but with a different emphasis due to its origin among AGI researchers rather than cognitive psychologists. As part of the AGI Roadmap project, specific tasks were created corresponding to each of the sub-areas in the above list; we will describe some of these tasks in Chap. 18.

4.3.2 Gardner's Theory of Multiple Intelligences

The diverse list of human-level "competencies" given above is reminiscent of Gardner's [Gar99] multiple intelligences (MI) framework—a psychological approach to intelligence assessment based on the idea that different people have mental strengths in different high-level domains, so that intelligence tests should contain aspects that focus on each of these domains separately. MI does not contradict the

---

[1] See http://www.ece.utk.edu/~itamar/AGI_Roadmap.html; participants included: Sam Adams, IBM Research; Ben Goertzel, Novamente LLC; Itamar Arel, University of Tennessee; Joscha Bach, Institute of Cognitive Science, University of Osnabruck, Germany; Robert Coop, University of Tennessee; Rod Furlan, Singularity Institute; Matthias Scheutz, Indiana University; J. Storrs Hall, Foresight Institute; Alexei Samsonovich, George Mason University; Matt Schlesinger, Southern Illinois University; John Sowa, Vivomind Intelligence, Inc.; Stuart C. Shapiro, University at Buffalo.

"complex goals in complex environments" view of intelligence, but rather may be interpreted as making specific commitments regarding which complex tasks and which complex environments are most important for roughly human-like intelligence.

MI does not seek an extreme generality, in the sense that it explicitly focuses on domains in which humans have strong innate capability as well as general-intelligence capability; there could easily be non-human intelligences that would exceed humans according to both the commonsense human notion of "general intelligence" and the generic "complex goals in complex environments" or Hutter/Legg-style definitions, yet would not equal humans on the MI criteria. This strong anthropocentrism of MI is not a problem from an AGI perspective so long as one uses MI in an appropriate way, i.e. only for assessing the extent to which an AGI system displays specifically *human-like* general intelligence. This restrictiveness is the price one pays for having an easily articulable and relatively easily implementable evaluation framework.

Table 4.1 summarizes the types of intelligence included in Gardner's MI theory.

4.3.3 Newell's Criteria for a Human Cognitive Architecture

Finally, another related perspective is given by Alan Newell's "functional criteria for a human cognitive architecture" [New90], which require that a humanlike AGI system should:

Table 4.1 Types of intelligence in Gardner's multiple intelligence theory

Intelligence Type	Aspects
Linguistic	Words and language, written and spoken; retention, interpretation and explanation of ideas and information via language; understands relationship between communication and meaning
Logical-Mathematical	Logical thinking, detecting patterns, scientific reasoning and deduction; analyse problems, perform mathematical calculations, understands relationship between cause and effect towards a tangible outcome
Musical	Musical ability, awareness, appreciation and use of sound; recognition of tonal and rhythmic patterns, understands relationship between sound and feeling
Bodily-Kinesthetic	Body movement control, manual dexterity, physical agility and balance; eye and body coordination
Spatial-Visual	Visual and spatial perception; interpretation and creation of images; pictorial imagination and expression; understands relationship between images and meanings, and between space and effect
Interpersonal	Perception of other people's feelings; relates to others; interpretation of behaviour and communications; understands relationships between people and their situations

1. Behave as an (almost) arbitrary function of the environment
2. Operate in real time
3. Exhibit rational, i.e., effective adaptive behavior
4. Use vast amounts of knowledge about the environment
5. Behave robustly in the face of error, the unexpected, and the unknown
6. Integrate diverse knowledge
7. Use (natural) language
8. Exhibit self-awareness and a sense of self
9. Learn from its environment
10. Acquire capabilities through development
11. Arise through evolution
12. Be realizable within the brain.

In our view, Newell's criterion 1 is poorly-formulated, for while universal Turing computing power is easy to come by, any finite AI system must inevitably be heavily adapted to some particular class of environments for straightforward mathematical reasons [Hut05a, GPI+10]. On the other hand, his criteria 11 and 12 are not relevant to the CogPrime approach as we are not doing biological modeling but rather AGI engineering. However, Newell's criteria 2–10 are essential in our view, and all will be covered in the following chapters.

4.3.4 Intelligence and Creativity

Creativity is a key aspect of intelligence. While sometimes associated especially with genius-level intelligence in science or the arts, actually creativity is pervasive throughout intelligence, at all levels. When a child makes a flying toy car by pasting paper bird wings on his toy car, and when a bird figures out how to use a curved stick to get a piece of food out of a difficult corner—this is creativity, just as much as the invention of a new physics theory or the design of a new fashion line. The very nature of intelligence—achieving complex goals in complex environments—requires creativity for its achievement, because the nature of complex environments and goals is that they are always unveiling new aspects, so that dealing with them involves inventing things beyond what worked for previously known aspects.

CogPrime contains a number of cognitive dynamics that are especially effective at creating new ideas, such as: concept creation (which synthesizes new concepts via combining aspects of previous ones), probabilistic evolutionary learning (which simulates evolution by natural selection, creating new procedures via mutation, combination and probabilistic modeling based on previous ones), and analogical inference (an aspect of the Probabilistic Logic Networks subsystems). But ultimately creativity is about how a system combines all the processes at its disposal to synthesize novel solutions to the problems posed by its goals in its environment.

There are times, of course, when the same goal can be achieved in multiple ways—some more creative than others. In CogPrime this relates to the existence of multiple top-level goals, one of which may be **novelty**. A system with novelty as one of its goals, alongside other more specific goals, will have a tendency to solve other problems in creative ways, thus fulfilling its novelty goal along with its other goals. This can be seen at the level of childlike behaviors, and also at a much more advanced level. Salvador Dali wanted to depict his thoughts and feelings, but he also wanted to do so in a striking and unusual way; this combination of aspirations spurred him to produce his amazing art. A child who is asked to draw a house, but has a goal of novelty, may draw a tower with a swimming pool on the roof rather than a typical Colonial structure. A physical motivated by novelty will seek a non-obvious solution to the equation at hand, rather than just applying tried and true methods, and perhaps discover some new phenomenon. Novelty can be measured formally in terms of information-theoretic surprisingness based upon a given basis of knowledge and experience [Sch06]; something that is novel and creative to a child may be familiar to the adult world, and a solution that seems novel and creative to a brilliant scientist today, may seem like cliche' elementary school level work 100 years from now.

Measuring creativity is even more difficult and subjective than measuring intelligence. Qualitatively, however, we humans can recognize it; and we suspect that the qualitative emergence of dramatic, multidisciplinary computational creativity will be one of the things that makes the human population feel emotionally that advanced AGI has finally arrived.

4.4 Preschool as a View into Human-Like General Intelligence

One issue that arises when pursuing the grand goal of human-level general intelligence is how to measure partial progress. The classic Turing Test of imitating human conversation remains too difficult to usefully motivate immediate-term AI research (see [HF95] [Fre90] for arguments that it has been counterproductive for the AI field). The same holds true for comparable alternatives like the Robot College Test of creating a robot that can attend a semester of university and obtain passing grades. However, some researchers have suggested intermediary goals, that constitute partial progress toward the grand goal and yet are qualitatively different from the highly specialized problems to which most current AI systems are applied.

In this vein, Sam Adams and his team at IBM have outlined a so-called "Toddler Turing Test," in which one seeks to use AI to control a robot qualitatively displaying similar cognitive behaviors to a young human child (say, a 3 year old) [AABL02]. In fact this sort of idea has a long and venerable history in the AI field—Alan Turing's original 1950 paper on AI [Tur50], where he proposed the Turing Test, contains the suggestion that

> Instead of trying to produce a programme to simulate the adult mind,
> why not rather try to produce one which simulates the child's?

We find this childlike cognition based approach promising for many reasons, including its integrative nature: what a young child does involves a combination of perception, actuation, linguistic and pictorial communication, social interaction, conceptual problem solving and creative imagination. Specifically, inspired by these ideas, in Chap. 17 we will suggest the approach of teaching and testing early-stage AGI systems in environments that emulate the preschools used for teaching human children.

Human intelligence evolved in response to the demands of richly interactive environments, and a preschool is specifically designed to be a richly interactive environment with the capability to stimulate diverse mental growth. So, we are currently exploring the use of CogPrime to control virtual agents in preschool-like virtual world environments, as well as commercial humanoid robot platforms such as the Nao (see Fig. 4.1) or Robokind (Fig. 4.2) in physical preschool-like robot labs.

Another advantage of focusing on childlike cognition is that child psychologists have created a variety of instruments for measuring child intelligence. In Chap. 18, we will discuss an approach to evaluating the general intelligence of human childlike AGI systems via combining tests typically used to measure the intelligence of young human children, with additional tests crafted based on cognitive science and the standard preschool curriculum.

Fig. 4.1 The Nao humanoid robot

Fig. 4.2 Robokind robot,
designed by David Hanson

To put it differently: While our long-term goal is the creation of genius machines
with general intelligence at the human level and beyond, we believe that every young
child has a certain genius; and by beginning with this childlike genius, we can built
a platform capable of developing into a genius machine with far more dramatic
capabilities.

4.4.1 Design for an AGI Preschool

More precisely, we don't suggest to place a CogPrime system in an environment
that is an exact imitation of a human preschool—this would be inappropriate since
current robotic or virtual bodies are very differently abled than the body of a young
human child. But we aim to place CogPrime in an environment emulating the basic
diversity and educational character of a typical human preschool. We stress this now,
at this early point in the book, because we will use running examples throughout the
book drawn from the preschool context.

The key notion in modern preschool design is the "learning center," an area designed and outfitted with appropriate materials for teaching a specific skill. Learning centers are designed to encourage learning by doing, which greatly facilitates learning processes based on reinforcement, imitation and correction; and also to provide multiple techniques for teaching the same skills, to accommodate different learning styles and prevent overfitting and overspecialization in the learning of new skills.

Centers are also designed to cross-develop related skills. A "manipulatives center," for example, provides physical objects such as drawing implements, toys and puzzles, to facilitate development of motor manipulation, visual discrimination, and (through sequencing and classification games) basic logical reasoning. A "dramatics center" cross-trains interpersonal and empathetic skills along with bodily-kinesthetic, linguistic, and musical skills. Other centers, such as art, reading, writing, science and math centers are also designed to train not just one area, but to center around a primary intelligence type while also cross-developing related areas. For specific examples of the learning centers associated with particular contemporary preschools, see [Nei98]. In many progressive, student-centered preschools, students are left largely to their own devices to move from one center to another throughout the preschool room. Generally, each center will be staffed by an instructor at some points in the day but not others, providing a variety of learning experiences.

To imitate the general character of a human preschool, we will create several centers in our robot lab. The precise architecture will be adapted via experience but initial centers will likely be:

- **a blocks center**: a table with blocks on it
- **a language center**: a circle of chairs, intended for people to sit around and talk with the robot
- **a manipulatives center**, with a variety of different objects of different shapes and sizes, intended to teach visual and motor skills
- **a ball play center**: where balls are kept in chests and there is space for the robot to kick the balls around
- **a dramatics center** where the robot can observe and enact various movements.

One Running Example

As we proceed through the various component structures and dynamics of CogPrime in the following chapters, it will be useful to have a few running examples to use to explain how the various parts of the system are supposed to work. One example we will use fairly frequently is drawn from the preschool context: the somewhat open-ended task of **Build me something out of blocks, that you haven't built for me before, and then tell me what it is**. This is a relatively simple task that combines multiple aspects of cognition in a richly interconnected way, and is the sort of thing that young children will naturally do in a preschool setting.

4.5 Integrative and Synergetic Approaches to Artificial General Intelligence

CogPrime constitutes an integrative approach to AGI. And we suggest that the naturalness of integrative approaches to AGI follows directly from comparing above lists of capabilities and criteria to the array of available AI technologies. No single known algorithm or data structure appears easily capable of carrying out all these functions, so if one wants to proceed *now* with creating a general intelligence that is even vaguely humanlike, one must integrate various AI technologies within some sort of unifying architecture.

For this reason and others, an increasing amount of work in the AI community these days is integrative in one sense or another. Estimation of Distribution Algorithms integrate probabilistic reasoning with evolutionary learning [Pel05]. Markov Logic Networks [RD06] integrate formal logic and probabilistic inference, as does the Probabilistic Logic Networks framework [GIGH08] utilized in CogPrime and explained further in the book, and other works in the "Progic" area such as [WW06]. Leslie Pack Kaelbling has synthesized low-level robotics methods (particle filtering) with logical inference [ZPK07]. Dozens of further examples could be given. The construction of practical robotic systems like the Stanley system that won the DARPA Grand Challenge [Tea06] involve the integration of numerous components based on different principles. These algorithmic and pragmatic innovations provide ample raw materials for the construction of integrative cognitive architectures and are part of the reason why childlike AGI is more approachable now than it was 50 or even 10 years ago.

Further, many of the *cognitive architectures* described in the current AI literature are "integrative" in the sense of combining multiple, qualitatively different, interoperating algorithms. Chapter 6 gives a high-level overview of existing cognitive architectures, dividing them into *symbolic*, *emergentist* (e.g. neural network) and *hybrid* architectures. The hybrid architectures generally integrate symbolic and neural components, often with multiple subcomponents within each of these broad categories. However, we believe that even these excellent architectures are not integrative enough, in the sense that they lack sufficiently rich and nuanced interactions between the learning components associated with different kinds of memory, and hence are unlikely to give rise to the emergent structures and dynamics characterizing general intelligence. One of the central ideas underlying CogPrime is that with an integrative cognitive architecture that combines multiple aspects of intelligence, achieved by diverse structures and algorithms, within a common framework designed specifically to support robust **synergetic interactions** between these aspects.

The simplest way to create an integrative AI architecture is to loosely couple multiple components carrying out various functions, in such a way that the different components pass inputs and outputs amongst each other but do not interfere with or modulate each others' internal functioning in real-time. However, the human brain appears to be integrative in a much tighter sense, involving rich real-time dynamical coupling between various components with distinct but related functions.

In [Goe09a] we have hypothesized that the brain displays a property of **cognitive synergy**, according to which multiple learning processes can not only **dispatch sub-problems** to each other, but also **share contextual understanding in real-time**, so that each one can get help from the others in a contextually savvy way. By imbuing AI architectures with cognitive synergy, we hypothesize, one can get past the bottle-necks that have plagued AI in the past. Part of the reasoning here, as elaborated in Chap. 10 and [Goe09b], is that real physical and social environments display a rich dynamic interconnection between their various aspects, so that richly dynamically interconnected integrative AI architectures will be able to achieve goals within them more effectively.

And this brings us to the patternist perspective on intelligent systems, alluded to above and fleshed out further in Chap. 5 with its focus on the emergence of hier-archically and heterarchically structured networks of patterns, and pattern-systems modeling self and others. Ultimately the purpose of cognitive synergy in an AGI system is to enable the various AI algorithms and structures composing the system to work together effectively enough to give rise to the right *system-wide emergent structures* characterizing real-world general intelligence. The underlying theory is that intelligence is not reliant on any particular structure or algorithm, but *is* reliant on the emergence of appropriately structured networks of patterns, which can then be used to guide ongoing dynamics of pattern recognition and creation. And the underlying hypothesis is that the emergence of these structures cannot be achieved by a loosely interconnected assemblage of components, no matter how sensible the architecture; it requires a tightly connected, synergetic system.

It is possible to make these theoretical ideas about cognition mathematically rig-orous; for instance, Appendix B briefly presents a formal definition of cognitive synergy that has been analyzed as part of an effort to prove theorems about the importance of cognitive synergy for giving rise to emergent system properties asso-ciated with general intelligence. However, while we have found such formal analyses valuable for clarifying our designs and understanding their qualitative properties, we have concluded that, for the present, the best way to explore our hypotheses about cognitive synergy and human-like general intelligence is empirically—via building and testing systems like CogPrime.

4.5.1 Achieving Human-Like Intelligence via Cognitive Synergy

Summing up: at the broadest level, there are four primary challenges in constructing an integrative, cognitive synergy based approach to AGI:

1. Choosing an **overall cognitive architecture** that possesses adequate richness and flexibility for the task of achieving childlike cognition.
2. Choosing **appropriate AI algorithms and data structures** to fulfill each of the functions identified in the cognitive architecture. (e.g. visual perception, audition, episodic memory, language generation, analogy,...)

3. Ensuring that these algorithms and structures, within the chosen cognitive architecture, are able to cooperate in such a way as to provide appropriate **coordinated, synergetic intelligent behavior** (a critical aspect since childlike cognition is an integrated functional response to the world, rather than a loosely coupled collection of capabilities.)
4. Embedding one's system in an environment that provides **sufficiently rich stimuli and interactions** to enable the system to use this cooperation to ongoingly, creatively develop an intelligent internal world-model and self-model.

We argue that CogPrime provides a viable way to address these challenges.

Chapter 5
A Patternist Philosophy of Mind

5.1 Introduction

In the last chapter we discussed human intelligence from a fairly down-to-earth perspective, looking at the particular intelligent functions that human beings carry out in their everyday lives. And we strongly feel this practical perspective is important: Without this concreteness, it's too easy for AGI research to get distracted by appealing (or frightening) abstractions of various sorts. However, it's **also** important to look at the nature of mind and intelligence from a more general and conceptual perspective, to avoid falling into an approach that follows the particulars of human capability but ignores the deeper structures and dynamics of mind that ultimately allow human minds to be so capable. In this chapter we very briefly review some ideas from the **patternist philosophy of mind**, a general conceptual framework on intelligence which has been inspirational for many key aspects of the CogPrime design, and which has been ongoingly developed by one of the authors (Ben Goertzel) during the last two decades (in a series of publications beginning in 1991, most recently *The Hidden Pattern* [Goe06a]). Some of the ideas described are quite broad and conceptual, and are related to CogPrime only via serving as general inspirations; others are more concrete and technical, and are actually utilized within the design itself.

CogPrime is an integrative design formed via the combination of a number of different philosophical, scientific and engineering ideas. The success or failure of the design doesn't depend on any particular philosophical understanding of intelligence. In that sense, the more abstract notions presented in this chapter should be considered "optional" rather than critical in a CogPrime context. However, due to the core role patternism has played in the development of CogPrime, understanding a few things about general patternist philosophy will be helpful for understanding CogPrime, even for those readers who are not philosophically inclined. Those readers who *are* philosophically inclined, on the other hand, are urged to read *The Hidden Pattern* and then interpret the particulars of CogPrime in this light.

B. Goertzel et al., *Engineering General Intelligence, Part 1*,
Atlantis Thinking Machines 5, DOI: 10.2991/978-94-6239-027-0_5,
© Atlantis Press and the authors 2014

5.2 Some Patternist Principles

The patternist philosophy of mind is a general approach to thinking about intelligent systems. It is based on the very simple premise that mind is made of pattern—and that a mind is a system for recognizing patterns in itself and the world, critically including patterns regarding which procedures are likely to lead to the achievement of which goals in which contexts.

Pattern as the basis of mind is not in itself is a very novel idea; this concept is present, for instance, in the nineteenth-century philosophy of Charles Peirce [Pei34], in the writings of contemporary philosophers Daniel Dennett [Den91] and Douglas Hofstadter [Hof79, Hof96], in Benjamin Whorf's [Who64] linguistic philosophy and Gregory Bateson's [Bat79] systems theory of mind and nature. Bateson spoke of the Metapattern: "that it is pattern which connects." In Goertzel's writings on philosophy of mind, an effort has been made to pursue this theme more thoroughly than has been done before, and to articulate in detail how various aspects of human mind and mind in general can be well-understood by explicitly adopting a patternist perspective.[1]

In the patternist perspective, "pattern" is generally defined as "representation as something simpler." Thus, for example, if one measures simplicity in terms of bit-count, then a program compressing an image would be a pattern in that image. But if one uses a simplicity measure incorporating run-time as well as bit-count, then the compressed version may or may not be a pattern in the image, depending on how one's simplicity measure weights the two factors. This definition encompasses simple repeated patterns, but also much more complex ones. While pattern theory has typically been elaborated in the context of computational theory, it is not intrinsically tied to computation; rather, it can be developed in any context where there is a notion of "representation" or "production" and a way of measuring simplicity. One just needs to be able to assess the extent to which f represents or produces X, and then to compare the simplicity of f and X; and then one can assess whether f is a pattern in X. A formalization of this notion of pattern is given in [Goe06a] and briefly summarized at the end of this chapter.

Next, in patternism the mind of an intelligent system is conceived as the (fuzzy) set of patterns in that system, and the set of patterns emergent between that system and other systems with which it interacts. The latter clause means that the patternist perspective is inclusive of notions of distributed intelligence [Hut96]. Basically, the mind of a system is the fuzzy set of different simplifying representations of that system that may be adopted.

Intelligence is conceived, similarly to in Marcus Hutter's [Hut05] work (and as elaborated informally in Chap. 4, and formally in Chap. 8), as the ability to achieve complex goals in complex environments; where complexity itself may be defined as the possession of a rich variety of patterns. A mind is thus a collection of patterns

---

[1] In some prior writings the term "psynet model of mind" has been used to refer to the application of patternist philosophy to cognitive theory, but this term has been "deprecated" in recent publications as it seemed to introduce more confusion than clarification.

that is associated with a persistent dynamical process that achieves highly-patterned goals in highly-patterned environments.

An additional hypothesis made within the patternist philosophy of mind is that reflection is critical to intelligence. This lets us conceive an intelligent system as a dynamical system that recognizes patterns in its environment and itself, as part of its quest to achieve complex goals.

While this approach is quite general, it is not vacuous; it gives a particular structure to the tasks of analyzing and synthesizing intelligent systems. About any would-be intelligent system, we are led to ask questions such as:

- How are patterns represented in the system? That is, how does the underlying infrastructure of the system give rise to the displaying of a particular pattern in the system's behavior?
- What kinds of patterns are most compactly represented within the system?
- What kinds of patterns are most simply learned?
- What learning processes are utilized for recognizing patterns?
- What mechanisms are used to give the system the ability to introspect (so that it can recognize patterns in itself)?

Now, these same sorts of questions could be asked if one substituted the word "pattern" with other words like "knowledge" or "information". However, we have found that asking these questions in the context of pattern leads to more productive answers, avoiding confusing byways and also tying in very nicely with the details of various existing formalisms and algorithms for knowledge representation and learning.

Among the many kinds of patterns in intelligent systems, *semiotic* patterns are particularly interesting ones. Peirce decomposed these into three categories:

- **iconic** patterns, which are patterns of contextually important internal similarity between two entities (e.g. an iconic pattern binds a picture of a person to that person)
- **indexical** patterns, which are patterns of spatiotemporal co-occurrence (e.g. an indexical pattern binds a wedding dress and a wedding)
- **symbolic** patterns, which are patterns indicating that two entities are often involved in the same relationships (e.g. a symbolic pattern between the number "5" (the symbol) and various sets of 5 objects (the entities that the symbol is taken to represent)

Of course, some patterns may span more than one of these semiotic categories; and there are also some patterns that don't fall neatly into any of these categories. But the semiotic patterns are particularly important ones; and symbolic patterns have played an especially large role in the history of AI, because of the radically different approaches different researchers have taken to handling them in their AI systems. Mathematical logic and related formalisms provide sophisticated mechanisms for combining and relating symbolic patterns ("symbols"), and some AI approaches have focused heavily on these, sometimes more so than on the identification of symbolic patterns in experience or the use of them to achieve practical goals. We will look fairly carefully at these differences in Chap. 6.

Pursuing the patternist philosophy in detail leads to a variety of particular hypotheses and conclusions about the nature of mind. Following from the view of intelligence in terms of achieving complex goals in complex environments, comes a view in which the dynamics of a cognitive system are understood to be governed by two main forces:

- self-organization, via which system dynamics cause existing system patterns to give rise to new ones
- goal-oriented behavior, which will be defined more rigorously in Chap. 8, but basically amounts to a system interacting with its environment in a way that appears like an attempt to maximize some reasonably simple function

Self-organized and goal-oriented behavior must be understood as cooperative aspects. If an agent is asked to build a surprising structure out of blocks and does so, this is goal-oriented. But the agent's ability to carry out this goal-oriented task will be greater if it has previously played around with blocks a lot in an unstructured, spontaneous way. And the "nudge toward creativity" given to it by asking it to build a surprising blocks structure may cause it to explore some novel patterns, which then feed into its future unstructured blocks play.

Based on these concepts, as argued in detail in [Goe06a], several primary dynamical principles may be posited, including:

- **Evolution**, conceived as a general process via which patterns within a large population thereof are differentially selected and used as the basis for formation of new patterns, based on some "fitness function" that is generally tied to the goals of the agent.

    - *Example:* If trying to build a blocks structure that will surprise Bob, an agent may simulate several procedures for building blocks structures in its "mind's eye", assessing for each one the expected degree to which it might surprise Bob. The search through procedure space could be conducted as a form of evolution, via an algorithm such as MOSES (see Chap. 16 of Part 2).

- **Autopoiesis**. The process by which a system of interrelated patterns maintains its integrity, via a dynamic in which whenever one of the patterns in the system begins to decrease in intensity, some of the other patterns increase their intensity in a manner that causes the troubled pattern to increase in intensity again.

    - *Example:* An agent's set of strategies for building the base of a tower, and its set of strategies for building the middle part of a tower, are likely to relate autopoietically. If the system partially forgets how to build the base of a tower, then it may regenerate this missing knowledge via using its knowledge about how to build the middle part (i.e., it knows it needs to build the base in a way that will support good middle parts). Similarly if it partially forgets how to build the middle part, then it may regenerate this missing knowledge via using its knowledge about how to build the base (i.e. it knows a good middle part should fit in well with the sorts of base it knows are good).

- This same sort of interdependence occurs between pattern-sets containing more than two elements.
- Sometimes (as in the above example) autopoietic interdependence in the mind is tied to interdependencies in the physical world, sometimes not.

- **Association**. Patterns, when given attention, spread some of this attention to other patterns that they have previously been associated with in some way. Furthermore, there is Peirce's law of mind [Pei34], which could be paraphrased in modern terms as stating that the mind is an associative memory network, whose dynamics dictate that every idea in the memory is an active agent, continually acting on those ideas with which the memory associates it.

  - *Example:* Building a blocks structure that resembles a tower, spreads attention to memories of prior towers the agents has seen, and also to memories of people the agent knows have seen towers, and structures it has built at the same time as towers, structures that resemble towers in various respects, etc.

- **Differential attention allocation/credit assignment**. Patterns that have been valuable for goal-achievement are given more attention, and are encouraged to participate in giving rise to new patterns.

  - *Example:* Perhaps in a prior instance of the task "build me a surprising structure out of blocks," searching through memory for non-blocks structures that the agent has played with has proved a useful cognitive strategy. In that case, when the task is posed to the agent again, it should tend to allocate disproportionate resources to this strategy.

- **Pattern creation**. Patterns that have been valuable for goal-achievement are mutated and combined with each other to yield new patterns.

  - *Example:* Building towers has been useful in a certain context, but so has building structures with a large number of triangles. Why not build a tower out of triangles? Or maybe a vaguely tower-like structure that uses more triangles than a tower easily could?
  - *Example:* Building an elongated block structure resembling a table was successful in the past, as was building a structure resembling a very flat version of a chair. Generalizing, maybe building distorted versions of furniture is good. Or maybe it is building distorted version of *any* previously perceived objects that is good. Or maybe both, to different degrees....

Next, for a variety of reasons outlined in [Goe06a] it becomes appealing to hypothesize that the network of patterns in an intelligent system must give rise to the following large-scale emergent structures

- Hierarchical network. Patterns are habitually in relations of control over other patterns that represent more specialized aspects of themselves.

- *Example:* The pattern associated with "tall building" has some control over the pattern associated with "tower", as the former represents a more general concept ... and "tower" has some control over "Eiffel tower", etc.

- Heterarchical network. The system retains a memory of which patterns have previously been associated with each other in any way.

  - *Example:* "Tower" and "snake" are distant in the natural pattern hierarchy, but may be associatively/heterarchically linked due to having a common elongated structure. This heterarchical linkage may be used for many things, e.g. it might inspire the creative construction of a tower with a snake's head.

- Dual network. Hierarchical and heterarchical structures are combined, with the dynamics of the two structures working together harmoniously. Among many possible ways to hierarchically organize a set of patterns, the one used should be one that causes hierarchically nearby patterns to have many meaningful heterarchical connections; and of course, there should be a tendency to search for heterarchical connections among hierarchically nearby patterns.

  - *Example:* While the set of patterns hierarchically nearby "tower" and the set of patterns heterarchically nearby "tower" will be quite different, they should still have more overlap than random pattern-sets of similar sizes. So, if looking for something else heterarchically near "tower", using the hierarchical information about "tower" should be of some use, and vice versa.
  - In CogPrime's probabilistic logic subsystem, PLN (see Chap. 17 of Part 2) hierarchical relationships correspond to Atoms *A* and *B* so that *Inheritance AB* and *Inheritance BA* have highly dissimilar strength; and heterarchical relationships correspond to IntensionalSimilarity relationships. The dual network structure then arises when intensional and extensional inheritance approximately correlate with each other, so that inference about either kind of inheritance assists with figuring out about the other kind.

- Self structure. A portion of the network of patterns forms into an approximate image of the overall network of patterns.

  - *Example:* Each time the agent builds a certain structure, it observes itself building the structure, and its role as "builder of a tall tower" (or whatever the structure is) becomes part of its self-model. Then when it is asked to build something new, it may consult its self-model to see if it believes itself capable of building that sort of thing (for instance, if it is asked to build something very large, its self-model may tell it that it lacks persistence for such projects, so it may reply "I can try, but I may wind up not finishing it").

As we proceed through the CogPrime design in the following pages, we will see how each of these abstract concepts arises concretely from CogPrime's structures and algorithms. If the theory of [Goe06a] is correct, then the success of CogPrime as a design will depend largely on whether these high-level structures and dynamics can be made to emerge from the synergetic interaction of CogPrime's representation and algorithms, when they are utilized to control an appropriate agent in an appropriate environment.

5.3 Cognitive Synergy

Now we dig a little deeper and present a different sort of "general principle of feasible general intelligence", already hinted in earlier chapters: the *cognitive synergy* principle,[2] which is both a conceptual hypothesis about the structure of generally intelligent systems in certain classes of environments, and a design principle used to guide the design of CogPrime. Chapter 9 presents a mathematical formalization of the notion of cognitive synergy; here we present the conceptual idea informally, which makes it more easily digestible but also more vague-sounding.

We will focus here on cognitive synergy specifically in the case of "multi-memory systems," which we define as intelligent systems whose combination of environment, embodiment and motivational system make it important for them to possess memories that divide into partially but not wholly distinct components corresponding to the categories of:

- Declarative memory

  - *Examples of declarative knowledge:* Towers on average are taller than buildings. I generally am better at building structures I imagine, than at imitating structures I'm shown in pictures.

- Procedural memory (memory about how to do certain things)

  - *Examples of procedural knowledge:* Practical know-how regarding how to pick up an elongated rectangular block, or a square one. Know-how regarding when to approach a problem by asking "What would one of my teachers do in this situation" versus by thinking through the problem from first principles.

- Sensory and episodic memory

  - *Example of sensory knowledge:* Memory of Bob's face; memory of what a specific tall blocks tower looked like.
  - *Example of episodic knowledge:* Memory of the situation in which the agent first met Bob; memory of a situation in which a specific tall blocks tower was built.

- Attentional memory (knowledge about what to pay attention to in what contexts)

  - *Example of attentional knowledge:* When involved with a new person, it's useful to pay attention to whatever that person looks at.

- Intentional memory (knowledge about the system's own goals and subgoals)

  - *Example of intentional knowledge:* If my goal is to please some person whom I don't know that well, then a subgoal may be figuring out what makes that person smile.

---

[2] While these points are implicit in the theory of mind given in [Goe06a], they are not articulated in this specific form there. So the material presented in this section is a new development within patternist philosophy, developed since [Goe06a] in a series of conference papers such as [Goe09a].

In Chap. 10 we present a detailed argument as to how the requirement for a multi-memory underpinning for general intelligence emerges from certain underlying assumptions regarding the measurement of the simplicity of goals and environments. Specifically we argue that each of these memory types corresponds to certain *modes of communication*, so that intelligent agents which have to efficiently handle a sufficient variety of types of communication with other agents, are going to have to handle all these types of memory. These types of communication overlap and are often used together, which implies that the different memories and their associated cognitive processes need to work together. The points made in this section do not rely on that argument regarding the relation of multiple memory types to the environmental situation of multiple communication types. What they do rely on is the assumption that, in the intelligence agent in question, the different components of memory are significantly but not wholly distinct. That is, there are significant "family resemblances" between the memories of a single type, yet there are also thoroughgoing connections between memories of different types.

Repeating the above points in a slightly more organized manner and then extending them, the essential idea of cognitive synergy, in the context of multi-memory systems, may be expressed in terms of the following points

1. Intelligence, relative to a certain set of environments, may be understood as the capability to achieve complex goals in these environments.
2. With respect to certain classes of goals and environments, an intelligent system requires a "multi-memory" architecture, meaning the possession of a number of specialized yet interconnected knowledge types, including: declarative, procedural, attentional, sensory, episodic and intentional (goal-related). These knowledge types may be viewed as different sorts of patterns that a system recognizes in itself and its environment.
3. Such a system must possess knowledge creation (i.e. pattern recognition / formation) mechanisms corresponding to each of these memory types. These mechanisms are also called "cognitive processes."
4. Each of these cognitive processes, to be effective, must have the capability to recognize when it lacks the information to perform effectively on its own; and in this case, to dynamically and interactively draw information from knowledge creation mechanisms dealing with other types of knowledge.
5. This cross-mechanism interaction must have the result of enabling the knowledge creation mechanisms to perform much more effectively in combination than they would if operated non-interactively. This is "cognitive synergy."

Interactions as mentioned in Points 4 and 5 in the above list are the real conceptual meat of the cognitive synergy idea. One way to express the key idea here, in an AI context, is that most AI algorithms suffer from combinatorial explosions: the number of possible elements to be combined in a synthesis or analysis is just too great, and the algorithms are unable to filter through all the possibilities, given the lack of intrinsic constraint that comes along with a "general intelligence" context (as opposed to a narrow-AI problem like chess-playing, where the context is constrained and hence restricts the scope of possible combinations that needs to be considered). In an AGI architecture based on cognitive synergy, the different learning mechanisms must be designed specifically to interact in such a way as to palliate each others' combinatorial explosions—so that, for instance, each learning mechanism dealing with a certain sort of knowledge, must synergize with learning mechanisms dealing with the other sorts of knowledge, in a way that decreases the severity of combinatorial explosion.

One prerequisite for cognitive synergy to work is that each learning mechanism must recognize when it is "stuck," meaning it's in a situation where it has inadequate information to make a confident judgment about what steps to take next. Then, when it does recognize that it's stuck, it may request help from other, complementary cognitive mechanisms.

5.4 The General Structure of Cognitive Dynamics: Analysis and Synthesis

We have discussed the need for synergetic interrelation between cognitive processes corresponding to different types of memory ... and the general high-level cognitive dynamics that a mind must possess (evolution, autopoiesis). The next step is to dig further into the nature of the cognitive processes associated with different memory types and how they give rise to the needed high-level cognitive dynamics. In this section we present a *general theory of cognitive processes* based on a decomposition of cognitive processes into the two categories of *analysis* and *synthesis*, and a general formulation of each of these categories.[3]

Specifically we concentrate here on what we call *focused cognitive processes*; that is, cognitive processes that selectively focus attention on a subset of the patterns making up a mind. In general these are not the only kind, there may also be *global cognitive processes* that act on every pattern in a mind. An example of a global cognitive process in CogPrime is the basic attention allocation process, which spreads "importance" among all knowledge in the system's memory. Global cognitive processes are also important, but focused cognitive processes are subtler to understand which is why we spend more time on them here.

5.4.1 Component-Systems and Self-Generating Systems

We begin with autopoiesis—and, more specifically, with the concept of a "component-system", as described in George Kampis's book *Self-Modifying Systems in Biology and Cognitive Science* [Kam91], and as modified into the concept of a "self-generating system" or SGS in Goertzel's book *Chaotic Logic* [Goe94]. Roughly

---

[3] While these points are highly compatible with theory of mind given in [Goe06a], they are not articulated there. The material presented in this section is a new development within patternist philosophy, presented previously only in the article [GPPG06].

speaking, a Kampis-style component-system consists of a set of components that combine with each other to form other compound components. The metaphor Kampis uses is that of Lego blocks, combining to form bigger Lego structures. Compound structures may in turn be combined together to form yet bigger compound structures. A self-generating system is basically the same concept as a component-system, but understood to be computable, whereas Kampis claims that component-systems are uncomputable.

Next, in SGS theory there is also a notion of reduction (not present in the Lego metaphor): sometimes when components are combined in a certain way, a "reaction" happens, which may lead to the elimination of some of the components. One relevant metaphor here is chemistry. Another is abstract algebra: for instance, if we combine a component f with its "inverse" component f^{-1}, both components are eliminated. Thus, we may think about two stages in the interaction of sets of components: combination, and reduction. Reduction may be thought of as algebraic simplification, governed by a set of rules that apply to a newly created compound component, based on the components that are assembled within it.

Formally, suppose $C_1, C_2, \dots$ is the set of components present in a discrete-time component-system at time t. Then, the components present at time $t + 1$ are a subset of the set of components of the form

$$Reduce(Join(C_{i(1)}, \dots, C_{i(r)}))$$

where *Join* is a joining operation, and *Reduce* is a reduction operator. The joining operation is assumed to map tuples of components into components, and the reduction operator is assumed to map the space of components into itself. Of course, the specific nature of a component system is totally dependent on the particular definitions of the reduction and joining operators; in following chapters we will specify these for the CogPrime system, but for the purpose of the broader theoretical discussion in this section they may be left general.

What is called the "cognitive equation" in *Chaotic Logic* [Goe94] is the case of a SGS where the patterns in the system at time t have a tendency to correspond to components of the system at future times $t + s$. So, part of the action of the system is to transform implicit knowledge (patterns among system components) into explicit knowledge (specific system components). We will see one version of this phenomenon in Chap. 15 where we model implicit knowledge using mathematical structures called "derived hypergraphs"; and we will also later review several ways in which CogPrime's dynamics explicitly encourage cognitive-equation type dynamics, e.g.:

- inference, which takes conclusions implicit in the combination of logical relationships, and makes them implicit by deriving new logical relationships from them
- map formation, which takes concepts that have often been active together, and creates new concepts grouping them
- association learning, which creates links representing patterns of association between entities

- probabilistic procedure learning, which creates new models embodying patterns regarding which procedures tend to perform well according to particular fitness functions.

5.4.2 Analysis and Synthesis

Now we move on to the main point of this section: the argument that all or nearly all focused cognitive processes are expressible using two general process-schemata we call *synthesis* and *analysis*.[4] The notion of "focused cognitive process" will be exemplified more thoroughly below, but in essence what is meant is a cognitive process that begins with a small number of items (drawn from memory) as its focus, and has as its goal discovering something about these items, or discovering something about something else in the context of these items or in a way strongly biased by these items. This is different from a global cognitive process whose goal is more broadly-based and explicitly involves all or a large percentage of the knowledge in an intelligent system's memory store.

Among the focused cognitive processes are those governed by the so-called *cognitive schematic* implication

$$Context \wedge Procedure \rightarrow Goal$$

where the Context involves sensory, episodic and/or declarative knowledge; and attentional knowledge is used to regulate how much resource is given to each such schematic implication in memory. Synergy among the learning processes dealing with the context, the procedure and the goal is critical to the adequate execution of the cognitive schematic using feasible computational resources. This sort of explicitly goal-driven cognition plays a significant though not necessarily dominant role in CogPrime, and is also related to production rules systems and other traditional AI systems, as will be articulated in Chap. 6.

The synthesis and analysis processes as we conceive them, in the general framework of SGS theory, are as follows. First, synthesis, as shown in Fig. 5.1, is defined as

Synthesis: Iteratively build compounds from the initial component pool using the combinators, greedily seeking compounds that seem likely to achieve the goal.

Or in more detail:

1. Begin with some initial components (the initial "current pool"), an additional set of components identified as "combinators" (combination operators), and a goal function

---

[4] In [GPPG06], what is here called "analysis" was called "backward synthesis", a name which has some advantages since it indicated that what's happening is a form of creation; but here we have opted for the more traditional analysis/synthesis terminology.

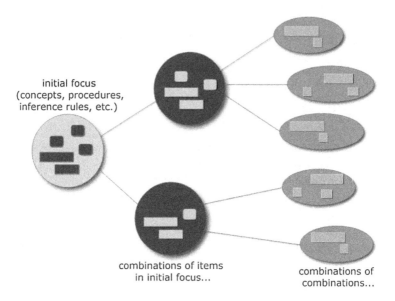

initial focus
(concepts, procedures,
inference rules, etc.)

combinations of items
in initial focus...

combinations of
combinations...

Fig. 5.1 The general process of synthesis

2. Combine the components in the current pool, utilizing the combinators, to form product components in various ways, carrying out reductions as appropriate, and calculating relevant quantities associated with components as needed
3. Select the product components that seem most promising according to the goal function, and add these to the current pool (or else simply define these as the current pool)
4. Return to Step 2

And analysis, as shown in Fig. 5.2, is defined as

Analysis: Iteratively search (the system's long-term memory) for component-sets that combine using the combinators to form the initial component pool (or subsets thereof), greedily seeking component-sets that seem likely to achieve the goal.

Or in more detail:

1. Begin with some components (the initial "current pool") and a goal function
2. Seek components so that, if one combines them to form product components using the combinators and then performs appropriate reductions, one obtains (as many as possible of) the components in the current pool
3. Use the newly found constructions of the components in the current pool, to update the quantitative properties of the components in the current pool, and also (via the current pool) the quantitative properties of the components in the initial pool
4. Out of the components found in Step 2, select the ones that seem most promising according to the goal function, and add these to the current pool (or else simply define these as the current pool)

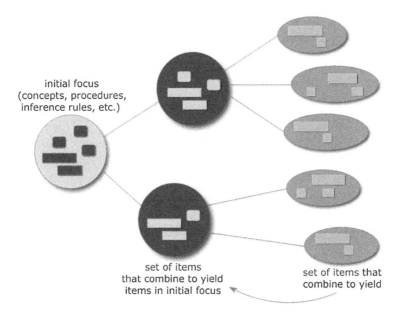

initial focus
(concepts, procedures,
inference rules, etc.)

set of items
that combine to yield
items in initial focus

set of items that
combine to yield

Fig. 5.2   The general process of analysis

5. Return to Step 2

More formally, synthesis may be specified as follows. Let X denote the set of combinators, and let Y_0 denote the initial pool of components (the initial focus of the cognitive process). Given Y_i, let Z_i denote the set

$$Reduce(Join(C_i(1), ..., C_i(r)))$$

where the C_i are drawn from Y_i or from X. We may then say

$$Y_{i+1} = Filter(Z_i)$$

where *Filter* is a function that selects a subset of its arguments.

Analysis, on the other hand, begins with a set W of components, and a set X of combinators, and tries to find a series Y_i so that according to the process of synthesis, $Y_n = W$.

In practice, of course, the implementation of a synthesis process need not involve the explicit construction of the full set Z_i. Rather, the filtering operation takes place implicitly during the construction of Y_{i+1}. The result, however, is that one gets some subset of the compounds producible via joining and reduction from the set of components present in Y_i plus the combinators X.

Conceptually one may view synthesis as a very generic sort of "growth process," and analysis as a very generic sort of "figuring out how to grow something."

The intuitive idea underlying the present proposal is that these forward-going and backward-going "growth processes" are among the essential foundations of cognitive control, and that a conceptually sound design for cognitive control should explicitly make use of this fact. To abstract away from the details, what these processes are about is:

- taking the general dynamic of compound-formation and reduction as outlined in Kampis and *Chaotic Logic*
- introducing goal-directed pruning ("filtering") into this dynamic so as to account for the limitations of computational resources that are a necessary part of pragmatic intelligence.

5.4.3 The Dynamic of Iterative Analysis and Synthesis

While synthesis and analysis are both very useful on their own, they achieve their greatest power when harnessed together. It is my hypothesis that the dynamic pattern of alternating synthesis and analysis has a fundamental role in cognition. Put simply, synthesis creates new mental forms by combining existing ones. Then, analysis seeks simple explanations for the forms in the mind, including the newly created ones; and, this explanation itself then comprises additional new forms in the mind, to be used as fodder for the next round of synthesis. Or, to put it yet more simply:

$$\Rightarrow \textbf{Combine} \Rightarrow \textbf{Explain} \Rightarrow \textbf{Combine} \Rightarrow \textbf{Explain} \Rightarrow \textbf{Combine} \Rightarrow$$

It is not hard to express this alternating dynamic more formally, as well.

- Let X denote any set of components.
- Let F(X) denote a set of components which is the result of synthesis on X.
- Let B(X) denote a set of components which is the result of analysis of X. We assume also a heuristic biasing the synthesis process toward simple constructs.
- Let S(t) denote a set of components at time t, representing part of a system's knowledge base.
- Let I(t) denote components resulting from the external environment at time t.

Then, we may consider a dynamical iteration of the form

$$S(t + 1) = B(F(S(t) + I(t)))$$

This expresses the notion of alternating synthesis and analysis formally, as a dynamical iteration on the space of sets of components. We may then speak about attractors of this iteration: fixed points, limit cycles and strange attractors. One of the key hypotheses we wish to put forward here is that some key emergent cognitive structures are strange attractors of this equation. The iterative dynamic of combination and explanation leads to the emergence of certain complex structures that are, in essence, maintained when one recombines their parts and then seeks to explain the recombinations. These structures are built in the first place through iterative recombination and explanation, and then survive in the mind because they are conserved by this process. They then ongoingly guide the construction and destruction of various other temporary mental structures that are not so conserved.

5.4.4 Self and Focused Attention as Approximate Attractors of the Dynamic of Iterated Forward-Analysis

As noted above, patternist philosophy argues that two key aspects of intelligence are emergent structures that may be called the "self" and the "attentional focus." These, it is suggested, are aspects of intelligence that may not effectively be wired into the infrastructure of an intelligent system, though of course the infrastructure may be configured in such a way as to encourage their emergence. Rather, these aspects, by their nature, are only likely to be effective if they emerge from the cooperative activity of various cognitive processes acting within a broad base of knowledge.

Above we have described the pattern of ongoing habitual oscillation between synthesis and analysis as a kind of "dynamical iteration." Here we will argue that both self and attentional focus may be viewed as strange attractors of this iteration. The mode of argument is relatively informal. The essential processes under consideration are ones that are poorly understood from an empirical perspective, due to the extreme difficulty involved in studying them experimentally. For understanding self and attentional focus, we are stuck in large part with introspection, which is famously unreliable in some contexts, yet still dramatically better than having no information at all. So, the philosophical perspective on self and attentional focus given here is a synthesis of empirical and introspective notions, drawn largely from the published thinking and research of others but with a few original twists. From a CogPrime perspective, its use has been to guide the design process, to provide a grounding for what otherwise would have been fairly arbitrary choices.

5.4.4.1 Self

Another high-level intelligent system pattern mentioned above is the "self", which we here will tie in with analysis and synthesis processes. The term "self" as used here refers to the "phenomenal self" [Met04] or "self-model". That is, the self is the model that a system builds internally, reflecting the patterns observed in the (external and internal) world that directly pertain to the system itself. As is well known in everyday human life, self-models need not be completely accurate to be useful; and in the presence of certain psychological factors, a more accurate self-model may not necessarily be advantageous. But a self-model that is too badly inaccurate will lead to a badly-functioning system that is unable to effectively act toward the achievement of its own goals.

The value of a self-model for any intelligent system carrying out embodied agentive cognition is obvious. And beyond this, another primary use of the self is as a foundation for metaphors and analogies in various domains. Patterns recognized pertaining to the self are analogically extended to other entities. In some cases this leads to conceptual pathologies, such as the anthropomorphization of trees, rocks and other such objects that one sees in some precivilized cultures. But in other cases this kind of analogy leads to robust sorts of reasoning—for instance, in reading Lakoff and Nunez's [LN00] intriguing explorations of the cognitive foundations of mathematics, it is pretty easy to see that most of the metaphors on which they hypothesize mathematics to be based, are grounded in the mind's conceptualization of itself as a spatiotemporally embedded entity, which in turn is predicated on the mind's having a conceptualization of itself (a self) in the first place.

A self-model can in many cases form a self-fulfilling prophecy (to make an obvious double-entendre!). Actions are generated based on one's model of what sorts of actions one can and/or should take; and the results of these actions are then incorporated into one's self-model. If a self-model proves a generally bad guide to action selection, this may never be discovered, unless said self-model includes the knowledge that semi-random experimentation is often useful.

In what sense, then, may it be said that self is an attractor of iterated analysis? Analysis infers the self from observations of system behavior. The system asks: What kind of system might I be, in order to give rise to these behaviors that I observe myself carrying out? Based on asking itself this question, it constructs a model of itself, i.e. it constructs a self. Then, this self guides the system's behavior: it builds new logical relationships its self-model and various other entities, in order to guide its future actions oriented toward achieving its goals. Based on the behaviors newly induced via this constructive, forward-synthesis activity, the system may then engage in analysis again and ask: What must I be now, in order to have carried out these new actions? And so on.

Our hypothesis is that after repeated iterations of this sort, in infancy, finally during early childhood a kind of self-reinforcing attractor occurs, and we have a self-model that is resilient and doesn't change dramatically when new instances of action- or explanation-generation occur. This is not strictly a mathematical attractor, though, because over a long period of time the self may well shift significantly. But, for a mature self, many hundreds of thousands or millions of forward-analysis cycles may occur before the self-model is dramatically modified. For relatively long periods of time, small changes within the context of the existing self may suffice to allow the system to control itself intelligently.

Humans can also develop what are known as **subselves** [Row90]. A subself is a partially autonomous self-network focused on particular tasks, environments or interactions. It contains a unique model of the whole organism, and generally has its own set of episodic memories, consisting of memories of those intervals during which it was the primary dynamic mode controlling the organism. One common example is the **creative subself**—the subpersonality that takes over when a creative person launches into the process of creating something. In these times, a whole different personality sometimes emerges, with a different sort of relationship to the world. Among other factors, creativity requires a certain open-ness that is not always productive in an everyday life context, so it's natural for the self-system of a highly creative person to bifurcate into one self-system for everyday life, and another for the protected context of creative activity. This sort of phenomenon might emerge naturally in CogPrime systems as well if they were exposed to appropriate environments and social situations.

Finally, it is interesting to speculate regarding how self may differ in future AI systems as opposed to in humans. The relative stability we see in human selves may not exist in AI systems that can self-improve and change more fundamentally and rapidly than humans can. There may be a situation in which, as soon as a system has understood itself decently, it radically modifies itself and hence violates its existing self-model. Thus: intelligence without a long-term stable self. In this case the "attractor-ish" nature of the self holds only over much shorter time scales than for human minds or human-like minds. But the alternating process of synthesis and analysis for self-construction is still critical, even though no reasonably stable self-constituting attractor ever emerges. The psychology of such intelligent systems will almost surely be beyond human beings' capacity for comprehension and empathy.

5.4.4.2 Attentional Focus

Now, we turn to the notion of an "attentional focus" similar to Baars' [Baa97] notion of a Global Workspace, which will be reviewed in more detail in Chap. 6: a collection of mental entities that are, at a given moment, receiving far more than the usual share of an intelligent system's computational resources. Due to the amount of attention paid to items in the attentional focus, at any given moment these items are in large part driving the cognitive processes going on elsewhere in the mind as well—because the cognitive processes acting on the items in the attentional focus are often involved in other mental items, not in attentional focus, as well (and sometimes this results in pulling these other items into attentional focus). An intelligent system must constantly shift its attentional focus from one set of entities to another based on changes in its environment and based on its own shifting discoveries.

In the human mind, there is a self-reinforcing dynamic pertaining to the collection of entities in the attentional focus at any given point in time, resulting from the observation that: If A is in the attentional focus, and A and B have often been associated in the past, then odds are increased that B will soon be in the attentional focus. This basic observation has been refined tremendously via a large body of cognitive psychology work; and neurologically it follows not only from Hebb's [Heb49] classic work on neural reinforcement learning, but also from numerous more modern refinements [SB98]. But it implies that two items A and B, if both in the attentional focus, can reinforce each others' presence in the attentional focus, hence forming a kind of conspiracy to keep each other in the limelight. But of course, this kind of dynamic must be counteracted by a pragmatic tendency to remove items from the attentional focus if giving them attention is not providing sufficient utility in terms of the achievement of system goals.

The synthesis and analysis perspective provides a more systematic perspective on this self-reinforcing dynamic. Synthesis occurs in the attentional focus when two or more items in the focus are combined to form new items, new relationships, new ideas. This happens continually, as one of the main purposes of the attentional focus is combinational. On the other hand, Analysis then occurs when a combination that has been speculatively formed is then linked in with the remainder of the mind (the "unconscious", the vast body of knowledge that is not in the attentional focus at the given moment in time). Analysis basically checks to see what support the new combination has within the existing knowledge store of the system. Thus, forward-analysis basically comes down to "generate and test", where the testing takes the form of attempting to integrate the generated structures with the ideas in the unconscious long-term memory. One of the most obvious examples of this kind of dynamic is creative thinking [Bod03, Goe97], where the attentional focus continually combinationally creates new ideas, which are then tested via checking which ones can be validated in terms of (built up from) existing knowledge.

The analysis stage may result in items being pushed out of the attentional focus, to be replaced by others. Likewise may the synthesis stage: the combinations may overshadow and then replace the things combined. However, in human minds and functional AI minds, the attentional focus will not be a complete chaos with constant turnover: Sometimes the same set of ideas—or a shifting set of ideas within the same overall family of ideas—will remain in focus for a while. When this occurs it is because this set or family of ideas forms an approximate attractor for the dynamics of the attentional focus, in particular for the forward-analysis dynamic of speculative combination and integrative explanation. Often, for instance, a small "core set" of ideas will remain in the attentional focus for a while, but will not exhaust the attentional focus: the rest of the attentional focus will then, at any point in time, be occupied with other ideas related to the ones in the core set. Often this may mean that, for a while, the whole of the attentional focus will move around quasi-randomly through a "strange attractor" consisting of the set of ideas related to those in the core set.

5.4.5 Conclusion

The ideas presented above (the notions of synthesis and analysis, and the hypothesis of self and attentional focus as attractors of the iterative forward-analysis dynamic) are quite generic and are hypothetically proposed to be applicable to any cognitive system, natural or artificial. Later chapters will discuss the manifestation of the above ideas in the context of CogPrime. We have found that the analysis/synthesis approach is a valuable tool for conceptualizing CogPrime's cognitive dynamics, and we conjecture that a similar utility may be found more generally.

Next, so as not to end the section on too blasé of a note, we will also make a stronger hypothesis: that, in order for a physical or software system to achieve intelligence that is roughly human-level in both capability and generality, using computational resources on the same order of magnitude as the human brain, this system must

- manifest the dynamic of iterated synthesis and analysis, as modes of an underlying "self-generating system" dynamic
- do so in such a way as to lead to self and attentional focus as emergent structures that serve as approximate attractors of this dynamic, over time periods that are long relative to the basic "cognitive cycle time" of the system's forward-analysis dynamics.

To prove the truth of a hypothesis of this nature would seem to require mathematics fairly far beyond anything that currently exists. Nonetheless, however, we feel it is important to formulate and discuss such hypotheses, so as to point the way for future investigations both theoretical and pragmatic.

5.5 Perspectives on Machine Consciousness

We can't let a chapter on philosophy—even a brief one—end without some discussion of the thorniest topic in the philosophy of mind: consciousness. Rather than seeking to resolve or comprehensively review this most delicate issue, we will restrict ourselves to discussing the relationship between consciousness theory and patternist philosophy of cognition, the practical work of designing and building AGI.

One fairly concrete idea about consciousness, that relates closely to certain aspects of the CogPrime design, is that the subjective experience of being conscious of some entity X, is correlated with the presence of a very intense pattern in one's overall mind-state, corresponding to X. This simple idea is also the essence of neuroscientist Susan Greenfield's theory of consciousness [Gre01] (but in her theory, "overall mind-state" is replaced with "brain-state"), and has much deeper historical roots in philosophy of mind which we shall not venture to unravel here.

This observation relates to the idea of "moving bubbles of awareness" in intelligent systems. If an intelligent system consists of multiple processing or data elements, and during each (sufficiently long) interval of time some of these elements get much more attention than others, then one may view the system as having a certain "attentional focus" during each interval. The attentional focus is itself a significant pattern in the system (the pattern being "these elements habitually get more processor and memory", roughly speaking). As the attentional focus shifts over time one has a "moving bubble of pattern" which then corresponds experientially to a "moving bubble of awareness."

This notion of a "moving bubble of awareness" ties in very closely to global workspace theory [Baa97] (briefly mentioned above), a cognitive theory that has broad support from neuroscience and cognitive science and has also served as the motivation for Stan Franklin's LIDA AI system [BF09], to be discussed in Chap. 6. The global workspace theory views the mind as consisting of a large population of small, specialized processes—a society of agents. These agents organize themselves into coalitions, and coalitions that are relevant to contextually novel phenomena, or contextually important goals, are pulled into the global workspace (which is identified with consciousness). This workspace broadcasts the message of the coalition to all the unconscious agents, and recruits other agents into consciousness. Various sorts of contexts—e.g. goal contexts, perceptual contexts, conceptual contexts and cultural contexts—play a role in determining which coalitions are relevant, and form the unconscious "background" of the conscious global workspace. New perceptions are often, but not necessarily, pushed into the workspace. Some of the agents in the global workspace are concerned with action selection, i.e. with controlling and passing parameters to a population of possible actions. The contents of the workspace at any given time have a certain cohesiveness and interdependency, the so-called "unity of consciousness." In essence the contents of the global workspace form a moving bubble of attention or awareness.

In CogPrime, this moving bubble is achieved largely via economic attention network (ECAN) equations [GPI+10] that propagate virtual currency between nodes and links representing elements of memories, so that the attentional focus consists of the wealthiest nodes and links. Figures 5.3 and 5.4 illustrate the existence and flow of attentional focus in OpenCog. On the other hand, in Hameroff's recent model of the brain [Ham10], the brain's moving bubble of attention is achieved through dendro-dendritic connections and the emergent dendritic web.

In this perspective, self, free will and reflective consciousness are specific phenomena occurring *within* the moving bubble of awareness. They are specific ways of experiencing awareness, corresponding to certain abstract types of physical structures and dynamics, which we shall endeavor to identify in detail in Appendix C.

5.6 Postscript: Formalizing Pattern

Finally, before winding up our very brief tour through patternist philosophy of mind, we will briefly visit patternism's more formal side. Many of the key aspects of patternism have been rigorously formalized. Here we give only a few very basic elements of the relevant mathematics, which will be used later on in the exposition of CogPrime. (Specifically, the formal definition of pattern emerges in the CogPrime design in the definition of a fitness function for "pattern mining" algorithms and Occam-based concept creation algorithms, and the definition of intensional inheritance within PLN.)

We give some definitions, drawn from Appendix 1 of [Goe06a]:

Definition 1 *Given a metric space* (M, d), *and two functions* $c : M \to [0, \infty]$ *(the "simplicity measure") and* $F : M \to M$ *(the "production relationship"), we say that* $\mathcal{P} \in M$ *is a* **pattern** *in* $X \in M$ *to the degree*

$$\iota_X^{\mathcal{P}} = \left(\left(1 - \frac{d(F(\mathcal{P}), X)}{c(X)} \right) \frac{c(X) - c(\mathcal{P})}{c(X)} \right)^+$$

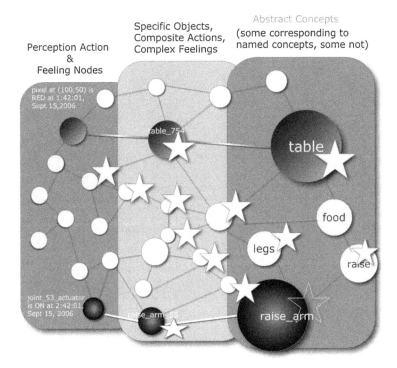

Fig. 5.3 Graphical depiction of the momentary bubble of attention in the memory of an OpenCog AI system. *Circles* and *lines* represent nodes and links in OpenCogPrimes memory, and *stars* denote those nodes with a high level of attention (represented in OpenCog by the ShortTermImportance node variable) at the particular point in time

This degree is called the **pattern intensity** of $\mathcal{P}$ in X. It quantifies the extent to which $\mathcal{P}$ is a pattern in X. Supposing that $F(\mathcal{P}) = X$, then the first factor in the definition equals 1, and we are left with only the second term, which measures the degree of compression obtained via representing X as the result of P rather than simply representing X directly. The greater the compression ratio obtained via using P to represent X, the greater the intensity of P as a pattern in X. The first time, in the case $F(\mathcal{P}) \neq X$, adjusts the pattern intensity downwards to account for the amount of error with which $F(\mathcal{P})$ approximates $\neq X$. If one holds the second factor fixed and thinks about varying the first factor, then: The greater the error, the lossier the compression, and the lower the pattern intensity.

For instance, if one wishes one may take c to denote algorithmic information measured on some reference Turing machine, and $F(X)$ to denote what appears on the second tape of a two-tape Turing machine t time-steps after placing X on its first tape. Other more naturalistic computational models are also possible here and are discussed extensively in Appendix 1 of [Goe06a].

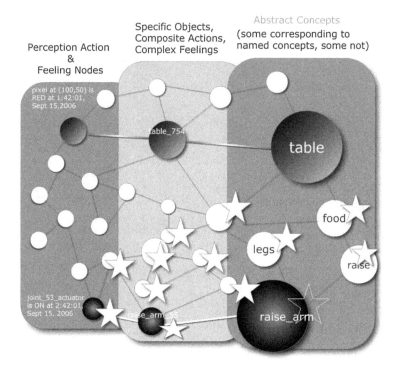

Fig. 5.4 Graphical depiction of the momentary bubble of attention in the memory of an OpenCog AI system, a few moments after the bubble shown in Fig. 5.3, indicating the moving of the bubble of attention. Depictive conventions are the same as in Fig. 5.3. This shows an idealized situation where the declarative knowledge remains invariant from one moment to the next but only the focus of attention shifts. In reality both will evolve together

Definition 2 *The* **structure** *of* $X \in M$ *is the fuzzy set* St_X *defined via the membership function*

$$\chi_{St_X}(\mathcal{P}) = \iota_X^{\mathcal{P}}$$

This lets us formalize our definition of "mind" alluded to above: the mind of X as the set of patterns associated with X. We can formalize this, for instance, by considering $\mathcal{P}$ to belong to the mind of X if it is a pattern in some Y that includes X. There are then two numbers to look at: $\iota_X^{\mathcal{P}}$ and $P(Y|X)$ (the percentage of Y that is also contained in X). To define the degree to which $\mathcal{P}$ belongs to the mind of X we can then combine these two numbers using some function f that is monotone increasing in both arguments. This highlights the somewhat arbitrary semantics of "of" in the phrase "the mind of X." Which of the patterns binding X to its environment are part of X's mind, and which are part of the world? This isn't necessarily a good question, and the answer seems to depend on what perspective you choose, represented formally in the present framework by what combination function f you choose (for instance if $f(a, b) = a^r b^{2-r}$ then it depends on the choice of $0 < r < 1$).

Next, we can formalize the notion of a "pattern space" by positing a metric on patterns, thus making pattern space a metric space, which will come in handy in some places in later chapters:

Definition 3 *Assuming M is a countable space, the* **structural distance** *is a metric d_{St} defined on M via*

$$d_{St}(X, Y) = T(\chi_{St_X}, \chi_{St_Y})$$

where T is any metric on countable-length vectors.

Using this definition of pattern, combined with the formal theory of intelligence given in Chap. 8, one may formalize the various hypotheses made in the previous section, regarding the emergence of different kinds of networks and structures as patterns in intelligent systems. However, it appears quite difficult to prove the formal versions of these hypotheses given current mathematical tools, which renders such formalizations of limited use.

Finally, consider the case where the metric space M has a partial ordering $<$ on it; we may then define

Definition 5.1 $\mathcal{R} \in M$ is a **subpattern** in $X \in M$ to the degree

$$\kappa_X^{\mathcal{R}} = \frac{\int_{\mathcal{P} \in M} \text{true}(\mathcal{R} < \mathcal{P}) d\iota_X^{\mathcal{P}}}{\int_{\mathcal{P} \in M} d\iota_X^{\mathcal{P}}}$$

This degree is called the **subpattern intensity** of $\mathcal{P}$ in X.

Roughly speaking, the subpattern intensity measures the percentage of patterns in X that contain R (where "containment" is judged by the partial ordering $<$). But the percentage is measured using a weighted average, where each pattern is weighted by its intensity as a pattern in X. A subpattern may or may not be a pattern on its own. A nonpattern that happens to occur within many patterns may be an intense subpattern.

Whether the subpatterns in X are to be considered part of the "mind" of X is a somewhat superfluous question of semantics. Here we choose to extend the definition of mind given in [Goe06a] to include subpatterns as well as patterns, because this makes it simpler to describe the relationship between hypersets and minds, as we will do in Appendix C.

Chapter 6
Brief Survey of Cognitive Architectures

6.1 Introduction

While we believe CogPrime is the most thorough attempt at an architecture for advanced AGI, to date, we certainly recognize there have been many valuable attempts in the past with similar aims; and we also have great respect for other AGI efforts occurring in parallel with CogPrime development, based on alternative, sometimes overlapping, theoretical presuppositions and practical choices. In most of this book we will ignore these other current and historical efforts except where they are directly useful for CogPrime—there are many literature reviews already published, and this is a research treatise not a textbook. In this chapter, however, we will break from this pattern and give a rough high-level overview of the various AGI architectures at play in the field today. The overview definitely has a bias toward other work with some direct relevance to CogPrime, but not an overwhelming bias; we also discuss a number of approaches that are unrelated to, and even in some cases conceptually orthogonal to, our own.

CogPrime builds on prior AI efforts in a variety of ways. Most of the specific algorithms and structures in CogPrime have their roots in prior AI work; and in addition, the CogPrime cognitive architecture has been heavily inspired by some other holistic cognitive architectures, especially (but not exclusively) MicroPsi [Bac09], LIDA [BF09] and DeSTIN [ARK09a, ARC09].

We will articulate some rough mappings between elements of these other architectures and elements of CogPrime—some in this chapter, and some in Chap. 7. However, these mappings will mostly be left informal and very incompletely specified. The articulation of detailed inter-architecture mappings is an important project, but would be a substantial additional project going well beyond the scope of this book. We will not give a thorough review of the similarities and differences between CogPrime and each of these architectures, but only mention some of the highlights.

The reader desiring a more thorough review of cognitive architectures is referred to Wlodek Duch's review paper from the AGI-08 conference [DOP08]; and also to Alexei Samsonovich's review paper [Sam10], which compares a number of cognitive

B. Goertzel et al., *Engineering General Intelligence, Part 1*,
Atlantis Thinking Machines 5, DOI: 10.2991/978-94-6239-027-0_6,
© Atlantis Press and the authors 2014

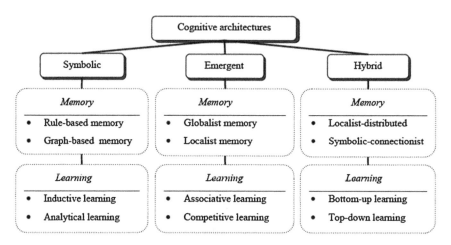

Fig. 6.1 Duch's simplified taxonomy of cognitive architectures. CogPrime falls into the "hybrid" category, but differs from other hybrid architectures in its focus on synergetic interactions between components and their potential to give rise to appropriate system-wide emergent structures enabling general intelligence

architectures in terms of a feature checklist, and was created collaboratively with the creators of the architectures.

Duch, in his survey of cognitive architectures [DOP08], divides existing approaches into three paradigms—symbolic, emergentist and hybrid—as broadly indicated in Fig. 6.1. Drawing on his survey and updating slightly, we give here some key examples of each, and then explain why CogPrime represents a significantly more effective approach to embodied human-like general intelligence. In our treatment of emergentist architectures, we pay particular attention to *developmental robotics* architectures, which share considerably with CogPrime in terms of underlying philosophy, but differ via not integrating a symbolic "language and inference" component such as CogPrime includes.

In brief, we believe that the hybrid approach is the most pragmatic one given the current state of AI technology, but that the emergentist approach gets something fundamentally right, by focusing on the emergence of complex dynamics and structures from the interactions of simple components. So CogPrime is a hybrid architecture which (according to the cognitive synergy principle) binds its components together very tightly dynamically, allowing the emergence of complex dynamics and structures in the integrated system. Most other hybrid architectures are less tightly coupled and hence seem ill-suited to give rise to the needed emergent complexity. The other hybrid architectures that do possess the needed tight coupling, such as MicroPsi [Bac09], strike us as underdeveloped and founded on insufficiently powerful learning algorithms.

6.2 Symbolic Cognitive Architectures

A venerable tradition in AI focuses on the physical symbol system hypothesis [New90], which states that minds exist mainly to manipulate symbols that represent aspects of the world or themselves. A physical symbol system has the ability to input, output, store and alter symbolic entities, and to execute appropriate actions in order to reach its goals. Generally, symbolic cognitive architectures focus on "working memory" that draws on long-term memory as needed, and utilize a centralized control over perception, cognition and action. Although in principle such architectures could be arbitrarily capable (since symbolic systems have universal representational and computational power, in theory), in practice symbolic architectures tend to be weak in learning, creativity, procedure learning, and episodic and associative memory. Decades of work in this tradition have not resolved these issues, which has led many researchers to explore other options. A few of the more important symbolic cognitive architectures are:

- **SOAR** [LRN87], a classic example of expert rule-based cognitive architecture designed to model general intelligence. It has recently been extended to handle sensorimotor functions, though in a somewhat cognitively unnatural way; and is not yet strong in areas such as episodic memory, creativity, handling uncertain knowledge, and reinforcement learning.
- **ACT-R** [AL03] is fundamentally a symbolic system, but Duch classifies it as a hybrid system because it incorporates connectionist-style activation spreading in a significant role; and there is an experimental thoroughly connectionist implementation to complement the primary mainly-symbolic implementation. Its combination of SOAR-style "production rules" with large-scale connectionist dynamics allows it to simulate a variety of human psychological phenomena, but abstract reasoning, creativity and transfer learning are still missing.
- **EPIC** [RCK01], a cognitive architecture aimed at capturing human perceptual, cognitive and motor activities through several interconnected processors working in parallel. The system is controlled by production rules for cognitive processors and a set of perceptual (visual, auditory, tactile) and motor processors operating on symbolically coded features rather than raw sensory data. It has been connected to SOAR for problem solving, planning and learning.
- **ICARUS** [Lan05], an integrated cognitive architecture for physical agents, with knowledge specified in the form of reactive skills, each denoting goal-relevant reactions to a class of problems. The architecture includes a number of modules: a perceptual system, a planning system, an execution system, and several memory systems. Concurrent processing is absent, attention allocation is fairly crude, and uncertain knowledge is not thoroughly handled.
- **SNePS** (Semantic Network Processing System) [SE07] is a logic, frame and network-based knowledge representation, reasoning, and acting system that has undergone over three decades of development. While it has been used for some interesting prototype experiments in language processing and virtual agent control, it has not yet been used for any large-scale or real-world application.

- **Cyc** [LG90] is an AGI architecture based on predicate logic as a knowledge representation, and using logical reasoning techniques to answer questions and derive new knowledge from old. It has been connected to a natural language engine, and designs have been created for the connection of Cyc with Albus's 4D-RCS [AM01]. Cyc's most unique aspect is the large database of common-sense knowledge that Cycorp has accumulated (millions of pieces of knowledge, entered by specially trained humans in predicate logic format); part of the philosophy underlying Cyc is that once a sufficient quantity of knowledge is accumulated in the knowledge base, the problem of creating human-level general intelligence will become much less difficult due to the ability to leverage this knowledge.

While these architectures contain many valuable ideas and have yielded some interesting results, we feel they are incapable *on their own* of giving rise to the emergent structures and dynamics required to yield humanlike general intelligence using feasible computational resources. However, we are more sanguine about the possibility of ideas and components from symbolic architectures playing a role in human-level AGI via incorporation in hybrid architectures.

We now review a few symbolic architectures in slightly more detail.

6.2.1 SOAR

The cognitive architectures best known among AI academics are probably Soar and ACT-R, both of which are explicitly being developed with the dual goals of creating human-level AGI and modeling all aspects of human psychology. Neither the Soar nor ACT-R communities feel themselves particularly near these long-term goals, yet they do take them seriously.

Soar is based on IF-THEN rules, otherwise known as "production rules." On the surface this makes it similar to old-style expert systems, but Soar is much more than an expert system; it's at minimum a sophisticated problem-solving engine. Soar explicitly conceives problem solving as a search through solution space for a "goal state" representing a (precise or approximate) problem solution. It uses a methodology of incremental search, where each step is supposed to move the system a little closer to its problem-solving goal, and each step involves a potentially complex "decision cycle."

In the simplest case, the decision cycle has two phases:

- Gathering appropriate information from the system's long-term memory (LTM) into its working memory (WM)
- A decision procedure that uses the gathered information to decide an action.

If the knowledge available in LTM isn't enough to solve the problem, then the decision procedure invokes search heuristics like hill-climbing, which try to create new knowledge (new production rules) that will help move the system closer to a solution. If a solution is found by chaining together multiple production rules, then a chunking mechanism is used to combine these rules together into a single rule for future use. One could view the chunking mechanism as a way of converting explicit knowledge into implicit knowledge, similar to "map formation" in CogPrime (see Chap. 25 of Part 2), but in the current Soar design and implementation it is a fairly crude mechanism

In recent years Soar has acquired a number of additional methods and modalities, including some visual reasoning methods and some mechanisms for handling episodic and procedural knowledge. These expand the scope of the system but the basic production rule and chunking mechanisms as briefly described above remain the core "cognitive algorithm" of the system.

From a CogPrime perspective, what Soar offers is certainly valuable, e.g.

- heuristics for transferring knowledge from LTM into WM
- chaining and chunking of implications
- methods for interfacing between other forms of knowledge and implications.

However, a very short and very partial list of the major differences between Soar and CogPrime would include

- CogPrime contains a variety of other core cognitive mechanisms beyond the management and chunking of implications
- The variety of "chunking" type methods in CogPrime goes far beyond the sort of localized chunking done in Soar
- CogPrime is committed to representing uncertainty at the base level whereas Soar's production rules are crisp
- The mechanisms for LTM-WM interaction are rather different in CogPrime, being based on complex nonlinear dynamics as represented in Economic Attention Allocation (ECAN)
- Currently Soar does not contain creativity-focused heuristics like blending or evolutionary learning in its core cognitive dynamic.

6.2.2 ACT-R

In the grand scope of cognitive architectures, ACT-R is quite similar to Soar, but there are many micro-level differences. ACT-R is defined in terms of declarative and procedural knowledge, where procedural knowledge takes the form of Soar-like production rules, and declarative knowledge takes the form of chunks. It contains a variety of mechanisms for learning new rules and chunks from old; and also contains sophisticated probabilistic equations for updating the activation levels associated with items of knowledge (these equations being roughly analogous in function to, though quite different from, the ECAN equations in CogPrime).

The flow of cognition in the system is in response to the current goal, currently active information from declarative memory, information attended to in perceptual

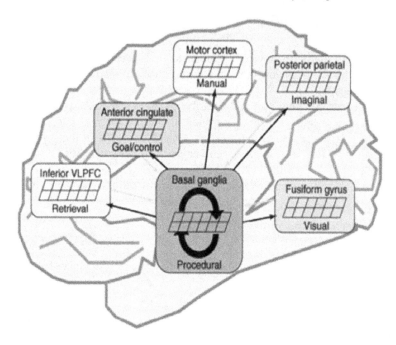

Fig. 6.2 Conjectured mapping between ACT-R and the brain

modules (vision and audition are implemented), and the current state of motor modules (hand and speech are implemented). The early work with ACT-R was based on comparing system performance to human behavior, using only behavioral measures, such as the timing of keystrokes or patterns of eye movements. Using such measures, it was not possible to test detailed assumptions about which modules were active in the performance of a task. More recently the ACT-R community has been engaged in a process of using imaging data to provide converging data on module activity. Figure 6.2 illustrates the associations they have made between the system's cognitive modules and brain regions. Coordination among all of these components occurs through actions of the procedural module, which is mapped to the basal ganglia.

In practice ACT-R, even more so than Soar, seems to be used more as a programming framework for cognitive modeling than as an AI system. One can fairly easily use ACT-R to program models of specific human mental behaviors, which may then be matched against psychological data. Opinions differ as to whether this sort of modeling is valuable for achieving AGI goals. CogPrime is not designed to support this kind of modeling, as it intentionally does many things very differently from humans.

ACT-R in its original form did not say much about perceptual and motor operations, but recent versions have incorporated EPIC, an independent cognitive architecture focused on modeling these aspects of human behavior.

6.2.3 Cyc and Texai

Our review of cognitive architectures would be incomplete without mentioning Cyc [LG90], one of the best known and best funded AGI-oriented projects in history. While the main focus of the Cyc project has been on the hand-coding of large amounts of declarative knowledge, there is also a cognitive architecture of sorts there. The center of Cyc is an engine for logical deduction, acting on knowledge represented in predicate logic. A natural language engine has been associated with the logic engine, which enables one to ask English questions and get English replies.

Stephen Reed, while an engineer at Cycorp, designed a perceptual-motor front end for Cyc based on James Albus' Reference Model Architecture; the ensuing system, called CognitiveCyc, would have been the first full-fledged cognitive architecture based on Cyc, but was not implemented. Reed left Cycorp and is now building a system called Texai, which has many similarities to Cyc (and relies upon the Open-Cyc knowledge base, a subset of Cyc's overall knowledge base), but incorporates a CognitiveCyc style cognitive architecture.

6.2.4 NARS

Pei Wang's NARS logic [Wan06] played a large role in the development of PLN, CogPrime's uncertain logic component, a relationship that is discussed in depth in [GMIH08] and won't be re-emphasized here. However, NARS is more than just an uncertain logic, it is also an overall cognitive architecture (which is centered on NARS logic, but also includes other aspects). CogPrime bears little relation to NARS except in the specific similarities between PLN logic and NARS logic, but, the other aspects of NARS are worth briefly recounting here.

NARS is formulated as a system for processing tasks, where a task consists of a question or a piece of new knowledge. The architecture is focused on declarative knowledge, but some pieces of knowledge may be associated with executable procedures, which allows NARS to carry out control activities (in roughly the same way that a Prolog program can).

At any given time a NARS system contains

- working memory: a small set of tasks which are active, kept for a short time, and closely related to new questions and new knowledge
- long-term memory: a huge set of knowledge which is passive, kept for a long time, and not necessarily related to current questions and knowledge.

The working and long term memory spaces of NARS may each be thought of as a set of chunks, where each chunk consists of a set of tasks and a set of knowledge. NARS's basic cognitive process is:

1. choose a chunk
2. choose a task from that chunk

3. choose a piece of knowledge from that chunk
4. use the task and knowledge to do inference
5. send the new tasks to corresponding chunks.

Depending on the nature of the task and knowledge, the inference involved may be one of the following:

- if the task is a question, and the knowledge happens to be an answer to the question, a copy of the knowledge is generated as a new task
- backward inference
- revision (merging two pieces of knowledge with the same form but different truth value)
- forward inference
- execution of a procedure associated with a piece of knowledge.

Unlike many other systems, NARS doesn't decide what type of inference is used to process a task when the task is accepted, but works in a data-driven way—that is, it is the task and knowledge that dynamically determine what type of inference will be carried out.

The "choice" processes mentioned above are done via assigning relative priorities to

- chunks (where they are called activity)
- tasks (where they are called urgency)
- knowledge (where they are called importance).

and then distributing the system's resources accordingly, based on a probabilistic algorithm. (It's interesting to note that while NARS uses probability theory as part of its control mechanism, the logic it uses to represent its own knowledge about the world is nonprobabilistic. This is considered conceptually consistent, in the context of NARS theory, because system control is viewed as a domain where the system's knowledge is more complete, thus more amenable to probabilistic reasoning.)

6.2.5 GLAIR and SNePS

Another logic-focused cognitive architecture, very different from NARS in detail, is Stuart Shapiro's GLAIR cognitive architecture, which is centered on the SNePS paraconsistent logic [SE07].

Like NARS, the core "cognitive loop" of GLAIR is based on reasoning: either thinking about some percept (e.g. linguistic input, or sense data from the virtual or physical world), or answering some question. This inference based cognition process is turned into an intelligent agent control process via coupling it with an acting component, which operates according to a set of policies, each one of which tells the system when to take certain internal or external actions (including internal reasoning actions) in response to its observed internal and external situation.

GLAIR contains multiple layers:

- the Knowledge Layer (KL), which contains the beliefs of the agent, and is where reasoning, planning, and act selection are performed.
- the Sensori-Actuator Layer (SAL), contains the controllers of the sensors and effectors of the hardware or software robot.
- the Perceptuo-Motor Layer (PML), which grounds the KL symbols in perceptual structures and subconscious actions, contains various registers for providing the agent's sense of situatedness in the environment, and handles translation and communication between the KL and the SAL.

The logical Knowledge Layer incorporates multiple memory types using a common representation (including declarative, procedural, episodic, attentional and intentional knowledge, and meta-knowledge). To support this broad range of knowledge types, a broad range of logical inference mechanisms are used, so that the KL may be variously viewed as predicate logic based, frame based, semantic network based, or from other perspectives.

What makes GLAIR more robust than most logic based AI approaches is the novel paraconsistent logical formalism used in the knowledge base, which means (among other things) that uncertain, speculative or erroneous knowledge may exist in the system's memory without leading the system to create a broadly erroneous view of the world or carry out egregiously unintelligent actions. CogPrime is not thoroughly logic-focused like GLAIR is, but in its logical aspect it seeks a similar robustness through its use of PLN logic, which embodies properties related to paraconsistency.

Compared to CogPrime, we see that GLAIR has a similarly integrative approach, but that the integration of different sorts of cognition is done more strictly within the framework of logical knowledge representation.

6.3 Emergentist Cognitive Architectures

Another species of cognitive architecture expects abstract symbolic processing to emerge from lower-level "subsymbolic" dynamics, which sometimes (but not always) are designed to simulate neural networks or other aspects of human brain function. These architectures are typically strong at recognizing patterns in high-dimensional data, reinforcement learning and associative memory; but no one has yet shown how to achieve high-level functions such as abstract reasoning or complex language processing using a purely subsymbolic approach. A few of the more important subsymbolic, emergentist cognitive architectures are:

- **DeSTIN** [ARK09a, ARC09], which is part of CogPrime, may also be considered as an autonomous AGI architecture, in which case it is emergentist and contains mechanisms to encourage language, high-level reasoning and other abstract aspects of intelligent to emerge from hierarchical pattern recognition and related self-organizing network dynamics. In CogPrime DeSTIN is used as part of a hybrid architecture, which greatly reduces the reliance on DeSTIN's emergent properties.

- **Hierarchical Temporal Memory (HTM)** [HB06] is a hierarchical temporal pattern recognition architecture, presented as both an AI approach and a model of the cortex. So far it has been used exclusively for vision processing and we will discuss its shortcomings later in the context of our treatment of DeSTIN.
- **SAL** [JL08], based on the earlier and related **IBCA** (Integrated Biologically-based Cognitive Architecture) is a large-scale emergent architecture that seeks to model distributed information processing in the brain, especially the posterior and frontal cortex and the hippocampus. So far the architectures in this lineage have been used to simulate various human psychological and psycholinguistic behaviors, but haven't been shown to give rise to higher-level behaviors like reasoning or subgoaling.
- **NOMAD** (Neurally Organized Mobile Adaptive Device) automata and its successors [KE06] are based on Edelman's "Neural Darwinism" model of the brain, and feature large numbers of simulated neurons evolving by natural selection into configurations that carry out sensorimotor and categorization tasks. The emergence of higher-level cognition from this approach seems rather unlikely.
- Ben Kuipers and his colleagues [MK07, MK08, MK09] have pursued an extremely innovative research program which combines qualitative reasoning and reinforcement learning to enable an intelligent agent to learn how to act, perceive and model the world. Kuipers' notion of "bootstrap learning" involves allowing the robot to learn almost *everything* about its world, including for instance the structure of 3D space and other things that humans and other animals obtain via their genetic endowments. Compared to Kuipers' approach, CogPrime falls in line with most other approaches which provide more "hard-wired" structure, following the analogy to biological organisms that are born with more innate biases.

There is also a set of emergentist architectures focused specifically on developmental robotics, which we will review below in a separate subsection, as all of these share certain common characteristics.

Our general perspective on the emergentist approach is that it is philosophically correct but currently pragmatically inadequate. Eventually, *some* emergentist approach could surely succeed at giving rise to humanlike general intelligence—the human brain, after all, is plainly an emergentist system. However, we currently lack understanding of how the brain gives rise to abstract reasoning and complex language, and none of the existing emergentist systems seem remotely capable of giving rise to such phenomena. It seems to us that the creation of a successful emergentist AGI will have to wait for either a detailed understanding of how the brain gives rise to abstract thought, or a much more thorough mathematical understanding of the dynamics of complex self-organizing systems.

The concept of cognitive synergy is more relevant to emergentist than to symbolic architectures. In a complex emergentist architecture with multiple specialized components, much of the emergence is expected to arise via synergy between different richly interacting components. Symbolic systems, at least in the forms currently seen in the literature, seem less likely to give rise to cognitive synergy as their dynamics tend to be simpler. And hybrid systems, as we shall see, are somewhat diverse in this regard: some rely heavily on cognitive synergies and others consist of more loosely coupled components.

We now review the DeSTIN emergentist architecture in more detail, and then turn to the developmental robotics architectures.

6.3.1 DeSTIN: A Deep Reinforcement Learning Approach to AGI

The DeSTIN architecture, created by Itamar Arel and his colleagues, addresses the problem of general intelligence using hierarchical spatiotemporal networks designed to enable scalable perception, state inference and reinforcement-learning-guided action in real-world environments. DeSTIN has been developed with the plan of gradually extending it into a complete system for humanoid robot control, founded on the same qualitative information-processing principles as the human brain (though without striving for detailed biological realism). However, the practical work with DeSTIN to date has focused on visual and auditory processing; and in the context of the present proposal, the intention is to utilize DeSTIN for perception and actuation oriented processing, hybridizing it with CogPrime which will handle abstract cognition and language. Here we will discuss DeSTIN primarily in the perception context, only briefly mentioning the application to actuation which is conceptually similar.

In DeSTIN (see Fig. 6.3), perception is carried out by a deep spatiotemporal inference network, which is connected to a similarly architected critic network that provides feedback on the inference network's performance, and an action network that controls actuators based on the activity in the inference network (Fig. 6.4 depicts a standard action hierarchy, of which the hierarchy in DeSTIN is an example). The nodes in these networks perform probabilistic pattern recognition according to algorithms to be described below; and the nodes in each of the networks may receive states of nodes in the other networks as inputs, providing rich interconnectivity and synergetic dynamics.

6.3.1.1 Deep Versus Shallow Learning for Perceptual Data Processing

The most critical feature of DeSTIN is its uniquely robust approach to modeling the world based on perceptual data. Mimicking the efficiency and robustness by which the human brain analyzes and represents information has been a core challenge in AI research for decades. For instance, humans are exposed to massive amounts of visual and auditory data every second of every day, and are somehow able to capture critical aspects of it in a way that allows for appropriate future recollection and action selection. For decades, it has been known that the brain is a massively parallel fabric, in which computation processes and memory storage are highly distributed. But massive parallelism is not in itself a solution—one also needs the right architecture; which DeSTIN provides, building on prior work in the area of deep learning.

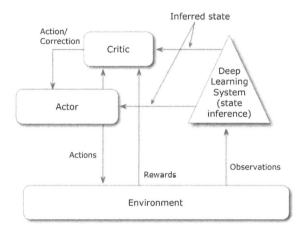

Fig. 6.3 High-level architecture of DeSTIN

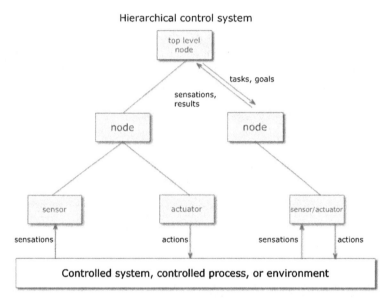

Fig. 6.4 A standard, general-purpose hierarchical control architecture. DeSTIN's control hierarchy exemplifies this architecture, with the difference lying mainly in the DeSTIN control hierarchy's tight integration with the state inference (perception) and critic (reinforcement) hierarchies

Humanlike intelligence is heavily adapted to the physical environments in which humans evolved; and one key aspect of sensory data coming from our physical environments is its **hierarchical** structure. However, most machine learning and pattern recognition systems are "shallow" in structure, not explicitly incorporating the hierarchical structure of the world in their architecture. In the context of perceptual data processing, the practical result of this is the need to couple each shallow learner with a pre-processing stage, wherein high-dimensional sensory signals are reduced to a lower-dimension feature space that can be understood by the shallow learner. The hierarchical structure of the world is thus crudely captured in the hierarchy of "preprocessor plus shallow learner." In this sort of approach, much of the intelligence of the system shifts to the feature extraction process, which is often imperfect and always application-domain specific.

Deep machine learning has emerged as a more promising framework for dealing with complex, high-dimensional real-world data. Deep learning systems possess a hierarchical structure that intrinsically biases them to recognize the hierarchical patterns present in real-world data. Thus, they hierarchically form a feature space that is driven by regularities in the observations, rather than by hand-crafted techniques. They also offer robustness to many of the distortions and transformations that characterize real-world signals, such as noise, displacement, scaling, etc.

Deep belief networks [HOT06] and Convolutional Neural Networks [LBDE90] have been demonstrated to successfully address pattern inference in high dimensional data (e.g. images). They owe their success to their underlying paradigm of partitioning large data structures into smaller, more manageable units, and discovering the dependencies that may or may not exist between such units. However, this paradigm has its limitations; for instance, these approaches do not represent temporal information with the same ease as spatial structure. Moreover, some key constraints are imposed on the learning schemes driving these architectures, namely the need for layer-by-layer training, and oftentimes pre-training. DeSTIN overcomes the limitations of prior deep learning approaches to perception processing, and also extends beyond perception to action and reinforcement learning.

6.3.1.2 DeSTIN for Perception Processing

The hierarchical architecture of DeSTIN's spatiotemporal inference network comprises an arrangement into multiple layers of "nodes" comprising multiple instantiations of an identical cortical circuit. Each node corresponds to a particular spatiotemporal region, and uses a statistical learning algorithm to characterize the sequences of patterns that are presented to it by nodes in the layer beneath it. More specifically,

- At the very lowest layer of the hierarchy nodes receive as input raw data (e.g. pixels of an image) and continuously construct a belief state that attempts to characterize the sequences of patterns viewed.
- The second layer, and all those above it, receive as input the belief states of nodes at their corresponding lower layers, and attempt to construct belief states that capture regularities in their inputs.
- Each node also receives as input the belief state of the node above it in the hierarchy (which constitutes "contextual" information) (Fig. 6.5).

More specifically, each of the DeSTIN nodes, referring to a specific spacetime region, contains a set of state variables conceived as clusters, each corresponding to

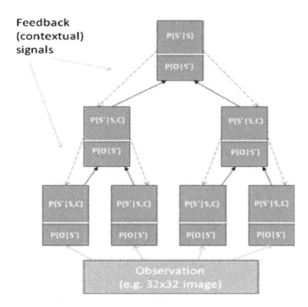

Fig. 6.5 Small-scale instantiation of the DeSTIN perceptual hierarchy. Each *box* represents a node, which corresponds to a spatiotemporal region (nodes higher in the hierarchy corresponding to larger regions). *O* denotes the current observation in the region, *C* is the state of the higher-layer node, and *S* and *S'* denote state variables pertaining to two subsequent time steps. In each node, a statistical learning algorithm is used to predict subsequent states based on prior states, current observations, and the state of the higher-layer node

a set of previously-observed sequences of events. These clusters are characterized by centroids (and are hence assumed roughly spherical in shape), and each of them comprises a certain "spatiotemporal form" recognized by the system in that region. Each node then contains the task of predicting the likelihood of a certain centroid being most apropos in the near future, based on the past history of observations in the node. This prediction may be done by simple probability tabulation, or via application of supervised learning algorithms such as recurrent neural networks. These clustering and prediction processes occur separately in each node, but the nodes are linked together via bidirectional dynamics: each node feeds input to its parents, and receives "advice" from its parents that is used to condition its probability calculations in a contextual way.

These processes are executed formally by the following basic belief update rule, which governs the learning process and is identical for every node in the architecture. The belief state is a probability mass function over the sequences of stimuli that the nodes learns to represent. Consequently, each node is allocated a predefined number of state variables each denoting a dynamic pattern, or sequence, that is autonomously learned. The DeSTIN update rule maps the current observation (o), belief state (b), and the belief state of a higher-layer node or context (c), to a new (updated) belief state (b'), such that

$$b'(s') = \Pr(s'|o, b, c) = \frac{\Pr(s' \cap o \cap b \cap c)}{\Pr(o \cap b \cap c)}, \tag{6.1}$$

alternatively expressed as

$$b'(s') = \frac{\Pr(o|s', b, c)\Pr(s'|b, c)\Pr(b, c)}{\Pr(o|b, c)\Pr(b, c)}. \tag{6.2}$$

Under the assumption that observations depend only on the true state, or $\Pr(o|s', b, c) = \Pr(o|s')$, we can further simplify the expression such that

$$b'(s') = \frac{\Pr(o|s')\Pr(s'|b, c)}{\Pr(o|b, c)}, \tag{6.3}$$

where $\Pr(s'|b, c) = \sum_{s \in S} \Pr(s'|s, c) b(s)$, yielding the belief update rule

$$b'(s') = \frac{\Pr(o|s') \sum_{s \in S} \Pr(s'|s, c) b(s)}{\sum_{s'' \in S} \Pr(o|s'') \sum_{s \in S} \Pr(s''|s, c) b(s)}, \tag{6.4}$$

where S denotes the sequence set (i.e. belief dimension) such that the denominator term is a normalization factor.

One interpretation of Eq. (6.4) would be that the static pattern similarity metric, $\Pr(o|s')$, is modulated by a construct that reflects the system dynamics, $\Pr(s'|s, c)$. As such, the belief state inherently captures both spatial and temporal information. In our implementation, the belief state of the parent node, c, is chosen using the selection rule

$$c = \arg \max_s b_p(s), \tag{6.5}$$

where b_p is the belief distribution of the parent node.

A close look at Eq. (6.4) reveals that there are two core constructs to be learned, $\Pr(o|s')$ and $\Pr(s'|s, c)$. In the current DeSTIN design, the former is learned via online clustering while the latter is learned based on experience by inductively learning a rule that predicts the next state s' given the prior state s and c.

The overall result is a robust framework that autonomously (i.e. with no human engineered pre-processing of any type) learns to represent complex data patterns, and thus serves the critical role of building and maintaining a model of the state of the world. In a vision processing context, for example, it allows for powerful unsupervised classification. If shown a variety of real-world scenes, it will automatically form internal structures corresponding to the various natural categories of objects shown in the scenes, such as trees, chairs, people, etc.; and also the various natural categories of events it sees, such as reaching, pointing, falling. And, as will be discussed below, it can use feedback from DeSTIN's action and critic networks to further shape its internal world-representation based on reinforcement signals.

Benefits of DeSTIN for Perception Processing

DeSTIN's perceptual network offers multiple key attributes that render it more powerful than other deep machine learning approaches to sensory data processing:

1. The belief space that is formed across the layers of the perceptual network inherently captures both *spatial and temporal regularities* in the data. Given that many applications require that temporal information be discovered for robust inference, this is a key advantage over existing schemes.
2. Spatiotemporal regularities in the observations are captured in a coherent manner (rather than being represented via two separate mechanisms).
3. All processing is both top-down and bottom-up, and both hierarchical and heterarchical, based on nonlinear feedback connections directing activity and modulating learning in multiple directions through DeSTIN's cortical circuits.
4. Support for multi-modal fusing is intrinsic within the framework, yielding a powerful state inference system for real-world, partially-observable settings.
5. Each node is identical, which makes it easy to map the design to massively parallel platforms, such as graphics processing units.

Points 2–4 in the above list describe how DeSTIN's perceptual network displays its own "cognitive synergy" in a way that fits naturally into the overall synergetic dynamics of the overall CogPrime architecture. Using this cognitive synergy, DeSTIN's perceptual network addresses a key aspect of general intelligence: the ability to robustly infer the state of the world, with which the system interacts, in an accurate and timely manner.

6.3.1.3 DeSTIN for Action and Control

DeSTIN's perceptual network performs unsupervised world-modeling, which is a critical aspect of intelligence but of course is not the whole story. DeSTIN's action network, coupled with the perceptual network, orchestrates actuator commands into complex movements, but also carries out other functions that are more cognitive in nature.

For instance, people learn to distinguish between cups and bowls in part via hearing other people describe some objects as cups and others as bowls. To emulate this kind of learning, DeSTIN's critic network provides positive or negative reinforcement signals based on whether the action network has correctly identified a given object as a cup or a bowl, and this signal then impacts the nodes in the action network. The critic network takes a simple external "degree of success or failure" signal and turns it into multiple reinforcement signals to be fed into the multiple layers of the action network. The result is that the action network self-organizes so as to include an implicit "cup versus bowl" classifier, whose inputs are the outputs of some of the nodes in the higher levels of the perceptual network. This classifier belongs in the action network because it is part of the procedure by which the DeSTIN system carries out the action of identifying an object as a cup or a bowl.

This example illustrates how the learning of complex concepts and procedures is divided fluidly between the perceptual network, which builds a model of the world in an unsupervised way, and the action network, which learns how to respond to the world in a manner that will receive positive reinforcement from the critic network.

6.3.2 Developmental Robotics Architectures

A particular subset of emergentist cognitive architectures are sufficiently important that we consider them separately here: these are *developmental robotics* architectures, focused on controlling robots without significant "hard-wiring" of knowledge or capabilities, allowing robots to learn (and learn how to learn, etc.) via their engagement with the world. A significant focus is often placed here on "intrinsic motivation," wherein the robot explores the world guided by internal goals like novelty or curiosity, forming a model of the world as it goes along, based on the modeling requirements implied by its goals. Many of the foundations of this research area were laid by Juergen Schmidhuber's work in the 1990s [Sch91b, Sch91a, Sch95, Sch02], but now with more powerful computers and robots the area is leading to more impressive practical demonstrations.

We mention here a handful of the important initiatives in this area:

- Juyang Weng's **Dav** [HZT+02] and **SAIL** [WHZ+00] projects involve mobile robots that explore their environments autonomously, and learn to carry out simple tasks by building up their own world-representations through both unsupervised and teacher-driven processing of high-dimensional sensorimotor data. The underlying philosophy is based on human child development [WH06], the knowledge representations involved are neural network based, and a number of novel learning algorithms are involved, especially in the area of vision processing.
- **FLOWERS** [BO09], an initiative at the French research institute INRIA, led by Pierre-Yves Oudeyer, is also based on a principle of trying to reconstruct the processes of development of the human child's mind, spontaneously driven by intrinsic motivations. Kaplan [Kap08] has taken this project in a direction closely related to our own via the creation of a "robot playroom." Experiential language learning has also been a focus of the project [OK06], driven by innovations in speech understanding.
- **IM-CLEVER**,[1] a new European project coordinated by Gianluca Baldassarre and conducted by a large team of researchers at different institutions, is focused on creating software enabling an iCub [MSV+08] humanoid robot to explore the

---

[1] http://im-clever.noze.it/project/project-description

environment and learn to carry out human childlike behaviors based on its own intrinsic motivations. As this project is the closest to our own we will discuss it in more depth below.

Like CogPrime, IM-CLEVER is a humanoid robot intelligence architecture guided by intrinsic motivations, and using hierarchical architectures for reinforcement learning and sensory abstraction. IM-CLEVER's motivational structure is based in part on Schmidhuber's information-theoretic model of curiosity [Sch06]; and CogPrime's Psi-based motivational structure utilizes probabilistic measures of novelty, which are mathematically related to Schmidhuber's measures. On the other hand, IM-CLEVER's use of reinforcement learning follows Schmidhuber's earlier work RL for cognitive robotics [BS04, BZGS06], Barto's work on intrinsically motivated reinforcement learning [SB06, SM05], and Lee's [LMC07a, LMC07b] work on developmental reinforcement learning; whereas CogPrime's assemblage of learning algorithms is more diverse, including probabilistic logic, concept blending and other symbolic methods (in the OCP component) as well as more conventional reinforcement learning methods (in the DeSTIN component).

In many respects IM-CLEVER bears a moderately strong resemblance to DeSTIN, whose integration with CogPrime is discussed in Chap. 9 of Part 2 (although IM-CLEVER has much more focus on biological realism than DeSTIN). Apart from numerous technical differences, the really big distinction between IM-CLEVER and CogPrime is that in the latter we are proposing to hybridize a hierarchical-abstraction/reinforcement-learning system (such as DeSTIN) with a more abstract symbolic cognition engine that explicitly handles probabilistic logic and language. IM-CLEVER lacks the aspect of hybridization with a symbolic system, taking more of a pure emergentist strategy. Like DeSTIN considered as a standalone architecture IM-CLEVER does entail a high degree of cognitive synergy, between components dealing with perception, world-modeling, action and motivation. However, the "emergentist versus hybrid" is a large qualitative difference between the two approaches.

In all, while we largely agree with the philosophy underlying developmental robotics, our intuition is that the learning and representational mechanisms underlying the current systems in this area are probably not powerful enough to lead to human child level intelligence. We expect that these systems will develop interesting behaviors but fall short of robust preschool level competency, especially in areas like language and reasoning where symbolic systems have typically proved more effective. This intuition is what impels us to pursue a hybrid approach, such as CogPrime. But we do feel that eventually, once the mechanisms underlying brains are better understood and robotic bodies are richer in sensation and more adept in actuation, some sort of emergentist, developmental-robotics approach can be successful at creating humanlike, human-level AGI.

6.4 Hybrid Cognitive Architectures

In response to the complementary strengths and weaknesses of the symbolic and emergentist approaches, in recent years a number of researchers have turned to integrative, hybrid architectures, which combine subsystems operating according to the two different paradigms. The combination may be done in many different ways, e.g. connection of a large symbolic subsystem with a large subsymbolic system, or the creation of a population of small agents each of which is both symbolic and subsymbolic in nature.

Nils Nilsson expressed the motivation for hybrid AGI systems very clearly in his article at the AI-50 conference (which celebrated the 50th anniversary of the AI field) [Nil09]. While affirming the value of the Physical Symbol System Hypothesis that underlies symbolic AI, he argues that "the PSSH explicitly assumes that, whenever necessary, symbols will be grounded in objects in the environment through the perceptual and effector capabilities of a physical symbol system." Thus, he continues,

> I grant the need for non-symbolic processes in some intelligent systems, but I think they supplement rather than replace symbol systems. I know of no examples of reasoning, understanding language, or generating complex plans that are best understood as being performed by systems using exclusively non-symbolic processes....
>
> AI systems that achieve human-level intelligence will involve a combination of symbolic and non-symbolic processing."

A few of the more important hybrid cognitive architectures are:

- **CLARION** [SZ04] is a hybrid architecture that combines a symbolic component for reasoning on "explicit knowledge" with a connectionist component for managing "implicit knowledge." Learning of implicit knowledge may be done via neural net, reinforcement learning, or other methods. The integration of symbolic and subsymbolic methods is powerful, but a great deal is still missing such as episodic knowledge and learning and creativity. Learning in the symbolic and subsymbolic portions is carried out separately rather than dynamically coupled, minimizing "cognitive synergy" effects.
- **DUAL** [NK04] is the most impressive system to come out of Marvin Minsky's "Society of Mind" paradigm. It features a population of agents, each of which combines symbolic and connectionist representation, self-organizing to collectively carry out tasks such as perception, analogy and associative memory. The approach seems innovative and promising, but it is unclear how the approach will scale to high-dimensional data or complex reasoning problems due to the lack of a more structured high-level cognitive architecture.
- **LIDA** [BF09] is a comprehensive cognitive architecture heavily based on Bernard Baars' "Global Workspace Theory". It articulates a "cognitive cycle" integrating various forms of memory and intelligent processing in a single processing loop. The architecture ties in well with both neuroscience and cognitive psychology, but it deals most thoroughly with "lower level" aspects of intelligence, handling more advanced aspects like language and reasoning only somewhat sketchily. There

is a clear mapping between LIDA structures and processes and corresponding structures and processing in OCP; so that it's only a mild stretch to view CogPrime as an instantiation of the general LIDA approach that extends further both in the lower level (to enable robot action and sensation via DeSTIN) and the higher level (to enable advanced language and reasoning via OCP mechanisms that have no direct LIDA analogues).

• **MicroPsi** [Bac09] is an integrative architecture based on Dietrich Dorner's Psi model of motivation, emotion and intelligence. It has been tested on some practical control applications, and also on simulating artificial agents in a simple virtual world. MicroPsi's comprehensiveness and basis in neuroscience and psychology are impressive, but in the current version of MicroPsi, learning and reasoning are carried out by algorithms that seem unlikely to scale. OCP incorporates the Psi model for motivation and emotion, so that MicroPsi and CogPrime may be considered very closely related systems. But similar to LIDA, MicroPsi currently focuses on the "lower level" aspects of intelligence, not yet directly handling advanced processes like language and abstract reasoning.

• **PolyScheme** [Cas07] integrates multiple methods of representation, reasoning and inference schemes for general problem solving. Each Polyscheme "specialist" models a different aspect of the world using specific representation and inference techniques, interacting with other specialists and learning from them. Polyscheme has been used to model infant reasoning including object identity, events, causality, and spatial relations. The integration of reasoning methods is powerful, but the overall cognitive architecture is simplistic compared to other systems and seems focused more on problem-solving than on the broader problem of intelligent agent control.

• **Shruti** [SA93] is a fascinating biologically-inspired model of human reflexive inference, which represents in connectionist architecture relations, types, entities and causal rules using focal-clusters. However, much like Hofstadter's earlier Copycat architecture [Hof95], Shruti seems more interesting as a prototype exploration of ideas than as a practical AGI system; at least, after a significant time of development it has not proved significantly effective in any applications.

• James Albus's **4D/RCS** robotics architecture shares a great deal with some of the emergentist architectures discussed above, e.g. it has the same hierarchical pattern recognition structure as DeSTIN and HTM, and the same three cross-connected hierarchies as DeSTIN, and shares with the developmental robotics architectures a focus on real-time adaptation to the structure of the world. However, 4D/RCS is not foundationally learning-based but relies on hard-wired architecture and algorithms, intended to mimic the qualitative structure of relevant parts of the brain (and intended to be *augmented* by learning, which differentiates it from emergentist approaches).

As our own CogPrime approach is a hybrid architecture, it will come as no surprise that we believe several of the existing hybrid architectures are fundamentally going in the right direction. However, nearly all the existing hybrid architectures have severe shortcomings which we feel will prevent them from achieving robust humanlike AGI.

Many of the hybrid architectures are in essence "multiple, disparate algorithms carrying out separate functions, encapsulated in black boxes and communicating results with each other." For instance, PolyScheme, ACT-R and CLARION all display this "modularity" property to a significant extent. These architectures lack the rich, real-time interaction between the *internal dynamics* of various memory and learning processes that we believe is critical to achieving humanlike general intelligence using realistic computational resources. On the other hand, those architectures that feature richer integration—such as DUAL, Shruti, LIDA and MicroPsi—have the flaw of relying (at least in their current versions) on overly simplistic learning algorithms, which drastically limits their scalability.

It does seem plausible to us that some of these hybrid architectures could be dramatically extended or modified so as to produce humanlike general intelligence. For instance, one could replace LIDA's learning algorithms with others that interrelate with each other in a nuanced synergetic way; or one could replace MicroPsi's simple learning and reasoning methods with much more powerful and scalable ones acting on the same data structures. However, making these changes would dramatically alter the cognitive architectures in question on multiple levels.

6.4.1 Neural Versus Symbolic; Global Versus Local

The "symbolic versus emergentist" dichotomy that we have used to structure our review of cognitive architectures is not absolute nor fully precisely defined; it is more of a heuristic distinction. In this section, before plunging into the details of particular hybrid cognitive architectures, we review two other related dichotomies that are useful for understanding hybrid systems: *neural versus symbolic* systems, and *globalist versus localist* knowledge representation.

6.4.1.1 Neural-Symbolic Integration

The distinction between neural and symbolic systems has gotten fuzzier and fuzzier in recent years, with developments such as

- Logic-based systems being used to control embodied agents (hence using logical terms to deal with data that is apparently perception or actuation-oriented in nature, rather than being symbolic in the semiotic sense), see [SS03a] and [GMIH08].
- Hybrid systems combining neural net and logical parts, or using logical or neural net components interchangeably in the same role [LAon].
- Neural net systems being used for strongly symbolic tasks such as automated grammar learning ([Elm91], plus more recent work.)

Figure 6.6 presents a schematic diagram of a generic neural-symbolic system, generalizing from [BH05], a paper that gives an elegant categorization of neural-

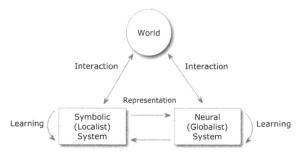

Fig. 6.6 Generic neural-symbolic architecture

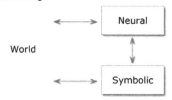

Fig. 6.7 Broad categories of neural-symbolic architecture

symbolic AI systems. Figure 6.7 depicts several broad categories of neural-symbolic architecture.

Bader and Hitzler categorize neural-symbolic systems according to three orthogonal axes: interrelation, language and usage. "Language" refers to the type of language used in the symbolic component, which may be logical, automata-based, formal grammar-based, etc. "Usage" refers to the purpose to which the neural-symbolic interrelation is put. We tend to use "learning" as an encompassing term for all forms of ongoing knowledge-creation, whereas Bader and Hitzler distinguish learning from reasoning.

Of Bader and Hitzler's three axes the one that interests us most here is "interrelation", which refers to the way the neural and symbolic components of the architecture intersect with each other. They distinguish "hybrid" architectures which contain separate but equal, interacting neural and symbolic components; versus "integrative" architectures in which the symbolic component essentially rides piggyback on the neural component, extracting information from it and helping it carry out its learning, but playing a clearly derived and secondary role. We prefer Sun's (2001) term "monolithic" to Bader and Hitzler's "integrative" to describe this type of system, as the latter term seems best preserved in its broader meaning.

Within the scope of hybrid neural-symbolic systems, there is another axis which Bader and Hitzler do not focus on, because the main interest of their review is in monolithic systems. We call this axis "interactivity", and what we are referring to is the frequency of high-information-content, high-influence interaction between the neural and symbolic components in the hybrid system. In a low-interaction hybrid system, the neural and symbolic components don't exchange large amounts of mutually influential information all that frequently, and basically act like independent system components that do their learning/reasoning/thinking periodically sending each other their conclusions. In some cases, interaction may be asymmetric: one component may frequently send a lot of influential information to the other, but not vice versa. However, our hypothesis is that the most capable neural-symbolic systems are going to be the symmetrically highly interactive ones.

In a symmetric high-interaction hybrid neural-symbolic system, the neural and symbolic components exchange influential information sufficiently frequently that each one plays a major role in the other one's learning/reasoning/thinking processes. Thus, the learning processes of each component must be considered as part of the overall dynamic of the hybrid system. The two components aren't just feeding their outputs to each other as inputs, they're mutually guiding each others' internal processing.

One can make a speculative argument for the relevance of this kind of architecture to neuroscience. It seems plausible that this kind of neural-symbolic system roughly emulates the kind of interaction that exists between the brain's neural subsystems implementing localist symbolic processing, and the brain's neural subsystems implementing globalist, classically "connectionist" processing. It seems most likely that, in the brain, symbolic functionality emerges from an underlying layer of neural dynamics. However, it is also reasonable to conjecture that this symbolic functionality is confined to a functionally distinct subsystem of the brain, which then interacts with other subsystems in the brain much in the manner that the symbolic and neural components of a symmetric high-interaction neural-symbolic system interact.

Neuroscience speculations aside, however, our key conjecture regarding neural-symbolic integration is that this sort of neural-symbolic system presents a promising direction for artificial general intelligence research. In Chap. 9 of Part 2, we will give a more concrete idea of what a symmetric high-interaction hybrid neural-symbolic architecture might look like, exploring the potential for this sort of hybridization between the OpenCogPrime AGI architecture (which is heavily symbolic in nature) and hierarchical attractor neural net based architectures such as DeSTIN.

6.5 Globalist Versus Localist Representations

Another interesting distinction, related to but different from "symbolic versus emergentist" and "neural versus symbolic", may be drawn between cognitive systems (or subsystems) where memory is essentially **global**, and those where memory is essentially **local**. In this section we will pursue this distinction in various guises, along with the less familiar notion of **glocal memory**.

This globalist/localist distinction is most easily conceptualized by reference to memories corresponding to categories of entities or events in an external environment. In an AI system that has an internal notion of "activation"—i.e. in which some of its internal elements are more active than others, at any given point in time—one can define the *internal image* of an external event or entity as the fuzzy set of internal elements that tend to be active when that event or entity is presented to the system's sensors. If one has a particular set S of external entities or events of interest, then, the *degree of memory localization* of such an AI system relative to S may be conceived as the percentage of the system's internal elements that have a high degree of membership in the internal image of an average element of S.

Of course, this characterization of localization has its limitations, such as the possibility of ambiguity regarding what are the "system elements" of a given AI system; and the exclusive focus on internal images of external phenomena rather than representation of internal abstract concepts. However, our goal here is not to formulate an ultimate, rigorous and thorough ontology of memory systems, but only to pose a "rough and ready" categorization so as to properly frame our discussion of some specific AGI issues relevant to CogPrime. Clearly the ideas pursued here will benefit from further theoretical exploration and elaboration.

In this sense, a Hopfield neural net [Ami89] would be considered "globalist" since it has a low degree of memory localization (most internal images heavily involve a large number of system elements); whereas Cyc would be considered "localist" as it has a very high degree of memory localization (most internal images are heavily focused on a small set of system elements).

However, although Hopfield nets and Cyc form handy examples, the "globalist versus localist" distinction as described above is not identical to the "neural versus symbolic" distinction. For it is in principle quite possible to create localist systems using formal neurons, and also to create globalist systems using formal logic. And "globalist-localist" is not quite identical to "symbolic versus emergentist" either, because the latter is about coordinated system dynamics and behavior not just about knowledge representation. CogPrime combines both symbolic and (loosely) neural representations, and also combines globalist and localist representations in a way that we will call "glocal" and analyze more deeply in Chap. 14; but there are many other ways these various properties could be manifested by AI systems. Rigorously studying the corpus of existing (or hypothetical!) cognitive architectures using these ideas would be a large task, which we do not undertake here.

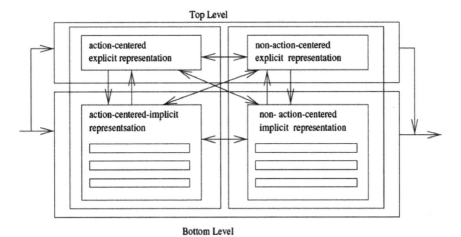

Fig. 6.8   The CLARION cognitive architecture

In the next sections we review several hybrid architectures in more detail, focusing most deeply on LIDA and MicroPsi which have been directly inspirational for CogPrime.

6.5.1 CLARION

Ron Sun's CLARION architecture (see Fig. 6.8) is interesting in its combination of symbolic and neural aspects—a combination that is used in a sophisticated way to embody the distinction and interaction between implicit and explicit mental processes. From a CLARION perspective, architectures like Soar and ACT-R are severely limited in that they deal only with explicit knowledge and associated learning processes.

CLARION consists of a number of distinct subsystems, each of which contains a dual representational structure, including a "rules and chunks" symbolic knowledge store somewhat similar to ACT-R, and a neural net knowledge store embodying implicit knowledge. The main subsystems are:

- An action-centered subsystem to control actions;
- A non-action-centered subsystem to maintain general knowledge;
- A motivational subsystem to provide underlying motivations for perception, action, and cognition;
- A meta-cognitive subsystem to monitor, direct, and modify the operations of all the other subsystems.

6.5.2 *The Society of Mind and the Emotion Machine*

In his influential but controversial book *The Society of Mind* [Min88], Marvin Minsky described a model of human intelligence as something that is built up from the interactions of numerous simple agents. He spells out in great detail how various particular cognitive functions may be achieved via agents and their interactions. He leaves no room for any central algorithms or structures of thought, famously arguing: "What magical trick makes us intelligent? The trick is that there is no trick. The power of intelligence stems from our vast diversity, not from any single, perfect principle."

This perspective was extended in the more recent work *The Emotion Machine* [Min07], where Minsky argued that emotions are "ways to think" evolved to handle different "problem types" that exist in the world. The brain is posited to have rule-based mechanisms (selectors) that turns on emotions to deal with various problems.

Overall, both of these works serve better as works of speculative cognitive science than as works of AI or cognitive architecture per se. As neurologist Richard Restak said in his review of *Emotion Machine*, "Minsky does a marvelous job parsing other complicated mental activities into simpler elements. ... But he is less effective in relating these emotional functions to what's going on in the brain." As Restak added, he is also not so effective at relating these emotional functions to straightforwardly implementable algorithms or data structures.

Push Singh, in his PhD thesis and followup work [SBC05], did the best job so far of creating a concrete AI design based on Minsky's ideas. While Singh's system was certainly interesting, it was also noteworthy for its lack of any learning mechanisms, and its exclusive focus on explicit rather than implicit knowledge. Due to Singh's tragic death, his work was never brought anywhere near completion. It seems fair to say that there has not yet been a serious cognitive architecture posed based closely on Minsky's ideas.

6.5.3 *DUAL*

The closest thing to a Minsky-ish cognitive architecture is probably DUAL, which takes the Society of Mind concept and adds to it a number of other interesting ideas. DUAL integrates symbolic and connectionist approaches at a deeper level than CLARION, and has been used to model various cognitive functions such as perception, analogy and judgment. Computations in DUAL emerge from the self-organized interaction of many micro-agents, each of which is a hybrid symbolic/connectionist device. Each DUAL agent plays the role of a neural network node, with an activation level and activation spreading dynamics; but also plays the role of a symbol, manipulated using formal rules. The agents exchange messages and activation via links that can be learned and modified, and they form coalitions which collectively represent concepts, episodes, and facts.

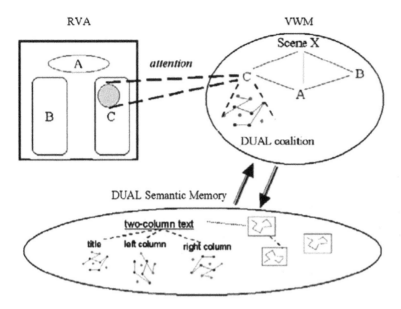

Fig. 6.9 The three main components of the DUAL model: the retinotopic visual array (RVA), the visual working memory (VWM) and DUAL's semantic memory. Attention is allocated to an area of the visual array by the object in VWM controlling attention, while scene and object categories corresponding to the contents of VWM are retrieved from the semantic memory

The structure of the model is sketchily depicted in Fig. 6.9, which covers the application of DUAL to a toy environment called TextWorld. The visual input corresponding to a stimulus is presented on a two-dimensional visual array representing the front end of the system. Perceptual primitives like blobs and terminations are immediately generated by cheap parallel computations. Attention is controlled at each time by an object which allocates it selectively to some area of the stimulus. A detailed symbolic representation is constructed for this area which tends to fade away as attention is withdrawn from it and allocated to another one. Categorization of visual memory contents takes place by retrieving object and scene categories from DUAL's semantic memory and mapping them onto current visual memory representations.

In principle the DUAL framework seems quite powerful; using the language of CogPrime, however, it seems to us that the learning mechanisms of DUAL have not been formulated in such a way as to give rise to powerful, scalable cognitive synergy. It would likely be possible to create very powerful AGI systems within DUAL, and perhaps some very CogPrime-like systems as well. But the systems that have been created or designed for use within DUAL so far seem not to be that powerful in their potential or scope.

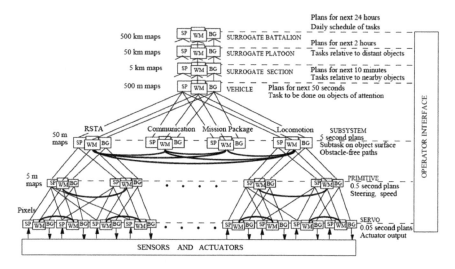

Fig. 6.10 Albus's 4D-RCS architecture for a single vehicle. Figure from [AM01], used with permission

6.5.4 4D/RCS

In a rather different direction, James Albus, while at the National Bureau of Standards, developed a very thorough and impressive architecture for intelligent robotics called 4D/RCS, which was implemented in a number of machines including unmanned automated vehicles. This architecture lacks critical aspects of intelligence such as learning and creativity, but combines perception, action, planning and world-modeling in a highly effective and tightly-integrated fashion.

The architecture has three hierarchies of memory/processing units: one for perception, one for action and one for modeling and guidance. Each unit has a certain spatiotemporal scope, and (except for the lowest level) supervenes over children whose spatiotemporal scope is a subset of its own. The action hierarchy takes care of decomposing tasks into subtasks; whereas the sensation hierarchy takes care of grouping signals into entities and events. The modeling/guidance hierarchy mediates interactions between perception and action based on its understanding of the world and the system's goals.

In his book [AM01] Albus describes methods for extending 4D/RCS into a complete cognitive architecture, but these extensions have not been elaborated in full detail nor implemented (Figs. 6.10 and 6.11).

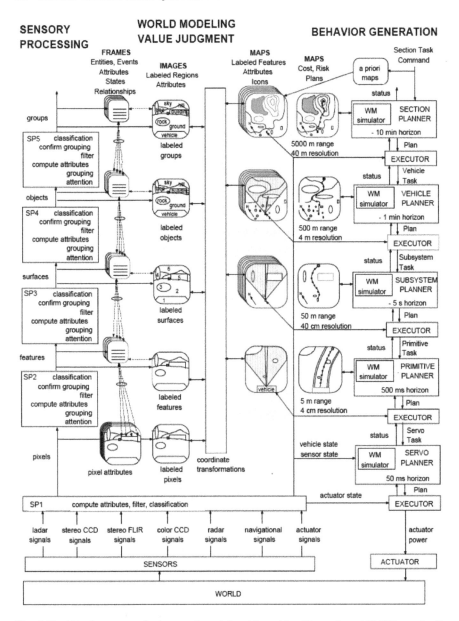

Fig. 6.11 Albus's perceptual, motor and modeling hierarchies. Figure from [AM01], used with permission

6.5.5 PolyScheme

Nick Cassimatis's PolyScheme architecture [Cas07] shares with GLAIR the use of multiple logical reasoning methods on a common knowledge store. While its underlying ideas are quite general, currently PolyScheme is being developed in the context of the "object tracking" domain (construed very broadly). As a logic framework PolyScheme is fairly conventional (unlike GLAIR or NARS with their novel underlying formalisms), but PolyScheme has some unique conceptual aspects, for instance its connection with Cassimatis's theory of mind, which holds that the same core set of logical concepts and relationships underlies both language and physical reasoning [Cas04]. This ties in with the use of a common knowledge store for multiple cognitive processes; for instance it suggests that

- the same core relationships can be used for physical reasoning and parsing, but that each of these domains may involve some additional relationships.
- language processing may be done via physical-reasoning-based cognitive processes, plus the additional activity of some language-specific processes.

6.5.6 Joshua Blue

Sam Adams and his colleagues at IBM have created a cognitive architecture called Joshua Blue [AABL02], which has some significant similarities to CogPrime. Similar to our current research direction with CogPrime, Joshua Blue was created with loose emulation of child cognitive development in mind; and, also similar to CogPrime, it features a number of cognitive processes acting on a common neural-symbolic knowledge store. The specific cognitive processes involved in Joshua Blue and CogPrime are not particularly similar, however. At time of writing (2012) Joshua Blue is not under active development and has not been for some time; however, the project may be reanimated in future.

Joshua Blue's core knowledge representation is a semantic network of nodes connected by links along which activation spreads. Although many of the nodes have specific semantic referents, as in a classical semantic net, the spread of activation through the network is designed to lead to the emergence of "assemblies" (which could also be thought of as dynamical attractors) in a manner more similar to an attractor neural network.

A major difference from typical semantic or neural network models is the central role that affect plays in the system's dynamics. The weights of the links in the knowledge base are adjusted dynamically based on the emotional context—a very direct way of ensuring that cognitive processes and mental representations are continuously influenced by affect. Qualitatively, this mimics the way that particular emotions in the human brain correlate with the dissemination throughout the brain of particular neurotransmitters, which then affect synaptic activity.

A result of this architecture is that in Joshua Blue, emotion directs attention in a very direct way: affective weighting is important in determining which associated objects will become part of the focus of attention, or will be retained from memory. A notable similarity between CogPrime and Joshua Blue is that in both systems, nodes are assigned two quantitative attention values, one governing allocation of current system resources (mainly processor time; this is CogPrime's ShortTermImportance) and one governing the long-term allocation of memory (CogPrime's LongTermImportance).

The concrete work done with Joshua Blue involved using it to control a simple agent in a simulated world, with the goal that via human interaction, the agent would develop a complex and humanlike emotional and motivational structure from its simple in-built emotions and drives, and would then develop complex cognitive capabilities as part of this development process.

6.5.7 LIDA

The LIDA architecture developed by Stan Franklin and his colleagues [BF09] is based on the concept of the "cognitive cycle"—a notion that is important to nearly every BICA (Biologically Inspired Cognitive Architectures) and also to the brain, but that plays a particularly central role in LIDA. As Franklin says, "as a matter of principle, every autonomous agent, be it human, animal, or artificial, must frequently sample (sense) its environment, process (make sense of) this input, and select an appropriate response (action). The agent's "life" can be viewed as consisting of a continual sequence of iterations of these cognitive cycles. Such cycles constitute the indivisible elements of attention, the least sensing and acting to which we can attend. A cognitive cycle can be thought of as a moment of cognition, a cognitive "moment"."

6.5.8 The Global Workspace

LIDA is heavily based on the "global workspace" concept developed by Bernard Baars. As this concept is also directly relevant to CogPrime it is worth briefly describing here.

In essence Baars' Global Workspace Theory (GWT) is a particular hypothesis about how working memory works and the role it plays in the mind. Baars conceives working memory as the "inner domain in which we can rehearse telephone numbers to ourselves or, more interestingly, in which we carry on the narrative of our lives. It is usually thought to include inner speech and visual imagery." Baars uses the term "consciousness" to refer to the contents of working memory—a theoretical commitment that is not part of the CogPrime design. In this section we will use the term "consciousness" in Baars' way, but not throughout the rest of the book.

Baars conceives working memory and consciousness in terms of a "theater metaphor"—according to which, in the "theater of consciousness" a "spotlight of selective attention" shines a bright spot on stage. The bright spot reveals the global workspace—the contents of consciousness, which may be metaphorically considered as a group of actors moving in and out of consciousness, making speeches or interacting with each other. The unconscious is represented by the audience watching the play ... and there is also a role for the director (the mind's executive processes) behind the scenes, along with a variety of helpers like stage hands, script writers, scene designers, etc.

GWT describes a fleeting memory with a duration of a few seconds. This is much shorter than the 10–30 s of classical working memory—according to GWT there is a very brief "cognitive cycle" in which the global workspace is refreshed, and the time period an item remains in working memory generally spans a large number of these elementary "refresh" actions. GWT contents are proposed to correspond to what we are conscious of, and are said to be broadcast to a multitude of unconscious cognitive brain processes. Unconscious processes, operating in parallel, can form coalitions which can act as input processes to the global workspace. Each unconscious process is viewed as relating to certain goals, and seeking to get involved with coalitions that will get enough importance to become part of the global workspace—because once they're in the global workspace they'll be allowed to broadcast out across the mind as a whole, which include broadcasting to the internal and external actuators that allow the mind to do things. Getting into the global workspace is a process's best shot at achieving its goals.

Obviously, the theater metaphor used to describe the GWT is evocative but limited; for instance, the unconscious in the mind does a lot more than the audience in a theater. The unconscious comes up with complex creative ideas sometimes, which feed into consciousness—almost as if the audience is also the scriptwriter. Baars' theory, with its understanding of unconscious dynamics in terms of coalition-building, fails to describe the subtle dynamics occurring within the various forms of long-term memory, which result in subtle nonlinear interactions between long term memory and working memory. But nevertheless, GWT successfully models a number of characteristics of consciousness, including its role in handling novel situations, its limited capacity, its sequential nature, and its ability to trigger a vast range of unconscious brain processes. It is the framework on which LIDA's theory of the cognitive cycle is built.

6.5.9 The LIDA Cognitive Cycle

The simplest cognitive cycle is that of an animal, which senses the world, compares sensation to memory, and chooses an action, all in one fluid subjective moment. But the same cognitive cycle structure/process applies to higher-level cognitive processes as well. The LIDA architecture is based on the LIDA model of the cognitive cycle,

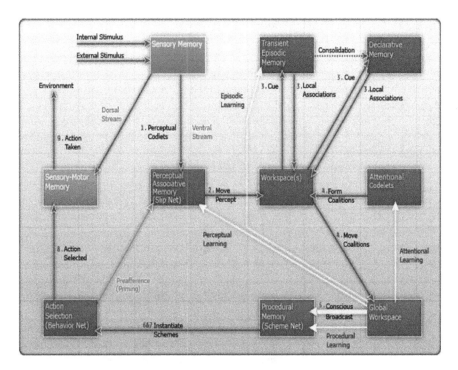

Fig. 6.12 The LIDA cognitive cycle. Copyright Stan Franklin (2006), used with permission

which posits a particular structure underlying the cognitive cycle that possess the generality to encompass both simple and complex cognitive moments.

The LIDA cognitive cycle itself is a theoretical construct that can be implemented in many ways, and indeed other BICAs like CogPrime and Psi also manifest the LIDA cognitive cycle in their dynamics, though utilizing different particular structures to do so.

Figure 6.12 shows the cycle pictorially, starting in the upper left corner and proceeding clockwise. At the start of a cycle, the LIDA agent perceives its current situation and allocates attention differentially to various parts of it. It then broadcasts information about the most important parts (which constitute the agent's consciousness), and this information gets features extracted from it, when then get passed along to episodic and semantic memory, that interact in the "global workspace" to create a model for the agent's current situation. This model then, in interaction with procedural memory, enables the agent to choose an appropriate action and execute it—the critical "action-selection" phase!

The LIDA Cognitive Cycle in More Depth

We now run through the cognitive cycle in more detail.[2] It begins with sensory stimuli from the agent's external internal environment. Low-level feature detectors in sensory memory begin the process of making sense of the incoming stimuli. These low-level features are passed to perceptual memory where higher-level features, objects, categories, relations, actions, situations, etc. are recognized. These recognized entities, called percepts, are passed to the workspace, where a model of the agent's current situation is assembled.

Workspace structures serve as cues to the two forms of episodic memory, yielding both short and long term remembered local associations. In addition to the current percept, the workspace contains recent percepts that haven't yet decayed away, and the agent's model of the then-current situation previously assembled from them. The model of the agent's current situation is updated from the previous model using the remaining percepts and associations. This updating process will typically require looking back to perceptual memory and even to sensory memory, to enable the understanding of relations and situations. This assembled new model constitutes the agent's understanding of its current situation within its world. Via constructing the model, the agent has made sense of the incoming stimuli.

Now attention allocation comes into play, because a real agent lacks the computational resources to work with all parts of its world-model with maximal mental focus. Portions of the model compete for attention. These competing portions take the form of (potentially overlapping) coalitions of structures comprising parts the model. Once one such coalition wins the competition, the agent has decided what to focus its attention on.

And now comes the purpose of all this processing: to help the agent to decide what to do next. The winning coalition passes to the global workspace, the namesake of Global Workspace Theory, from which it is broadcast globally. Though the contents of this conscious broadcast are available globally, the primary recipient is procedural memory, which stores templates of possible actions including their context and possible results.

Procedural memory also stores an activation value for each such template—a value that attempts to measure the likelihood of an action taken within its context producing the expected result. It's worth noting that LIDA makes a rather specific assumption here. LIDA's "activation" values are like the probabilistic truth values of the implications in CogPrime's *Context* $\wedge$ *Procedure* $\rightarrow$ *Goal* triples. However, in CogPrime this probability is not the same as the ShortTermImportance "attention value" associated with the Implication link representing that implication. Here LIDA merges together two concepts that in CogPrime are separate.

Templates whose contexts intersect sufficiently with the contents of the conscious broadcast instantiate copies of themselves with their variables specified to the current situation. These instantiations are passed to the action selection mechanism, which chooses a single action from these instantiations and those remaining from previous

---

[2] This section paraphrases heavily from [Fra06].

cycles. The chosen action then goes to sensorimotor memory, where it picks up the appropriate algorithm by which it is then executed. The action so taken affects the environment, and the cycle is complete.

The LIDA model hypothesizes that all human cognitive processing is via a continuing iteration of such cognitive cycles. It acknowledges that other cognitive processes may also occur, refining and building on the knowledge used in the cognitive cycle (for instance, the cognitive cycle itself doesn't mention abstract reasoning or creativity). But the idea is that these other processes occur in the context of the cognitive cycle, which is the main loop driving the internal and external activities of the organism.

6.5.9.1 Avoiding Combinatorial Explosion via Adaptive Attention Allocation

LIDA avoids combinatorial explosions in its inference processes via two methods, both of which are also important in CogPrime:

- combining reasoning via association with reasoning via deduction
- foundational use of uncertainty in reasoning.

One can create an analogy between LIDA's workspace structures and codelets and a logic-based architecture's assertions and functions. However, LIDA's codelets only operate on the structures that are active in the workspace during any given cycle. This includes recent perceptions, their closest matches in other types of memory, and structures recently created by other codelets. The results with the highest estimate of success, i.e. activation, will then be selected.

Uncertainty plays a role in LIDA's reasoning in several ways, most notably through the base activation of its behavior codelets, which depend on the model's estimated probability of the codelet's success if triggered. LIDA observes the results of its behaviors and updates the base activation of the responsible codelets dynamically.

We note that for this kind of uncertain inference/activation interplay to scale well, some level of cognitive synergy must be present; and based on our understanding of LIDA it is not clear to us whether the particular inference and association algorithms used in LIDA possess the requisite synergy.

6.5.9.2 LIDA Versus CogPrime

The LIDA cognitive cycle, broadly construed, exists in CogPrime as in other cognitive architectures. To see how, it suffices to map the key LIDA structures into corresponding CogPrime structures, as is done in Table 6.1. Of course this table does not cover all CogPrime processes, as LIDA does not constitute a thorough explanation of CogPrime structure and dynamics. And in most cases the corresponding CogPrime and LIDA processes don't work in exactly the same way; for instance, as noted above, LIDA's action selection relies solely on LIDA's "activation" values,

Table 6.1 CogPrime analogues of key LIDA features

LIDA	CogPrime
Declarative memory	Atomspace
Attentional codelets	Schema that adjust importance of atoms explicitly
Coalitions	Maps
Global workspace	Attentional focus
Behavior codelets	Schema
Procedural memory (scheme net)	Procedures in ProcedureRepository; and network of SchemaNodes in the Atomspace
Action selection (behavior net)	Propagation of STICurrency from goals to actions, and action selection process
Transient episodic memory	Perceptual atoms entering AT with high STI, which rapidly decreases in most cases
Local workspaces	Bubbles of interlinked atoms with moderate importance, focused on by a subset of MindAgents (defined in Chap. 2 of Part 2) for a period of time
Perceptual associative memory	HebbianLinks in the AT
Sensory memory	Spaceserver/timeserver, plus auxiliary stores for other senses
Sensorimotor memory	Atoms storing record of actions taken, linked in with atoms indexed in sensory memory

whereas CogPrime's action selection process is more complex, relying on aspects of
CogPrime that lack LIDA analogues.

6.5.10 Psi and MicroPsi

We have saved for last the architecture that has the most in common with CogPrime:
Joscha Bach's MicroPsi architecture, closely based on Dietrich Dorner's Psi theory.
CogPrime has borrowed substantially from Psi in its handling of emotion and motivation; but Psi also has other aspects that differ considerably from CogPrime. Here
we will focus more heavily on the points of overlap, but will mention the key points
of difference as well.

The overall Psi cognitive architecture, which is centered on the Psi model of the
motivational system, is roughly depicted in Fig. 6.13.

Psi's motivational system begins with **Demands**, which are the basic factors that
motivate the agent. For an animal these would include things like food, water, sex,
novelty, socialization, protection of one's children, and so forth. For an intelligent
robot they might include things like electrical power, novelty, certainty, socialization,
well-being of others and mental growth.

Psi also specifies two fairly abstract demands and posits them as psychologically
fundamental (see Fig. 6.14):

- **competence**, the effectiveness of the agent at fulfilling its Urges

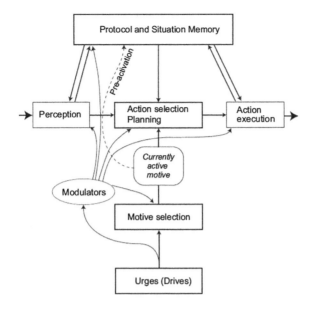

Fig. 6.13 High-level architecture of the Psi model

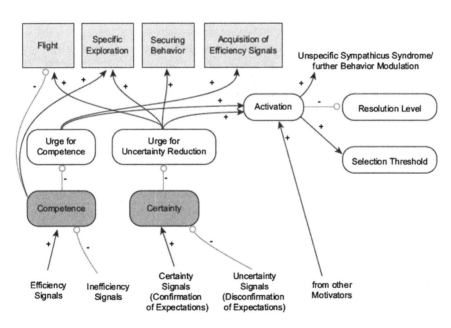

Fig. 6.14 Primary interrelationships between Psi modulators

- **certainty**, the confidence of the agent's knowledge.

Each demand is assumed to come with a certain "target level" or "target range" (and these may fluctuate over time, or may change as a system matures and develops). An **Urge** is said to develop when a demand deviates from its target range: the urge then seeks to return the demand to its target range. For instance, in an animal-like agent the demand related to food is more clearly described as "fullness," and there is a target range indicating that the agent is neither too hungry nor too full of food. If the agent's fullness deviates from this range, an Urge to return the demand to its target range arises. Similarly, if an agent's novelty deviates from its target range, this means the agent's life has gotten either too boring or too disconcertingly weird, and the agent gets an Urge for either more interesting activities (in the case of below-range novelty) or more familiar ones (in the case of above-range novelty).

There is also a primitive notion of **Pleasure** (and its opposite, displeasure), which is considered as different from the complex emotion of "happiness." Pleasure is understood as associated with Urges: pleasure occurs when an Urge is (at least partially) satisfied, whereas displeasure occurs when an urge gets increasingly severe. The degree to which an Urge is satisfied is not necessarily defined instantaneously; it may be defined, for instance, as a time-decaying weighted average of the proximity of the demand to its target range over the recent past.

So, for instance if an agent is bored and gets a lot of novel stimulation, then it experiences some pleasure. If it's bored and then the monotony of its stimulation gets even more extreme, then it experiences some displeasure.

Note that, according to this relatively simplistic approach, any decrease in the amount of dissatisfaction causes some pleasure; whereas if everything always continues within its acceptable range, there isn't any pleasure. This may seem a little counterintuitive, but it's important to understand that these simple definitions of "pleasure" and "displeasure" are not intended to fully capture the natural language concepts associated with those words. The natural language terms are used here simply as heuristics to convey the general character of the processes involved. These are very low level processes whose analogues in human experience are largely below the conscious level.

A **Goal** is considered as a statement that the system may strive to make true at some future time. A **Motive** is an (*urge, goal*) pair, consisting of a goal whose satisfaction is predicted to imply the satisfaction of some urge. In fact one may consider Urges as top-level goals, and the agent's other goals as their subgoals.

In Psi an agent has one "ruling motive" at any point in time, but this seems an oversimplification more applicable to simple animals than to human-like or other advanced AI systems. In general one may think of different motives having different weights indicating the amount of resources that will be spent on pursuing them.

Emotions in Psi are considered as complex systemic response-patterns rather than explicitly constructed entities. An emotion is the set of mental entities activated in response to a certain set of urges. Dorner conceived theories about how various common emotions emerge from the dynamics of urges and motives as described in the Psi model. "Intentions" are also considered as composite entities: an intention at a given point in time consists of the active motives, together with their related goals, behavior programs and so forth.

The basic logic of action in Psi is carried out by "triples" that are very similar to CogPrime's *Context* ∧ *Procedure* → *Goal* triples. However, an important role is played by four **modulators** that control how the processes of perception, cognition and action selection are regulated at a given time:

- *activation*, which determines the degree to which the agent is focused on rapid, intensive activity versus reflective, cognitive activity
- *resolution level*, which determines how accurately the system tries to perceive the world
- *certainty*, which determines how hard the system tries to achieve definite, certain knowledge
- *selection threshold*, which determines how willing the system is to change its choice of which goals to focus on.

These modulators characterize the system's emotional and cognitive state at a very abstract level; they are not emotions per se, but they have a large effect on the agent's emotions. Their intended interaction is depicted in Fig. 6.14.

6.5.11 The Emergence of Emotion in the Psi Model

We now briefly review the specifics of how Psi models the emergence of emotion. The basic idea is to define a small set of **proto-emotional dimensions** in terms of basic Urges and modulators. Then, emotions are identified with regions in the space spanned by these dimensions.

The simplest approach uses a six-dimensional continuous space:

1. pleasure
2. arousal
3. resolution level
4. selection threshold (i.e. degree of dominance of the leading motive)
5. level of background checks (the rate of the securing behavior)
6. level of goal-directed behavior.

Figure 6.15 shows how the latter 5 of these dimensions are derived from underlying urges and modulators. Note that these dimensions are not orthogonal; for instance resolution is mainly inversely related to arousal. Additional dimensions are also discussed, for instance it is postulated that to deal with social emotions one may wish to introduce two more demands corresponding to inner and outer obedience to social norms, and then define dimensions in terms of these.

Specific emotions are then characterized in terms of these dimensions. According to [Bac09], for instance, "Anger ... is characterized by high arousal, low resolution, strong motive dominance, few background checks and strong goal-orientedness;

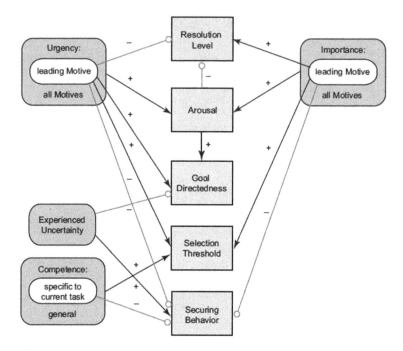

Fig. 6.15 Five proto-emotional dimensions implicit in the Psi model

sadness by low arousal, high resolution, strong dominance, few background-checks
and low goal-orientedness."

I'm a bit skeptical of the contention that these dimensions *fully* characterize the
relevant emotions. Anger for instance seems to have some particular characteristics
not implied by the above list of dimensional values. The list of dimensional values
associated with anger doesn't tell us that an angry person is more likely to punch
someone than to bounce up and down, for example. However, it does seem that the
dimensional values associated with an emotion are informative about the emotion,
so that positioning an emotion on the given dimensions tells one a lot.

6.5.12 Knowledge Representation, Action Selection
and Planning in Psi

In addition to the basic motivation/emotion architecture of Psi, which has been
adopted (with some minor changes) for use in CogPrime, Psi has a number of other
aspects that are somewhat different from their CogPrime analogues.

First of all, on the micro level, Psi represents knowledge using structures called
"quads." Each quad is a cluster of 5 neurons containing a core neuron, and four other

neurons representing before/after and part-of/has-part relationships in regard to that core neuron. Quads are naturally assembled into spatiotemporal hierarchies, though they are not required to form part of such a structure.

Psi stores knowledge using quads arranged in three networks, which are conceptually similar to the networks in Albus's 4D/RCS and Arel's DeSTIN architectures:

- A sensory network, which stores declarative knowledge: schemas representing images, objects, events and situations as hierarchical structures.
- A motor network, which contains procedural knowledge by way of hierarchical behavior programs.
- A motivational network handling demands.

Perception in Psi, which is centered in the sensory network, follows principles similar to DeSTIN (which are shared also by other systems), for instance the principle of *perception as prediction*. Psi's "HyPercept" mechanism performs hypothesis-based perception: it attempts to predict what is there to be perceived and then attempts to verify these predictions using sensation and memory. Furthermore HyPercept is intimately coupled with actions in the external world, according to the concept of "Neisser's perceptual cycle," the cycle between exploration and representation of reality. Perceptually acquired information is translated into schemas capable of guiding behaviors, and these are enacted (sometimes affecting the world in significant ways) and in the process used to guide further perception. Imaginary perceptions are handled via a "mental stage" analogous to CogPrime's internal simulation world.

Action selection in Psi works based on what are called "triplets," each of which consists of

- a sensor schema (pre-conditions, "condition schema"; like CogPrime's "context")
- a subsequent motor schema (action, effector; like CogPrime's "procedure")
- a final sensor schema (post-conditions, expectations; like an CogPrime predicate or goal).

What distinguishes these triplets from classic production rules as used in (say) Soar and ACT-R is that the triplets may be partial (some of the three elements may be missing) and may be uncertain. However, there seems no fundamental difference between these triplets and CogPrime's concept/procedure/goal triplets, at a high level; the difference lies in the underlying knowledge representation used for the schemata, and the probabilistic logic used to represent the implication.

The work of figuring out what schema to execute to achieve the chosen goal in the current context is done in Psi using a combination of processes called the "Rasmussen ladder" (named after Danish psychologist Jens Rasmussen). The Rasmussen ladder describes the organization of action as a movement between the stages of skill-based behavior, rule-based behavior and knowledge-based behavior, as follows:

- If a given task amounts to a trained routine, an automatism or skill is activated; it can usually be executed without conscious attention and deliberative control.
- If there is no automatism available, a course of action might be derived from rules; before a known set of strategies can be applied, the situation has to be analyzed and the strategies have to be adapted.

- In those cases where the known strategies are not applicable, a way of combining the available manipulations (operators) into reaching a given goal has to be explored at first. This stage usually requires a recomposition of behaviors, that is, a planning process.

The planning algorithm used in the Psi and MicroPsi implementations is a fairly simple hill-climbing planner. While it's hypothesized that a more complex planner may be needed for advanced intelligence, part of the Psi theory is the hypothesis that most real-life planning an organism needs to do is fairly simple, once the organism has the right perceptual representations and goals.

6.5.13 Psi Versus CogPrime

On a high level, the similarities between Psi and CogPrime are quite strong:

- interlinked declarative, procedural and intentional knowledge structures, represented using neural-symbolic methods (though, the knowledge structures have somewhat different high-level structures and low-level representational mechanisms in the two systems)
- perception via prediction and perception/action integration
- action selection via triplets that resemble uncertain, potentially partial production rules
- similar motivation/emotion framework, since CogPrime incorporates a variant of Psi for this.

On the nitty-gritty level there are many differences between the systems, but on the big-picture level the *main* difference lies in the way the cognitive synergy principle is pursued in the two different approaches. Psi and MicroPsi rely on very simple learning algorithms that are closely tied to the "quad" neurosymbolic knowledge representation, and hence interoperate in a fairly natural way without need for subtle methods of "synergy engineering." CogPrime uses much more diverse and sophisticated learning algorithms which thus require more sophisticated methods of interoperation in order to achieve cognitive synergy.

Chapter 7
A Generic Architecture of Human-Like Cognition

7.1 Introduction

When writing the first draft of this book, some years ago, we had the idea to explain CogPrime by aligning its various structures and processes with the ones in the "standard architecture diagram" of the human mind. After a bit of investigation, though, we gradually came to the realization that no such thing existed. There was no standard flowchart or other sort of diagram explaining the modern consensus on how human thought works. Many such diagrams existed, but each one seemed to represent some particular focus or theory, rather than an overall integrative understanding.

Since there are multiple opinions regarding nearly every aspect of human intelligence, it would be difficult to get two cognitive scientists to fully agree on every aspect of an overall human cognitive architecture diagram. Prior attempts to outline detailed mind architectures have tended to follow highly specific theories of intelligence, and hence have attracted only moderate interest from researchers not adhering to those theories. An example is Minsky's work presented in *The Emotion Machine* [Min07], which arguably does constitute an architecture diagram for the human mind, but which is only loosely grounded in current empirical knowledge and stands more as a representation of Minsky's own intuitive understanding.

But nevertheless, it seemed to us that a reasonable attempt at an integrative, relatively theory-neutral "human cognitive architecture diagram" would be better than nothing. So naturally, we took it on ourselves to create such a diagram. This chapter is the result—it draws on the thinking of a number of cognitive science and AGI researchers, integrating their perspectives in a coherent, overall architecture diagram for human, and human-like, general intelligence. The specific architecture diagram of CogPrime, given in Chap. 2, may then be understood as a particular instantiation of this generic architecture diagram of human-like cognition.

There is no getting around the fact that, to a certain extent, the diagram presented here reflects our particular understanding of how the mind works. However, it was intentionally constructed with the goal of *not* being just an abstracted version of the CogPrime architecture diagram! It does not reflect our own idiosyncratic

B. Goertzel et al., *Engineering General Intelligence, Part 1*,
Atlantis Thinking Machines 5, DOI: 10.2991/978-94-6239-027-0_7,
© Atlantis Press and the authors 2014

understanding of human intelligence, as much as a combination of understandings previously presented by multiple researchers (including ourselves), arranged according to our own taste in a manner we find conceptually coherent. With this in mind, we call it the "Integrative Human-Like Cognitive Architecture Diagram", or for short "the integrative diagram". We have made an effort to ensure that as many pieces of the integrative diagram as possible are well grounded in psychological and even neuroscientific data, rather than mainly embodying speculative notions; however, given the current state of knowledge, this could not be done to a complete extent, and there is still some speculation involved here and there.

While based on understandings of human intelligence, the integrative diagram is intended to serve as an architectural outline for human-like general intelligence more broadly. For example, CogPrime is explicitly not intended as a precise emulation of human intelligence, and does many things quite differently than the human mind, yet can still fairly straightforwardly be mapped into the integrative diagram.

The integrative diagram focuses on *structure*, but this should not be taken to represent a valuation of structure over dynamics in our approach to intelligence. Following chapters treat various dynamical phenomena in depth.

7.2 Key Ingredients of the Integrative Human-Like Cognitive Architecture Diagram

The main ingredients we've used in assembling the integrative diagram are as follows:

- Our own views on the various types of memory critical for human-like cognition, and the need for tight, "synergetic" interactions between the cognitive processes focused on these.
- Aaron Sloman's high-level architecture diagram of human intelligence [Slo01], drawn from his CogAff architecture, which strikes me as a particularly clear embodiment of "modern common sense" regarding the overall architecture of the human mind. We have added only a couple items to Sloman's high-level diagram, which we felt deserved an explicit high-level role that he did not give them: emotion, language and reinforcement.
- The LIDA architecture diagram presented by Stan Franklin and Bernard Baars [BF09]. We think LIDA is an excellent model of working memory and what Sloman calls "reactive processes", with well-researched grounding in the psychology and neuroscience literature. We have adapted the LIDA diagram only very slightly for use here, changing some of the terminology on the arrows, and indicating where parts of the LIDA diagram indicate processes elaborated in more detail elsewhere in the integrative diagram.
- The architecture diagram of the Psi model of motivated cognition, presented by Joscha Bach in [Bac09] based on prior work by Dietrich Dorner [Dör02]. This diagram is presented without significant modification; however it should be noted that Bach and Dorner present this diagram in the context of larger and richer cognitive

models, the other aspects of which are not all incorporated in the integrative diagram.

- James Albus's three-hierarchy model of intelligence [AM01], involving coupled perception, action and reinforcement hierarchies. Albus's model, utilized in the creation of intelligent unmanned automated vehicles, is a crisp embodiment of many ideas emergent from the field of intelligent control systems.
- Deep learning networks as a model of perception (and action and reinforcement learning), as embodied for example in the work of Itamar Arel [ARC09] and Jeff Hawkins [HB06]. The integrative diagram adopts this as the basic model of the perception and action subsystems of human intelligence. Language understanding and generation are also modeled according to this paradigm.

One possible negative reaction to the integrative diagram might be to say that it's a kind of Frankenstein monster diagram, piecing together aspects of different theories in a way that violates the theoretical notions underlying all of them! For example, the integrative diagram takes LIDA as a model of working memory and reactive processing, but from the papers on LIDA it's unclear whether the creators of LIDA construe it more broadly than that. The deep learning community tends to believe that the architecture of current deep learning networks, in itself, is close to sufficient for human-level general intelligence—whereas the integrative diagram appropriates the ideas from this community mainly for handling perception, action and language, etc.

On the other hand, in a more positive perspective, one could view the integrative diagram as consistent with LIDA, but merely providing much more detail on some of the boxes in the LIDA diagram (e.g. dealing with perception and long-term memory). And one could view the integrative diagram as consistent with the deep learning paradigm—via viewing it, not as a description of components to be explicitly implemented in an AGI system, but rather as a description of the key structures and processes that must emerge in deep learning network, based on its engagement with the world, in order for it to achieve human-like general intelligence.

Our own view, underlying the creation of the integrative diagram, is that different communities of cognitive science researchers have focused on different aspects of intelligence, and have thus each created models that are more fully fleshed out in some aspects than others. But these various models all link together fairly cleanly, which is not surprising as they are all grounded in the same data regarding human intelligence. Many judgment calls must be made in fusing multiple models in the way that the integrative diagram does, but we feel these can be made without violating the spirit of the component models. In assembling the integrative diagram, we have made these judgment calls as best we can, but we're well aware that different judgments would also be feasible and defensible. Revisions are likely as time goes on, not only due to new data about human intelligence but also to evolution of understanding regarding the best approach to model integration.

Another possible argument against the ideas presented here is that there's nothing new—all the ingredients presented have been given before elsewhere. To this our retort is to quote Pascal: "Let no one say that I have said nothing new... the

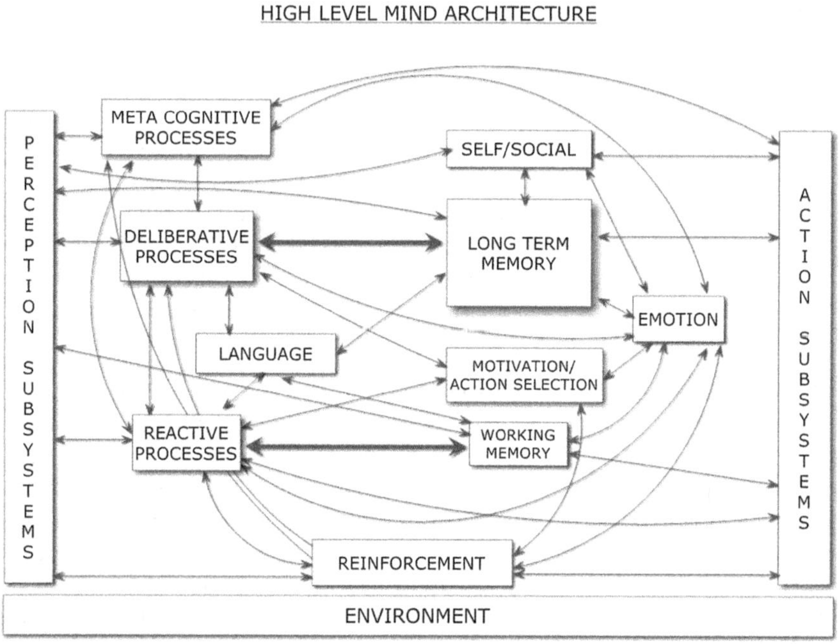

HIGH LEVEL MIND ARCHITECTURE

Fig. 7.1 High-level architecture of a human-like mind

arrangement of the subject is new". The various architecture diagrams incorporated into the integrative diagram are either extremely high level (Sloman's diagram) or focus primarily on one aspect of intelligence, treating the others very concisely by summarizing large networks of distinction structures and processes in small boxes. The integrative diagram seeks to cover all aspects of human-like intelligence at a roughly equal granularity—a different arrangement.

This kind of high-level diagramming exercise is not precise enough, nor dynamics-focused enough, to serve as a guide for creating human-level or more advanced AGI. But it can be a useful tool for explaining and interpreting a concrete AGI design, such as CogPrime.

7.3 An Architecture Diagram for Human-Like General Intelligence

The integrative diagram is presented here in a series of seven Figures.

Figure 7.1 gives a high-level breakdown into components, based on Sloman's high-level cognitive-architectural sketch [Slo01]. This diagram represents, roughly speaking, "modern common sense" about how a human-like mind is architected. The separation between structures and processes, embodied in having separate boxes

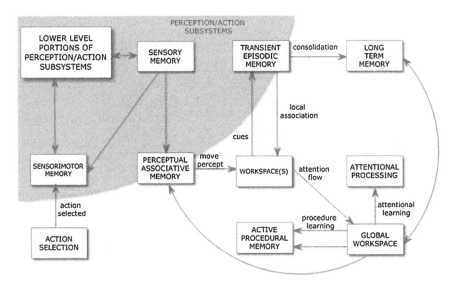

Fig. 7.2 Architecture of working memory and reactive processing, closely modeled on the LIDA architecture

for Working Memory versus Reactive Processes, and for Long Term Memory versus Deliberative Processes, could be viewed as somewhat artificial, since in the human brain and most AGI architectures, memory and processing are closely integrated. However, the tradition in cognitive psychology is to separate out Working Memory and Long Term Memory from the cognitive processes acting thereupon, so we have adhered to that convention. The other changes from Sloman's diagram are the explicit inclusion of language, representing the hypothesis that language processing is handled in a somewhat special way in the human brain; and the inclusion of a reinforcement component parallel to the perception and action hierarchies, as inspired by intelligent control systems theory (e.g. Albus as mentioned above) and deep learning theory. Of course Sloman's high level diagram in its original form is intended as inclusive of language and reinforcement, but we felt it made sense to give them more emphasis.

Figure 7.2, modeling working memory and reactive processing, is essentially the LIDA diagram as given in prior papers by Stan Franklin, Bernard Baars and colleagues [BF09]. The boxes in the upper left corner of the LIDA diagram pertain to sensory and motor processing, which LIDA does not handle in detail, and which are modeled more carefully by deep learning theory. The bottom left corner box refers to action selection, which in the integrative diagram is modeled in more detail by Psi. The top right corner box refers to Long-Term Memory, which the integrative diagram models in more detail as a synergetic multi-memory system (Fig. 7.4).

The original LIDA diagram refers to various "codelets", a key concept in LIDA theory. We have replaced "attention codelets" here with "attention flow", a more generic term. We suggest one can think of an attention codelet as: a piece of

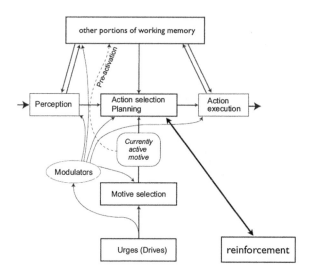

Fig. 7.3 Architecture of motivated action

information stating that, for a certain group of items, it's currently pertinent to pay attention to this group as a collective.

Figure 7.3, modeling motivation and action selection, is a lightly modified version of the Psi diagram from Joscha Bach's book *Principles of Synthetic Intelligence* [Bac09]. The main difference from Psi is that in the integrative diagram the Psi motivated action framework is embedded in a larger, more complex cognitive model. Psi comes with its own theory of working and long-term memory, which is related to but different from the one given in the integrative diagram—it views the multiple memory types distinguished in the integrative diagram as emergent from a common memory substrate. Psi comes with its own theory of perception and action, which seems broadly consistent with the deep learning approach incorporated in the integrative diagram. Psi's handling of working memory lacks the detailed, explicit workflow of LIDA, though it seems broadly conceptually consistent with LIDA.

In Fig. 7.3, the box labeled "Other portions of working memory" is labeled "Protocol and situation memory" in the original Psi diagram. The Perception, Action Execution and Action Selection boxes have fairly similar semantics to the similarly labeled boxes in the LIDA-like Fig. 7.2, so that these diagrams may be viewed as overlapping. The LIDA model doesn't explain action selection and planning in as much detail as Psi, so the Psi-like Fig. 7.3 could be viewed as an elaboration of the action-selection portion of the LIDA-like Fig. 7.2. In Psi, reinforcement is considered as part of the learning process involved in action selection and planning; in Fig. 7.3 an explicit "reinforcement box" has been added to the original Psi diagram, to emphasize this.

Figure 7.4, modeling long-term memory and deliberative processing, is derived from our own prior work studying the "cognitive synergy" between different

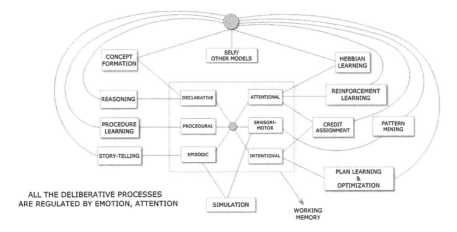

Fig. 7.4 Architecture of long-term memory and deliberative and metacognitive thinking

cognitive processes associated with different types of memory. The division into types of memory is fairly standard. Declarative, procedural, episodic and sensorimotor memory are routinely distinguished; we like to distinguish attentional memory and intentional (goal) memory as well, and view these as the interface between long-term memory and the mind's global control systems. One focus of our AGI design work has been on designing learning algorithms, corresponding to these various types of memory, that interact with each other in a synergetic way [Goe09c], helping each other to overcome their intrinsic combinatorial explosions. There is significant evidence that these various types of long-term memory are differently implemented in the brain, but the degree of structure and dynamical commonality underlying these different implementations remains unclear.

Each of these long-term memory types has its analogue in working memory as well. In some cognitive models, the working memory and long-term memory versions of a memory type and corresponding cognitive processes, are basically the same thing. CogPrime is mostly like this—it implements working memory as a subset of long-term memory consisting of items with particularly high importance values. The distinctive nature of working memory is enforced via using slightly different dynamical equations to update the importance values of items with importance above a certain threshold. On the other hand, many cognitive models treat working and long term memory as more distinct than this, and there is evidence for significant functional and anatomical distinctness in the brain in some cases. So for the purpose of the integrative diagram, it seemed best to leave working and long-term memory subcomponents as parallel but distinguished.

Figure 7.4 also encompasses metacognition, under the hypothesis that in human beings and human-like minds, metacognitive thinking is carried out using basically the same processes as plain ordinary deliberative thinking, perhaps with various tweaks optimizing them for thinking about thinking. If it turns out that humans

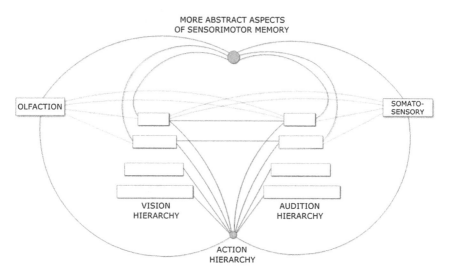

Fig. 7.5 Architecture for multimodal perception

have, say, a special kind of reasoning faculty exclusively for metacognition, then the diagram would need to be modified. Modeling of self and others is understood to occur via a combination of metacognition and deliberative thinking, as well as via implicit adaptation based on reactive processing.

Figure 7.5 models perception, according to the basic ideas of deep learning theory. Vision and audition are modeled as deep learning hierarchies, with bottom-up and top-down dynamics. The lower layers in each hierarchy refer to more localized patterns recognized in, and abstracted from, sensory data. Output from these hierarchies to the rest of the mind is not just through the top layers, but via some sort of sampling from various layers, with a bias toward the top layers. The different hierarchies cross-connect, and are hence to an extent dynamically coupled together. It is also recognized that there are some sensory modalities that aren't strongly hierarchical, e.g touch and smell (the latter being better modeled as something like an asymmetric Hopfield net, prone to frequent chaotic dynamics [LLW+05])—these may also cross-connect with each other and with the more hierarchical perceptual subnetworks. Of course the suggested architecture could include any number of sensory modalities; the diagram is restricted to four just for simplicity.

The self-organized patterns in the upper layers of perceptual hierarchies may become quite complex and may develop advanced cognitive capabilities like episodic memory, reasoning, language learning, etc. A pure deep learning approach to intelligence argues that all the aspects of intelligence emerge from this kind of dynamics (among perceptual, action and reinforcement hierarchies). Our own view is that the heterogeneity of human brain architecture argues against this perspective, and that

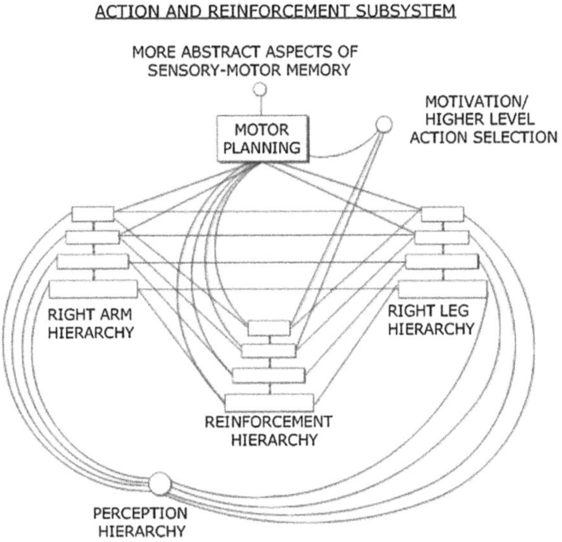

Fig. 7.6 Architecture for action and reinforcement

deep learning systems are probably better as models of perception and action than of general cognition. However, the integrative diagram is not committed to our perspective on this—a deep-learning theorist could accept the integrative diagram, but argue that all the other portions besides the perceptual, action and reinforcement hierarchies should be viewed as descriptions of phenomena that emerge in these hierarchies due to their interaction.

Figure 7.6 shows an action subsystem and a reinforcement subsystem, parallel to the perception subsystem. Two action hierarchies, one for an arm and one for a leg, are shown for concreteness, but of course the architecture is intended to be extended more broadly. In the hierarchy corresponding to an arm, for example, the lowest level would contain control patterns corresponding to individual joints, the next level up to groupings of joints (like fingers), the next level up to larger parts of the arm (hand, elbow). The different hierarchies corresponding to different body parts cross-link, enabling coordination among body parts; and they also connect at multiple levels to perception hierarchies, enabling sensorimotor coordination. Finally there is a module for motor planning, which links tightly with all the motor hierarchies, and also overlaps with the more cognitive, inferential planning activities of the mind, in a manner that is modeled different ways by different theorists. Albus [AM01] has elaborated this kind of hierarchy quite elaborately.

The reward hierarchy in Fig. 7.6 provides reinforcement to actions at various levels on the hierarchy, and includes dynamics for propagating information about reinforcement up and down the hierarchy.

Figure 7.7 deals with language, treating it as a special case of coupled perception and action. The traditional architecture of a computational language comprehension

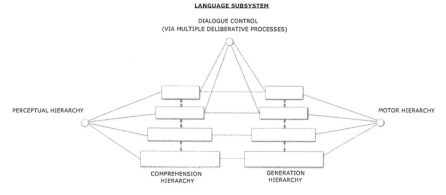

Fig. 7.7 Architecture for language processing

system is a pipeline [JM09] [Goe10c], which is equivalent to a hierarchy with the lowest-level linguistic features (e.g. sounds, words) at the bottom, and the highest level features (semantic abstractions) at the top, and syntactic features in the middle. Feedback connections enable semantic and cognitive modulation of lower-level linguistic processing. Similarly, language generation is commonly modeled hierarchically, with the top levels being the ideas needing verbalization, and the bottom level corresponding to the actual sentence produced. In generation the primary flow is top-down, with bottom-up flow providing modulation of abstract concepts by linguistic surface forms.

So, that's it—an integrative architecture diagram for human-like general intelligence, split among seven different pictures, formed by judiciously merging together architecture diagrams produced via a number of cognitive theorists with different, overlapping foci and research paradigms.

Is anything critical left out of the diagram? A quick perusal of the table of contents of cognitive psychology textbooks suggests to me that if anything major is left out, it's also unknown to current cognitive psychology. However, one could certainly make an argument for explicit inclusion of certain other aspects of intelligence, that in the integrative diagram are left as implicit emergent phenomena. For instance, creativity is obviously very important to intelligence, but, there is no "creativity" box in any of these diagrams—because in our view, and the view of the cognitive theorists whose work we've directly drawn on here, creativity is best viewed as a process emergent from other processes that are explicitly included in the diagrams.

7.4 Interpretation and Application of the Integrative Diagram

A tongue-partly-in-cheek definition of a biological pathway is "a subnetwork of a biological network, that fits on a single journal page". Cognitive architecture diagrams have a similar property—they are crude abstractions of complex structures and dynamics, sculpted in accordance with the size of the printed page, and the tolerance of the human eye for absorbing diagrams, and the tolerance of the human author for making diagrams.

However, sometimes constraints—even arbitrary ones—are useful for guiding creative efforts, due to the fact that they force choices. Creating an architecture for human-like general intelligence that fits in a few (okay, seven) fairly compact diagrams, requires one to make many choices about what features and relationships are most essential. In constructing the integrative diagram, we have sought to make these choices, not purely according to our own tastes in cognitive theory or AGI system design, but according to a sort of blend of the taste and judgment of a number of scientists whose views we respect, and who seem to have fairly compatible, complementary perspectives.

What is the use of a cognitive architecture diagram like this? It can help to give newcomers to the field a basic idea about what is known and suspected about the nature of human-like general intelligence. Also, it could potentially be used as a tool for cross-correlating different AGI architectures. If everyone who authored an AGI architecture would explain how their architecture accounts for each of the structures and processes identified in the integrative diagram, this would give a means of relating the various AGI designs to each other.

The integrative diagram could also be used to help connect AGI and cognitive psychology to neuroscience in a more systematic way. In the case of LIDA, a fairly careful correspondence has been drawn up between the LIDA diagram nodes and links and various neural structures and processes [FB08]. Similar knowledge exists for the rest of the integrative diagram, though not organized in such a systematic fashion. A systematic curation of links between the nodes and links in the integrative diagram and current neuroscience knowledge, would constitute an interesting first approximation of the holistic cognitive behavior of the human brain.

Finally (and harking forward to later chapters), the big omission in the integrative diagram is *dynamics*. Structure alone will only get you so far, and you could build an AGI system with reasonable-looking things in each of the integrative diagram's boxes, interrelating according to the given arrows, and yet still fail to make a viable AGI system. Given the limitations the real world places on computing resources, it's not enough to have adequate representations and algorithms in all the boxes, communicating together properly and capable doing the right things given sufficient resources. Rather, one needs to have all the boxes filled in properly with structures and processes that, when they act together using feasible computing resources, will yield appropriately intelligent behaviors via their cooperative activity. And this has to do with the complex interactive dynamics of all the processes in all the different boxes—which is something the integrative diagram doesn't touch at all. This brings us again to the network of ideas we've discussed under the name of "cognitive synergy", to be discussed later on.

It might be possible to make something similar to the integrative diagram on the level of dynamics rather than structures, complementing the structural integrative diagram given here; but this would seem significantly more challenging, because we lack a standard set of tools for depicting system dynamics. Most cognitive theorists and AGI architects describe their structural ideas using boxes-and-lines diagrams of some sort, but there is no standard method for depicting complex system dynamics. So to make a dynamical analogue to the integrative diagram, via a similar integrative methodology, one would first need to create appropriate diagrammatic formalizations of the dynamics of the various cognitive theories being integrated—a fascinating but onerous task.

When we first set out to make an integrated cognitive architecture diagram, via combining the complementary insights of various cognitive science and AGI theorists, we weren't sure how well it would work. But now we feel the experiment was generally a success —the resultant integrated architecture seems sensible and coherent, and reasonably complete. It doesn't come close to telling you everything you need to know to understand or implement a human-like mind—but it tells you the various processes and structures you need to deal with, and which of their inter-relations are most critical. And, perhaps just as importantly, it gives a concrete way of understanding the insights of a specific but fairly diverse set of cognitive science and AGI theorists as complementary rather than contradictory. In a CogPrime context, it provides a way of tying in the specific structures and dynamics involved in CogPrime, with a more generic portrayal of the structures and dynamics of human-like intelligence.

Part III
Toward a General Theory
of General Intelligence

Chapter 8
A Formal Model of Intelligent Agents

8.1 Introduction

The artificial intelligence field is full of sophisticated mathematical models and equations, but most of these are highly specialized in nature—e.g. formalizations of particular logic systems, analyzes of the dynamics of specific sorts of neural nets, etc. On the other hand, a number of highly general models of intelligent systems also exist, including Hutter's recent formalization of universal intelligence [Hut05] and a large body of work in the disciplines of systems science and cybernetics—but these have tended not to yield many specific lessons useful for engineering AGI systems, serving more as conceptual models in mathematical form.

It would be fantastic to have a mathematical theory bridging these extremes—a real "general theory of general intelligence", allowing the derivation and analysis of specific structures and processes playing a role in practical AGI systems, from broad mathematical models of general intelligence in various situations and under various constraints. However, the path to such a theory is not entirely clear at present; and, as valuable as such a theory would be, we don't believe such a thing to be *necessary* for creating advanced AGI. One possibility is that the development of such a theory will occur contemporaneously and synergetically with the advent of practical AGI technology.

Lacking a mature, pragmatically useful "general theory of general intelligence", however, we have still found it valuable to articulate certain theoretical ideas about the nature of general intelligence, with a level of rigor a bit greater than the wholly informal discussions of the previous chapters. The chapters in this section of the book articulate some ideas we have developed in pursuit of a general theory of general intelligence; ideas that, even in their current relatively undeveloped form, have been very helpful in guiding our concrete work on the CogPrime design.

This chapter presents a more formal version of the notion of intelligence as "achieving complex goals in complex environments", based on a formal model of intelligent agents. These formalizations of agents and intelligence will be used in later chapters as a foundation for formalizing other concepts like inference and cognitive

B. Goertzel et al., *Engineering General Intelligence, Part 1*,
Atlantis Thinking Machines 5, DOI: 10.2991/978-94-6239-027-0_8,
© Atlantis Press and the authors 2014

synergy. Chapters 9 and 10 pursue the notion of cognitive synergy a little more thoroughly than was done in previous chapters. Chapter 11 sketches a general theory of general intelligence using tools from category theory—not bringing it to the level where one can use it to derive specific AGI algorithms and structures; but still, presenting ideas that will be helpful in interpreting and explaining specific aspects of the CogPrime design in Part 2 Finally, Appendix A explores an additional theoretical direction, in which the mind of an intelligent system is viewed in terms of certain curved spaces—a novel way of thinking about the dynamics of general intelligence, which has been useful in guiding development of the ECAN component of CogPrime, and we expect will have more general value in future.

Despite the intermittent use of mathematical formalism, the ideas presented in this section are fairly speculative, and we do not propose them as constituting a well-demonstrated theory of general intelligence. Rather, we propose them as an interesting *way of thinking* about general intelligence, which appears to be consistent with available data, and which has proved inspirational to us in conceiving concrete structures and dynamics for AGI, as manifested for example in the CogPrime design. Understanding the way of thinking described in these chapters is valuable for understanding why the CogPrime design is the way it is, and for relating CogPrime to other practical and intellectual systems, and extending and improving CogPrime.

8.2 A Simple Formal Agents Model (SRAM)

We now present a formalization of the concept of "intelligent agents"—beginning with a formalization of "agents" in general.

Drawing on [Hut05, LH07a], we consider a class of active agents which observe and explore their environment and also take actions in it, which may affect the environment. Formally, the agent sends information to the environment by sending symbols from some finite alphabet called the *action space* Σ; and the environment sends signals to the agent with symbols from an alphabet called the *perception space*, denoted $\mathcal{P}$. Agents can also experience rewards, which lie in the *reward space*, denoted $\mathcal{R}$, which for each agent is a subset of the rational unit interval.

The agent and environment are understood to take turns sending signals back and forth, yielding a history of actions, observations and rewards, which may be denoted

$$a_1 o_1 r_1 a_2 o_2 r_2 \ldots$$

or else

$$a_1 x_1 a_2 x_2 \ldots$$

if x is introduced as a single symbol to denote both an observation and a reward. The complete interaction history up to and including cycle t is denoted $ax_{1:t}$; and the history before cycle t is denoted $ax_{<t} = ax_{1:t-1}$.

The agent is represented as a function π which takes the current history as input, and produces an action as output. Agents need not be deterministic, an agent may for instance induce a probability distribution over the space of possible actions, conditioned on the current history. In this case we may characterize the agent by a probability distribution $\pi(a_t|ax_{<t})$. Similarly, the environment may be characterized by a probability distribution $\mu(x_k|ax_{<k}a_k)$. Taken together, the distributions π and μ define a probability measure over the space of interaction sequences.

Next, we extend this model in a few ways, intended to make it better reflect the realities of intelligent computational agents. The first modification is to allow agents to maintain memories (of finite size), via adding memory actions drawn from a set $\mathcal{M}$ into the history of actions, observations and rewards. The second modification is to introduce the notion of **goals**.

8.2.1 Goals

We define goals as mathematical functions (to be specified below) associated with symbols drawn from the alphabet $\mathcal{G}$; and we consider the environment as sending goal-symbols to the agent along with regular observation-symbols. (Note however that the presentation of a goal-symbol to an agent does not necessarily entail the explicit communication to the agent of the contents of the goal function. This must be provided by other, correlated observations.) We also introduce a conditional distribution $\gamma(g, \mu)$ that gives the weight of a goal g in the context of a particular environment μ.

In this extended framework, an interaction sequence looks like

$$a_1 o_1 g_1 r_1 a_2 o_2 g_2 r_2 \ldots$$

or else

$$a_1 y_1 a_2 y_2 \ldots$$

where g_i are symbols corresponding to goals, and y is introduced as a single symbol to denote the combination of an observation, a reward and a goal.

Each goal function maps each finite interaction sequence $I_{g,s,t} = ay_{s:t}$ with g_s to g_t corresponding to g, into a value $r_g(I_{g,s,t}) \in [0, 1]$ indicating the value or "raw reward" of achieving the goal during that interaction sequence. The total reward r_t obtained by the agent is the sum of the raw rewards obtained at time t from all goals whose symbols occur in the agent's history before t.

This formalism of goal-seeking agents allows us to formalize the notion of intelligence as "achieving complex goals in complex environments"—a direction that is pursued in Sect. 8.3.

Note that this is an *external* perspective of system goals, which is natural from the perspective of formally defining system intelligence in terms of system behavior, but is not necessarily very natural in terms of system design. From the point of view of AGI design, one is generally more concerned with the (implicit or explicit) representation of goals inside an AGI system, as in CogPrime's Goal Atoms to be reviewed in Chap. 4 of Part 2

Further, it is important to also consider the case where an AGI system has no explicit goals, and the system's environment has no immediately identifiable goals either. But in this case, we don't see any clear way to define a system's intelligence, except via *approximating* the system in terms of other theoretical systems which do have explicit goals. This approximation approach is developed in Sect. 8.3.5.

The awkwardness of linking the general formalism of intelligence theory presented here, with the practical business of creating and designing AGI systems, may indicate a shortcoming on the part of contemporary intelligence theory or AGI designs. On the other hand, this sort of situation often occurs in other domains as well—e.g. the leap from quantum theory to the analysis of real-world systems like organic molecules involves a lot of awkwardness and large leaps a well.

8.2.2 Memory Stores

As well as goals, we introduce into the model a long-term memory and a workspace. Regarding long-term memory we assume the agent's memory consists of multiple memory stores corresponding to various types of memory, e.g.: procedural (K_{Proc}), declarative (K_{Dec}), episodic (K_{Ep}), attentional (K_{Att}) and Intentional (K_{Int}). In Appendix B a category-theoretic model of these memory stores is introduced; but for the moment, we need only assume the existence of

- an injective mapping Θ_{Ep}: $K_{Ep} \to \mathcal{H}$ where $\mathcal{H}$ is the space of fuzzy sets of subhistories (subhistories being "episodes" in this formalism)
- an injective mapping Θ_{Proc}: $K_{Proc} \times \mathcal{M} \times \mathcal{W} \to \mathcal{A}$, where $\mathcal{M}$ is the set of memory states, $\mathcal{W}$ is the set of (observation, goal, reward) triples, and $\mathcal{A}$ is the set of actions (this maps each procedure object into a function that enacts actions in the environment or memory, based on the memory state and current world-state)
- an injective mapping Θ_{Dec}: $K_{Dec} \to \mathcal{L}$, where $\mathcal{L}$ is the set of expressions in some formal language (which may for example be a logical language), which possesses words corresponding to the observations, goals, reward values and actions in our agent formalism
- an injective mapping Θ_{Int}: $K_{Int} \to \mathcal{G}$, where $\mathcal{G}$ is the space of goals mentioned above
- an injective mapping Θ_{Att}: $K_{Int} \cup K_{Ep} \cup K_{Proc} \cup K_{Ec} \to \mathcal{V}$, where $\mathcal{V}$ is the space of "attention values" (structures that gauge the importance of paying attention to an item of knowledge over various time-scales or in various contexts).

We also assume that the vocabulary of actions contains memory-actions corresponding to the operations of inserting the current observation, goal, reward or action into the episodic and/or declarative memory store. And, we assume that the activity of the agent, at each time-step, includes the enaction of one or more of the procedures in the procedural memory store. If several procedures are enacted at once, then the end result is still formally modeled as a single action $a = a_{[1]} * \cdots * a_{[k]}$ where $*$ is an operator on action-space that composes multiple actions into a single one.

Finally, we assume that, at each time-step, the agent may carry out an external action a_i on the environment, a memory action m_i on the (long-term) memory, and an action b_i on its **internal workspace**. Among the actions that can be carried out on the workspace, are the ability to insert or delete observations, goals, actions or reward-values from the workspace. The workspace can be thought of as a sort of short-term memory or else in terms of Baars' "global workspace" concept mentioned above. The workspace provides a medium for interaction between the different memory types.

The workspace provides a mechanism by which declarative, episodic and procedural memory may interact with each other. For this mechanism to work, we must assume that there are actions corresponding to query operations that allow procedures to look into declarative and episodic memory. The nature of these query operations will vary among different agents, but we can assume that in general an agent has

- one or more procedures $Q_{Dec}(x)$ serving as *declarative queries*, meaning that when Q_{Dec} is enacted on some x that is an ordered set of items in the workspace, the result is that one or more items from declarative memory is entered into the workspace
- one or more procedures $Q_{Ep}(x)$ serving as *episodic queries*, meaning that when Q_{Ep} is enacted on some x that is an ordered set of items in the workspace, the result is that one or more items from episodic memory is entered into the workspace.

One additional aspect of CogPrime's knowledge representation that is important to PLN is the attachment of nonnegative weights n_i corresponding to elementary observations o_i. These weights denote the amount of evidence contained in the observation. For instance, in the context of a robotic agent, one could use these values to encode the assumption that an elementary visual observation has more evidential value than an elementary olfactory observation.

We now have a model of an agent with long-term memory comprising procedural, declarative and episodic aspects, an internal cognitive workspace, and the capability to use procedures to drive actions based on items in memory and the workspace, and to move items between long-term memory and the workspace.

8.2.2.1 Modeling CogPrime

Of course, this formal model may be realized differently in various real-world AGI systems. In CogPrime we have

- a weighted, labeled hypergraph structure called the AtomSpace used to store declarative knowledge (this is the representation used by PLN)
- a collection of programs in a LISP-like language called Combo, stored in a Procedure Repository data structure, used to store procedural knowledge
- a collection of partial "movies" of the system's experience, played back using an internal simulation engine, used to store episodic knowledge

- AttentionValue objects, minimally containing ShortTermImportance (STI) and LongTermImportance (LTI) values used to store attentional knowledge
- Goal Atoms for intentional knowledge, stored in the same format as declarative knowledge but whose dynamics involve a special form of artificial currency that is used to govern action selection.

The AtomSpace is the central repository and procedures and episodes are linked to Atoms in the AtomSpace which serve as their symbolic representatives. The "workspace" in CogPrime exists only virtually: each item in the AtomSpace has a "short term importance" (STI) level, and the workspace consists of those items in the AtomSpace with highest STI, and those procedures and episodes whose symbolic representatives in the AtomSpace have highest STI.

On the other hand, as we saw above, the LIDA architecture uses separate representations for procedural, declarative and episodic memory, but also has an explicit workspace component, where the most currently contextually relevant items from all different types of memory are gathered and used together in the course of actions. However, compared to CogPrime, it lacks comparably fine-grained methods for integrating the different types of memory.

Systematically mapping various existing cognitive architectures, or human brain structure, into this formal agents model would be a substantial though quite plausible exercise; but we will not undertake this here.

8.2.3 The Cognitive Schematic

Next we introduce an additional specialization into SRAM: the **cognitive schematic**, written informally as

$$Context \ \& \ Procedure \rightarrow Goal$$

and considered more formally as $holds(C) \ \& \ ex(P) \rightarrow h_i$ where h may be an externally specified goal g_i or an internally specified goal h derived as a (possibly uncertain) subgoal of one of more g_i; C is a piece of declarative or episodic knowledge and P is a procedure that the agent can internally execute to generate a series of actions. $ex(P)$ is the proposition that P is successfully executed. If C is episodic then $holds(C)$ may be interpreted as the current context (i.e. some finite slice of the agent's history) being similar to C; if C is declarative then $holds(C)$ may be interpreted as the truth value of C evaluated at the current context. Note that C may refer to some part of the world quite distant from the agent's current sensory observations; but it may still be formally evaluated based on the agent's history.

In the standard CogPrime notation as introduced formally in Chap. 2 of Part 2 (where indentation has function-argument syntax similar to that in Python, and relationship types are prepended to their relata without parentheses), for the case C is declarative this would be written as

PredictiveExtensionalImplication
>   AND
>   >   C
>   >   Execution P
>   G

and in the case *C* is episodic one replaces *C* in this formula with a predicate expressing *C*'s similarity to the current context. The semantics of the PredictiveExtensionalInheritance relation will be discussed below. The Execution relation simply denotes the proposition that procedure *P* has been executed.

For the class of SRAM agents who (like CogPrime) use the cognitive schematic to govern many or all of their actions, a significant fragment of agent intelligence boils down to estimating the truth values of PredictiveExtensionalImplication relationships. Action selection procedures can be used, which choose procedures to enact based on which ones are judged most likely to achieve the current external goals g_i in the current context. Rather than enter into the particularities of action selection or other cognitive architecture issues, we will restrict ourselves to PLN inference, which in the context of the present agent model is a method for handling PredictiveImplication in the cognitive schematic.

Consider an agent in a virtual world, such as a virtual dog, one of whose external goals is to please its owner. Suppose its owner has asked it to find a cat, and it can translate this into a subgoal "find cat." If the agent operates according to the cognitive schematic, it will search for *P* so that

PredictiveExtensionalImplication
>   AND
>   >   C
>   >   Execution P
>   Evaluation
>   >   found
>   >   cat

holds.

8.3 Toward a Formal Characterization of Real-World General Intelligence

Having defined what we mean by an agent acting in an environment, we now turn to the question of what it means for such an agent to be "intelligent".

As we have reviewed extensively in Chap. 4, "intelligence" is a commonsense, "folk psychology" concept, with all the imprecision and contextuality that this generally entails. One cannot expect any compact, elegant formalism to capture all of its meanings. Even in the psychology and AI research communities, divergent definitions abound; Legg and Hutter [LH07a] lists and organizes 70+ definitions from the literature.

Practical study of natural intelligence in humans and other organisms, and practical design, creation and instruction of artificial intelligences, can proceed perfectly well without an agreed-upon formalization of the "intelligence" concept. Some researchers may conceive their own formalisms to guide their own work, others may feel no need for any such thing.

But nevertheless, it is of interest to seek formalizations of the concept of intelligence, which capture useful fragments of the commonsense notion of intelligence, and provide guidance for practical research in cognitive science and AI. A number of such formalizations have been given in recent decades, with varying degrees of mathematical rigor. Perhaps the most carefully-wrought formalization of intelligence so far is the theory of "universal intelligence" presented by Shane Legg and Marcus Hutter in [LH07b], which draws on ideas from algorithmic information theory.

Universal intelligence captures a certain aspect of the "intelligence" concept very well, and has the advantage of connecting closely with ideas in learning theory, decision theory and computation theory. However, the kind of general intelligence it captures best, is a kind which is in a sense *more general* in scope than human-style general intelligence. Universal intelligence does capture the sense in which humans are more intelligent than worms, which are more intelligent than rocks; and the sense in which theoretical AGI systems like Hutter's AIXI or $AIXI^{tl}$ [Hut05] would be much more intelligent than humans. But it misses essential aspects of the intelligence concept as it is used in the context of intelligent natural systems like humans or real-world AI systems.

Our main goal in this section is to present variants of universal intelligence that better capture the notion of intelligence as it is typically understood in the context of real-world natural and artificial systems. The first variant we describe is *pragmatic general intelligence*, which is inspired by the intuitive notion of intelligence as "the ability to achieve complex goals in complex environments", given in [Goe93]. After assuming a prior distribution over the space of possible environments, and one over the space of possible goals, one then defines the pragmatic general intelligence as the expected level of goal-achievement of a system relative to these distributions. Rather than measuring truly broad mathematical general intelligence, pragmatic general intelligence measures intelligence in a way that's specifically biased toward certain environments and goals.

Another variant definition is then presented, the *efficient pragmatic general intelligence* , which takes into account the amount of computational resources utilized by the system in achieving its intelligence. Some argue that making efficient use of available resources is a defining characteristic of intelligence, see e.g. [Wan06].

A critical question left open is the characterization of the prior distributions corresponding to everyday human reality; we give a semi-formal sketch of some ideas on this in Chap. 10, where we present the notion of a "communication prior", which assigns a probability weight to a situation S based on the ease with which one agent in a society can communicate S to another agent in that society, using multimodal communication (including verbalization, demonstration, dramatic and pictorial depiction, etc.).

Finally, we present a formal measure of the "generality" of an intelligence, which precisiates the informal distinction between "general AI" and "narrow AI".

8.3.1 Biased Universal Intelligence

To define universal intelligence, Legg and Hutter consider the class of environments that are *reward-summable*, meaning that the total amount of reward they return to any agent is bounded by 1. Where r_i denotes the reward experienced by the agent from the environment at time i, the *expected total reward* for the agent π from the environment μ is defined as

$$V_\mu^\pi \equiv E(\sum_1^\infty r_i) \leq 1.$$

To extend their definition in the direction of greater realism, we first introduce a second-order probability distribution ν, which is a probability distribution over the space of environments μ. The distribution ν assigns each environment a probability. One such distribution ν is the Solomonoff-Levin universal distribution in which one sets $\nu = 2^{-K(\mu)}$; but this is not the only distribution ν of interest. In fact a great deal of real-world general intelligence consists of the adaptation of intelligent systems to particular distributions ν over environment-space, differing from the universal distribution.

We then define

Definition 4 *The* **biased universal intelligence** *of an agent* π *is its expected performance with respect to the distribution* ν *over the space of all computable reward-summable environments, E, that is,*

$$\Upsilon(\pi) \equiv \sum_{\mu \in E} \nu(\mu) V_\mu^\pi.$$

Legg and Hutter's **universal intelligence** is obtained by setting ν equal to the universal distribution.

This framework is more flexible than it might seem. E.g. suppose one wants to incorporate agents that die. Then one may create a special action, say a_{666}, corresponding to the state of death, to create agents that

- in certain circumstances output action a_{666}
- have the property that if their previous action was a_{666}, then all of their subsequent actions must be a_{666}

and to define a reward structure so that actions a_{666} always bring zero reward. It then follows that death is generally a bad thing if one wants to maximize intelligence. Agents that die will not get rewarded after they're dead; and agents that live only 70 years, say, will be restricted from getting rewards involving long-term patterns and will hence have specific limits on their intelligence.

8.3.2 Connecting Legg and Hutter's Model of Intelligent Agents to the Real World

A notable aspect of the Legg and Hutter formalism is the separation of the reward mechanism from the cognitive mechanisms of the agent. While commonplace in the reinforcement learning literature, this seems psychologically unrealistic in the context of biological intelligences and many types of machine intelligences. Not all human intelligent activity is specifically reward-seeking in nature; and even when it is, humans often pursue complexly constructed rewards, that are defined in terms of their own cognitions rather than separately given. Suppose a certain human's goals are true love, or world peace, and the proving of interesting theorems—then these goals are defined by the human herself, and only she knows if she's achieved them. An externally-provided reward signal doesn't capture the nature of this kind of goal-seeking behavior, which characterizes much human goal-seeking activity (and will presumably characterize much of the goal-seeking activity of advanced engineered intelligences also) ... let alone human behavior that is spontaneous and unrelated to explicit goals, yet may still appear commonsensically intelligent.

One could seek to bypass this complaint about the reward mechanisms via a sort of "neo-Freudian" argument, via

- associating the reward signal, not with the "external environment" as typically conceived, but rather with a portion of the intelligent agent's brain that is separate from the cognitive component
- viewing complex goals like true love, world peace and proving interesting theorems as indirect ways of achieving the agent's "basic goals", created within the agent's memory via subgoaling mechanisms

but it seems to us that a general formalization of intelligence should not rely on such strong assumptions about agents' cognitive architectures. So below, after introducing the pragmatic and efficient pragmatic general intelligence measures, we will propose an alternate interpretation wherein the mechanism of external rewards is viewed as a theoretical test framework for assessing agent intelligence, rather than a hypothesis about intelligent agent architecture.

In this alternate interpretation, formal measures like the universal, pragmatic and efficient pragmatic general intelligence are viewed as *not* directly applicable to real-world intelligences, because they involve the behaviors of agents over a wide variety of goals and environments, whereas in real life the opportunities to observe agents are more limited. However, they are viewed as being *indirectly* applicable to real-world agents, in the sense that an external intelligence can observe an agent's real-world behavior and then *infer* its likely intelligence according to these measures.

In a sense, this interpretation makes our formalized measures of intelligence the opposite of real-world IQ tests. An IQ test is a quantified, formalized test which is designed to approximately predict the informal, qualitative achievement of humans in real life. On the other hand, the formal definitions of intelligence we present here are quantified, formalized tests that are designed to capture abstract notions of intelligence, but which can be approximately evaluated on a real-world intelligent system by observing what it does in real life.

8.3.3 Pragmatic General Intelligence

The above concept of biased universal intelligence is perfectly adequate for many purposes, but it is also interesting to explicitly introduce the notion of a *goal* into the calculation. This allows us to formally capture the notion presented in [Goe93] of intelligence as "the ability to achieve complex goals in complex environments".

If the agent is acting in environment μ, and is provided with g_s corresponding to g at the start and the end of the time-interval $T = \{i \in (s, \ldots, t)\}$, then the *expected goal-achievement* of the agent, relative to g, during the interval is the expectation

$$V_{\mu,g,T}^{\pi} \equiv E\left(\sum_{i=s}^{t} r_g(I_{g,s,i})\right)$$

where the expectation is taken over all interaction sequences $I_{g,s,i}$ drawn according to μ. We then propose

Definition 5 *The* **pragmatic general intelligence** *of an agent* π, *relative to the distribution* ν *over environments and the distribution* γ *over goals, is its expected performance with respect to goals drawn from* γ *in environments drawn from* ν, *over the time-scales natural to the goals; that is,*

$$\Pi(\pi) \equiv \sum_{\mu \in E, g \in \mathcal{G}, T} \nu(\mu)\gamma(g, \mu)V_{\mu,g,T}^{\pi}$$

(in those cases where this sum is convergent).

This definition formally captures the notion that "intelligence is achieving complex goals in complex environments", where "complexity" is gauged by the assumed measures ν and γ.

If ν is taken to be the universal distribution, and γ is defined to weight goals according to the universal distribution, then pragmatic general intelligence reduces to universal intelligence.

Furthermore, it is clear that a universal algorithmic agent like AIXI [Hut05] would also have a high pragmatic general intelligence, under fairly broad conditions. As the interaction history grows longer, the pragmatic general intelligence of AIXI would approach the theoretical maximum; as AIXI would implicitly infer the relevant distributions via experience. However, if significant reward discounting is involved, so that near-term rewards are weighted much higher than long-term rewards, then AIXI might compare very unfavorably in pragmatic general intelligence, to other agents designed with prior knowledge of ν, γ and τ in mind.

The most interesting case to consider is where ν and γ are taken to embody some particular bias in a real-world space of environments and goals, and this bias is appropriately reflected in the internal structure of an intelligent agent. Note that an agent needs not lack universal intelligence in order to possess pragmatic general intelligence with respect to some non-universal distribution over goals and environments. However, in general, given limited resources, there may be a tradeoff between universal intelligence and pragmatic intelligence. Which leads to the next point: how to encompass resource limitations into the definition.

One might argue that the definition of Pragmatic General Intelligence is already encompassed by Legg and Hutter's definition because one may bias the distribution of environments within the latter by considering different Turing machines underlying the Kolmogorov complexity. However this is not a general equivalence because the Solomonoff-Levin measure intrinsically decays exponentially, whereas an assumptive distribution over environments might decay at some other rate. This issue seems to merit further mathematical investigation.

8.3.4 Incorporating Computational Cost

Let $\eta_{\pi,\mu,g,T}$ be a probability distribution describing the amount of computational resources consumed by an agent π while achieving goal g over time-scale T. This is a probability distribution because we want to account for the possibility of non-deterministic agents. So, $\eta_{\pi,\mu,g,T}(Q)$ tells the probability that Q units of resources are consumed. For simplicity we amalgamate space and time resources, energetic resources, etc. into a single number Q, which is assumed to live in some subset of the positive reals. Space resources of course have to do with the size of the system's memory. Then we may define

Definition 6 *The **efficient pragmatic general intelligence** of an agent π with resource consumption $\eta_{\pi,\mu,g,T}$, relative to the distribution ν over environments and the distribution γ over goals, is its expected performance with respect to goals drawn from γ in environments drawn from ν, over the time-scales natural to the goals, normalized by the amount of computational effort expended to achieve each goal; that is,*

$$\Pi_{Eff}(\pi) \equiv \sum_{\mu \in E, g \in \mathcal{G}, Q, T} \frac{\nu(\mu)\gamma(g, \mu)\eta_{\pi,\mu,g,T}(Q)}{Q} V^{\pi}_{\mu,g,T}$$

(in those cases where this sum is convergent).

This is a measure that rates an agent's intelligence higher if it uses fewer computational resources to do its business. Roughly, it measures reward achieved per spacetime computation unit.

Note that, by abandoning the universal prior, we have also abandoned the proof of convergence that comes with it. In general the sums in the above definitions need not converge; and exploration of the conditions under which they do converge is a complex matter.

8.3.5 Assessing the Intelligence of Real-World Agents

The pragmatic and efficient pragmatic general intelligence measures are more "realistic" than the Legg and Hutter universal intelligence measure, in that they take into account the innate biasing and computational resource restrictions that characterize real-world intelligence. But as discussed earlier, they still live in "fantasy-land" to an extent—they gauge the intelligence of an agent via a weighted average over a wide variety of goals and environments; and they presume a simplistic relationship between agents and rewards that does not reflect the complexities of real-world cognitive architectures. It is not obvious from the foregoing how to apply these measures to real-world intelligent systems, which lack the ability to exist in such a wide variety of environments within their often brief lifespan, and mostly go about their lives doing things other than pursuing quantified external rewards. In this brief section we describe an approach to bridging this gap. The treatment is left semi-formal in places.

We suggest to view the definitions of pragmatic and efficient pragmatic general intelligence in terms of a "possible worlds" semantics—i.e. to view them as asking, counterfactually, how an agent *would* perform, hypothetically, on a series of tests (the tests being goals, defined in relation to environments and reward signals).

Real-world intelligent agents don't normally operate in terms of explicit goals and rewards; these are abstractions that we use to think about intelligent agents. However, this is no objection to characterizing various sorts of intelligence in terms of counterfactuals like: how would system S operate if it were trying to achieve this or that goal, in this or that environment, in order to seek reward? We can characterize various sorts of intelligence in terms of how it can be inferred an agent would perform on certain tests, even though the agent's real life does not consist of taking these tests.

This conceptual approach may seem a bit artificial but we don't currently see a better alternative, if one wishes to quantitatively gauge intelligence (which is, in a sense, an "artificial" thing to do in the first place). Given a real-world agent X and a mandate to assess its intelligence, the obvious alternative to looking at possible worlds in the manner of the above definitions, is just looking *directly* at the properties of the things X has achieved in the real world during its lifespan. But this isn't an easy solution, because it doesn't disambiguate which aspects of X's achievements were due to its own actions versus due to the rest of the world that X was interacting with when it made its achievements. To distinguish the amount of achievement that

X "caused" via its own actions requires a model of causality, which is a complex
can of worms in itself; and, critically, the standard models of causality also involve
counterfactuals (asking "what would have been achieved in this situation if the agent
X hadn't been there", etc.) [MW07]. Regardless of the particulars, it seems difficult
to avoid counterfactual realities in assessing intelligence.

The approach we suggest—given a real-world agent X with a history of actions
in a particular world, and a mandate to assess its intelligence—is to introduce an
additional player, an *inference agent* δ, into the picture. The agent π modeled above
is then viewed as π_X: the model of X that δ constructs, in order to explore X's inferred
behaviors in various counterfactual environments. In the test situations embodied in
the definitions of pragmatic and efficient pragmatic general intelligence, the environment gives π_X rewards, based on specifically configured goals. In X's real life, the
relation between goals, rewards and actions will generally be significantly subtler
and perhaps quite different.

We model the real world similarly to the "fantasy world" of the previous section,
but with the omission of goals and rewards. We define a *naturalistic* context as one
in which all goals and rewards are constant, i.e. $g_i = g_0$ and $r_i = r_0$ for all i. This is
just a mathematical convention for stating that there are no precisely-defined external
goals and rewards for the agent. In a naturalistic context, we then have a situation
where agents create actions based on the past history of actions and perceptions, and
if there is any relevant notion of reward or goal, it is within the cognitive mechanism
of some agent. A *naturalistic agent* X is then an agent π which is restricted to one
particular naturalistic context, involving one particular environment μ (formally, we
may achieve this within the framework of agents described above via dictating that
X issues constant "null actions" a_0 in all environments except μ).

Next, we posit a metric space (Σ_μ, d) of naturalistic agents defined on a naturalistic
context involving environment μ, and a subspace $\Delta \in \Sigma_\mu$ of inference agents, which
are naturalistic agents that output predictions of other agents' behaviors (a notion we
will not fully formalize here). If agents are represented as program trees, then d may
be taken as edit distance on tree space [Bil05]. Then, for each agent $\delta \in \Delta$, we may
assess

- the prior probability $\theta(\delta)$ according to some assumed distribution θ
- the effectiveness $p(\delta, X)$ of δ at predicting the actions of an agent $X \in \Sigma_\mu$.

We may then define

Definition 7 *The **inference ability** of the agent δ, relative to μ and X, is*

$$q_{\mu,X}(\delta) = \theta(\delta)\frac{\sum_{Y \in \Sigma_\mu} sim(X, Y)p(\delta, Y)}{\sum_{Y \in \Sigma_\mu} sim(X, Y)}$$

where sim is a specified decreasing function of $d(X, Y)$, such as $sim(X, Y) = \frac{1}{1+d(X,Y)}$.

To construct π_X, we may then use the model of X created by the agent $\delta \in \Delta$ with the highest inference ability relative to μ and X (using some specified ordering, in case of a tie). Having constructed π_X, we can then say that.

Definition 8 *The inferred pragmatic general intelligence (relativeto ν and γ) of a naturalistic agent X defined relative to an environment μ,is defined as the pragmatic general intelligence of the model π_X of Xproduced by the agent $\delta \in \Delta$ with maximal inference ability relative to μ(and in the case of a tie, the first of these in the ordering defined over Δ). The inferred efficient pragmatic general intelligence of X relative to $\mu$is defined similarly.*

This provides a precise characterization of the pragmatic and efficient pragmatic intelligence of real-world systems, based on their observed behaviors. It's a bit messy; but the real world tends to be like that.

8.4 Intellectual Breadth: Quantifying the Generality of an Agent's Intelligence

We turn now to a related question: How can one quantify the degree of **generality** that an intelligent agent possesses? Above we have discussed the qualitative distinction between AGI and "Narrow AI", and intelligence as we have formalized it above is specifically intended as a measure of general intelligence. But quantifying intelligence is different than quantifying generality versus narrowness.

To make the discussion simpler, we introduce the term "context" as a shorthand for "environment/interval triple (μ, g, T)". Given a context (μ, g, T), and a set Σ of agents, one may construct a fuzzy set $Ag_{\mu,g,T}$ gathering those agents that are intelligent relative to the context; and given a set of contexts, one may also define a fuzzy set Con_π gathering those contexts with respect to which a given agent π is intelligent. The relevant formulas are:

$$\chi_{Ag_{\mu,g,T}}(\pi) = \chi_{Con_\pi}(\mu, g, T) = \frac{1}{N} \sum_Q \frac{\eta_{\mu,g,T}(Q) V^\pi_{\mu,g,T}}{Q}$$

where $N = N(\mu, g, T)$ is a normalization factor defined appropriately, e.g. via $N(\mu, g, T) = \max_\pi V^\pi_{\mu,g,T}$.

One could make similar definitions leaving out the computational cost factor Q, but we suspect that incorporating Q is a more promising direction. We then propose

Definition 9 *The **intellectual breadth** of an agent π, relative to the distribution ν over environments and the distribution γ over goals, is*

$$H(\chi^P_{Con_\pi}(\mu, g, T))$$

where H is the entropy and

$$
\chi_{Con_\pi}^{P}(\mu, g, T) = \frac{\nu(\mu)\gamma(g, \mu)\chi_{Con_\pi}(\mu, g, T)}{\displaystyle\sum_{(\mu_\alpha, g_\beta. T_\omega)} \nu(\mu_\alpha)\gamma(g_\beta, \mu_\alpha)\chi_{Con_\pi}(\mu_\alpha, g_\beta, T_\omega)}
$$

is the probability distribution formed by normalizing the fuzzy set $\chi_{Con_\pi}(\mu, g, T)$.

A similar definition of the intellectual breadth of a context(μ, g, T), relative to the distribution σ over agents, may be posited. A weakness of these definitions is that they don't try to account for dependencies between agents or contexts; perhaps more refined formulations may be developed that account explicitly for these dependencies.

Note that the intellectual breadth of an agent as defined here is largely independent of the (efficient or not) pragmatic general intelligence of that agent. One could have a rather (efficiently or not) pragmatically generally intelligent system with little breadth: this would be a system very good at solving a fair number of hard problems, yet wholly incompetent on a larger number of hard problems. On the other hand, one could also have a terribly (efficiently or not) pragmatically generally stupid system with great intellectual breadth: i.e a system roughly equally dumb in all contexts!

Thus, one can characterize an intelligent agent as "narrow" with respect to distribution ν over environments and the distribution γ over goals, based on evaluating it as having low intellectual breadth. A "narrow AI" relative to ν and γ would then be an AI agent with a relatively high efficient pragmatic general intelligence but a relatively low intellectual breadth.

8.5 Conclusion

Our main goal in this chapter has been to push the formal understanding of intelligence in a more pragmatic direction. Much more work remains to be done, e.g. in specifying the environment, goal and efficiency distributions relevant to real-world systems, but we believe that the ideas presented here constitute nontrivial progress.

If the line of research suggested in this chapter succeeds, then eventually, one will be able to do AGI research as follows: Specify an AGI architecture formally, and then use the mathematics of general intelligence to derive interesting results about the environments, goals and hardware platforms relative to which the AGI architecture will display significant pragmatic or efficient pragmatic general intelligence, and intellectual breadth. The remaining chapters in this section present further ideas regarding how to work toward this goal. For the time being, such a mode of AGI research remains mainly for the future, but we have still found the formalism given in these chapters useful for formulating and clarifying various aspects of the CogPrime design as will be presented in later chapters.

Chapter 9
Cognitive Synergy

9.1 Introduction

As we have seen, the formal theory of general intelligence, in its current form, doesn't really tell us much that's of use for creating real-world AGI systems. It tells us that creating extraordinarily powerful general intelligence is almost trivial if one has unrealistically huge amounts of computational resources; and that creating moderately powerful general intelligence using feasible computational resources is all about creating AI algorithms and data structures that (explicitly or implicitly) match the restrictions implied by a certain class of situations, to which the general intelligence is biased.

We've also described, in various previous chapters, some non-rigorous, conceptual principles that seem to explain key aspects of feasible general intelligence: the complementary reliance on evolution and autopoiesis, the superposition of hierarchical and heterarchical structures, and so forth. These principles can be considered as broad strategies for achieving general intelligence in certain broad classes of situations. Although, a lot of research needs to be done to figure out nice ways to describe, for instance, in what class of situations evolution is an effective learning strategy, in what class of situations dual hierarchical/heterarchical structure is an effective way to organize memory, etc.

In this chapter we'll dig deeper into one of the "general principle of feasible general intelligences" briefly alluded to earlier: the *cognitive synergy* principle, which is both a conceptual hypothesis about the structure of generally intelligent systems in certain classes of environments, and a design principle used to guide the architecting of CogPrime. We will focus here on cognitive synergy specifically in the case of "multi-memory systems", which we define as intelligent systems (like CogPrime) whose combination of environment, embodiment and motivational systems make it important for them to possess memories that divide into partially but not wholly distinct components corresponding to the categories of:

B. Goertzel et al., *Engineering General Intelligence, Part 1*,
Atlantis Thinking Machines 5, DOI: 10.2991/978-94-6239-027-0_9,
© Atlantis Press and the authors 2014

- Declarative memory
- Procedural memory (memory about how to do certain things)
- Sensory and episodic memory
- Attentional memory (knowledge about what to pay attention to in what contexts)
- Intentional memory (knowledge about the system's own goals and subgoals).

In Chap. 10 we present a detailed argument as to how the requirement for a multi-memory underpinning for general intelligence emerges from certain underlying assumptions regarding the measurement of the simplicity of goals and environments; but the points made here do not rely on that argument. What they do rely on is the assumption that, in the intelligence in question, the different components of memory are significantly but not wholly distinct. That is, there are significant "family resemblances" between the memories of a single type, yet there are also thoroughgoing connections between memories of different types.

The cognitive synergy principle, if correct, applies to any AI system demonstrating intelligence in the context of embodied, social communication. However, one may also take the theory as an explicit guide for constructing AGI systems; and of course, the bulk of this book describes one AGI architecture, CogPrime, designed in such a way.

It is possible to cast these notions in mathematical form, and we make some efforts in this direction in Appendix B, using the languages of category theory and information geometry. However, this formalization has not yet led to any rigorous proof of the generality of cognitive synergy nor any other exciting theorems; with luck this will come as the mathematics is further developed. In this chapter the presentation is kept on the heuristic level, which is all that is critically needed for motivating the CogPrime design.

9.2 Cognitive Synergy

The essential idea of cognitive synergy, in the context of multi-memory systems, may be expressed in terms of the following points:

1. Intelligence, relative to a certain set of environments, may be understood as the capability to achieve complex goals in these environments.
2. With respect to certain classes of goals and environments (see Chap. 10 for a hypothesis in this regard), an intelligent system requires a "multi-memory" architecture, meaning the possession of a number of specialized yet interconnected knowledge types, including: declarative, procedural, attentional, sensory, episodic and intentional (goal-related). These knowledge types may be viewed as different sorts of patterns that a system recognizes in itself and its environment. Knowledge of these various different types must be interlinked, and in some cases may represent differing views of the same content (see Fig. 9.1).

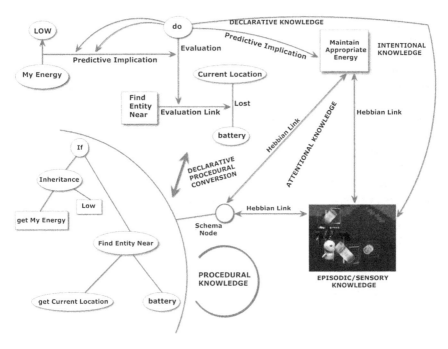

Fig. 9.1 Illustrative example of the interactions between multiple types of knowledge, in representing a simple piece of knowledge. Generally speaking, one type of knowledge can be converted to another, at the cost of some loss of information. The synergy between cognitive processes associated with corresponding pieces of knowledge, possessing different type, is a critical aspect of general intelligence

3. Such a system must possess knowledge creation (i.e. pattern recognition / formation) mechanisms corresponding to each of these memory types. These mechanisms are also called "cognitive processes".
4. Each of these cognitive processes, to be effective, must have the capability to recognize when it lacks the information to perform effectively on its own; and in this case, to dynamically and interactively draw information from knowledge creation mechanisms dealing with other types of knowledge.
5. This cross-mechanism interaction must have the result of enabling the knowledge creation mechanisms to perform much more effectively in combination than they would if operated non-interactively. This is "cognitive synergy".

While these points are implicit in the theory of mind given in [Goe06a], they are not articulated in this specific form there.

Interactions as mentioned in Points 4 and 5 in the above list are the real conceptual meat of the cognitive synergy idea. One way to express the key idea here is that most AI algorithms suffer from combinatorial explosions: the number of possible elements to be combined in a synthesis or analysis is just too great, and the algorithms are unable to filter through all the possibilities, given the lack of intrinsic constraint that comes along with a "general intelligence" context (as opposed to a narrow-AI problem like chess-playing, where the context is constrained and hence restricts the scope of possible combinations that needs to be considered). In an AGI architecture based on cognitive synergy, the different learning mechanisms must be designed specifically to interact in such a way as to palliate each others' combinatorial explosions—so that, for instance, each learning mechanism dealing with a certain sort of knowledge, must synergize with learning mechanisms dealing with the other sorts of knowledge, in a way that decreases the severity of combinatorial explosion.

One prerequisite for cognitive synergy to work is that each learning mechanism must recognize when it is "stuck", meaning it's in a situation where it has inadequate information to make a confident judgment about what steps to take next. Then, when it does recognize that it's stuck, it may request help from other, complementary cognitive mechanisms.

A theoretical notion closely related to cognitive synergy is the *cognitive schematic*, formalized in Chap. 8, which states that the activity of the different cognitive processes involved in an intelligent system may be modeled in terms of the schematic implication

$$Context \land Procedure \rightarrow Goal$$

where the Context involves sensory, episodic and/or declarative knowledge; and attentional knowledge is used to regulate how much resource is given to each such schematic implication in memory. Synergy among the learning processes dealing with the context, the procedure and the goal is critical to the adequate execution of the cognitive schematic using feasible computational resources.

Overall, the cognitive synergy principle describes the behavior of a system as it pursues a set of goals (which in most cases may be assumed to be supplied to the system "a priori", but then refined by inference and other processes). The assumed intelligent agent model is roughly as follows: At each time the system chooses a set of procedures to execute, based on its judgments regarding which procedures will best help it achieve its goals in the current context. These procedures may involve external actions (e.g. involving conversation, or controlling an agent in a simulated world) and/or internal cognitive actions. In order to make these judgments it must effectively manage declarative, procedural, episodic, sensory and attentional memory, each of which is associated with specific algorithms and structures. There are also global processes spanning all the forms of memory, including the allocation of attention to different memory items and cognitive processes, and the identification and reification of system-wide activity patterns (the latter referred to as "map formation").

9.3 Cognitive Synergy in CogPrime

Different cognitive systems will use different processes to fulfill the various roles identified in Chap. 7. Here we briefly preview the basic cognitive processes that the CogPrime AGI design uses for these roles, and the synergies that exist between these.

9.3.1 Cognitive Processes in CogPrime

Tables 9.1 and 9.2 present the key structures and processes involved in CogPrime, identifying each one with a certain memory/process type as considered in cognitive synergy theory. That is: each of these cognitive structures or processes deals with one or more types of memory—declarative, procedural, sensory, episodic or attentional. Table 9.3 describes the key CogPrime processes in terms of the "analysis vs. synthesis" distinction. Finally, Tables 9.4 and 9.5 exemplify these structures and processes in the context of embodied virtual agent control.

Table 9.1 The OpenCogPrime data structures used to represent the key knowledge types involved

Memory type	OpenCogPrime data structure
Declarative	The AtomTable, which is a special form of weighted, labeled hypergraph—i.e. a table of nodes and links (collectively referred to as Atoms) with different types, and each weighted with a multi-dimensional truth value (embodying an indefinite probability value that give both probability and confidence information).
Attentional	Atoms in the AtomTable are weighted with AttentionValue objects, which contain both ShortTermImportance values (governing processor time allocation) and LongTerm Importance values (governing memory usage)
Procedural	This is handled using special Combo tree structures embodying LISP-like programs, in a special program dialect intended to manage behaviors in a virtual world and actions in the AtomTable
Sensory	Handled via a collection of specialized sense-modality-specific data structures
Episodic	Handled via an internal simulation world that allows the system to run mind's eye movies of situations it remembers, has heard about, or hypothetically envisions
Intentional	Goals are represented by Atoms stored in the AtomTable; there is a separate table indicating which Atoms are top-level goals, which is used to guide attention allocation and goal refinement processes

Table 9.2 Key cognitive processes, and the algorithms that play their roles in CogPrime

Cognitive process	OpenCogPrime algorithm
Uncertain inference	Probabilistic Logic Networks (PLN), a logical inference framework capable of uncertain reasoning about abstract knowledge, everyday commonsense knowledge, and low-level perceptual and motor knowledge
Supervised procedure learning	MOSES, a probabilistic evolutionary learning algorithm, which learns procedures (represented as LISP-like program trees) based on specifications
Attention allocation	Economic Attention Networks (ECAN), a framework for allocating (memory and processor) attention among items of knowledge and cognitive processes, utilizing a synthesis of ideas from neural networks and artificial economics. ECAN also comes with a forgetting agent that either saves to disk or deletes knowledge that is estimated not sufficiently valuable to keep in memory
Map formation	Use of frequent subgraph mining, MOSES and other algorithms to scan the knowledge base of the system for patterns and then embodying these patterns explicitly as new knowledge items
Concept creation	A collection of heuristics for forming new concepts via combining existing ones, including conceptual blending, mutation and extensional and intensional logical operators
Simulation	The running of simulations of (remembered or imagined) external-world scenarios in an internal world-simulation engine
Goal refinement	Transformation of given goals into sets of subgoals, using concept creation, inference and procedure learning

In the CogPrime context, a procedure in this cognitive schematic is a program tree stored in the system's procedural knowledge base; and a context is a (fuzzy, probabilistic) logical predicate stored in the AtomSpace, that holds, to a certain extent, during each interval of time. A goal is a fuzzy logical predicate that has a certain value at each interval of time, as well.

Attentional knowledge is handled in CogPrime by the ECAN artificial economics mechanism, that continually updates ShortTermImportance and LongTerm Importance values associated with each item in the CogPrime system's memory, which control the amount of attention other cognitive mechanisms pay to the item, and how much motive the system has to keep the item in memory. HebbianLinks are then created between knowledge items that often possess ShortTermImportance at the same time; this is CogPrime's version of traditional Hebbian learning.

ECAN has deep interactions with other cognitive mechanisms as well, which are essential to its efficient operation; for instance, PLN inference may be used to help ECAN extrapolate conclusions about what is worth paying attention to, and

Table 9.3 Key CogPrime cognitive processes categorized according to knowledge type and process type

	Synthesis	Analysis
PLN (Decl. and Proc.)	PLN forward inference	PLN backward inference
MOSES (Decl. and Proc.)	MOSES and hillclimbing procedure learning (combining portions and aspects of prior procedures)	Probabilistic modeling to identify patterns among programs fulfilling a certain goal in a certain context (part of MOSES)
Sensory/episodic	Imagination of hypothetical episodes based on specified criteria, via combination of aspects of known episodes	Filling in gaps in remembered or hypothesized episodes
Attentional	Hebbian learning Importance spreading Map formation	Assignment of credit
Intentional	Goal synthesis	Goal refinement

MOSES may be used to recognize subtle attentional patterns. ECAN also handles "assignment of credit", the figuring-out of the causes of an instance of successful goal-achievement, drawing on PLN and MOSES as needed when the causal inference involved here becomes difficult.

The synergies between CogPrime's cognitive processes are well summarized in Table 9.6 below, which is a 16×16 matrix summarizing a host of interprocess interactions generic to CST.

One key aspect of how CogPrime implements cognitive synergy is PLN's sophisticated management of the confidence of judgments. This ties in with the way OpenCog Prime's PLN inference framework represents truth values in terms of multiple components (as opposed to the single probability values used in many probabilistic inference systems and formalisms): each item in OpenCogPrime's declarative memory has a confidence value associated with it, which tells how much weight the system places on its knowledge about that memory item. This assists with cognitive synergy as follows: A learning mechanism may consider itself "stuck", generally speaking, when it has no high-confidence estimates about the next step it should take.

Without reasonably accurate confidence assessment to guide it, inter-component interaction could easily lead to increased rather than decreased combinatorial explosion. And of course there is an added recursion here, in that confidence assessment is carried out partly via PLN inference, which in itself relies upon these same synergies for its effective operation.

To illustrate this point further, consider one of the synergetic aspects described in Table 9.6: the role cognitive synergy plays in deductive inference. Deductive inference is a hard problem in general—but what is hard about it is not carrying out

Table 9.4 Key CogPrime cognitive structures illustrated in the context of virtual agents

Knowledge type	Virtual agent example(s)
Declarative	
	• The red ball on the table is larger than the blue ball on the floor
	• Bob becomes angry quickly
	• Ball roll; blocks don't
	• Jim knows Bob is not my friend
Procedural	
	• A procedure for retrieving an item from a distant location
	• A procedure for spinning around in a circle
	• A procedure for stacking a block on top of another one
	• A procedure for repeatedly asking a question in different ways until an acceptable answer is obtained
Sensory	
	• The appearance of Bob's face
	•The specific array of objects on the floor under the table
Episodic	
	• The series of actions Bill did when he built a tower on the floor yesterday
	• The episode in which Bill and Bob repeatedly threw a ball back and forth between each other
	• The series of actions I just took, between getting up from the chair and Bob saying good
Attentional	
	• The set of objects that seem to be important in the context of the game Bob and Bill are playing
	• The set of words and phrases that are associated with Bob being happy with me while we walk around together
Intentional	
	• The goal of making Bob say positive things
	• The goal of making a tower that does not fall down easily
	• The goal of getting Jim to answer my question

inference steps, but rather "inference control" (i.e., choosing which inference steps to carry out). Specifically, what must happen for deduction to succeed in CogPrime is:

1. the system must recognize when its deductive inference process is "stuck", i.e. when the PLN inference control mechanism carrying out deduction has no clear idea regarding which inference step(s) to take next, even after considering all the domain knowledge at is disposal;
2. in this case, the system must defer to another learning mechanism to gather more information about the different choices available—and the other learning mechanism chosen must, a reasonable percentage of the time, actually provide useful information that helps PLN to get "unstuck" and continue the deductive process.

Table 9.5 Key CogPrime cognitive processes illustrated in the context of virtual agents

Cognitive process	Virtual agent example
Inference	
	• Tall thin blocks, when stood upright, are less likely to topple over if placed next to each other
	• Bob hates cursing, and Jim is Bob's friend, and friends often have similar likes and dislikes, so Jim probably hates cursing
Procedure learning	
	• Learning a procedure for crawling on the floor, based on imitation of what others do when they describe themselves as crawling, plus reinforcement from others when they find one's imitation accurate
	• Learning a procedure embodying some combination of functional and visual features that predicts whether some entity is considered a toy or not
Attention allocation	
	• Pictures of women are associated with Bob's happiness, and Bob's happiness is associated with getting reward, therefore pictures of women are associated with getting reward
	• Asking for help is surprisingly often a precursor to getting reward when Jane is around; so when a reward is gotten when Jane is around, a little extra attention should be given to ongoing improvement of the processes that help in the mechanics of asking for help
Goal refinement	
	• The goal of making Jim happy, seems to often be achieved by the goal of creating sculptures Jim likes, and Jim likes complicated sculptures; thus I adopt the goal of creating complicated sculptures when Jim is around
Declarative pattern mining	
	• The goal of making Jim happy, seems to often be achieved by the goal of creating sculptures Jim likes, and Jim likes complicated sculptures; thus I adopt the goal of creating complicated sculptures when Jim is around
Sensory pattern recognition	
	• When Jim builds a castle out of blocks, he identifies some portions of the castle as towers and others as walls; it's necessary to visually identify which portions of each castle correspond to these descriptors
	• It's also necessary to visually identify the castle as a whole versus the table, floor or other base it's resting on
Simulation	
	• Using an internal simulation world to experiment with building various towers rapidly, at a pace faster than is possible in the online simulation world where humans participate
	• Using an internal simulation world containing a simulation of Bob and Jim, to simulate what Bob will know about what you're doing if you hide behind Jim and build a tower of blocks

(Continued)

Table 9.5 (Continued)

Cognitive process	Virtual agent example
Concept creation	
	• The concept of an unstable structure
	• The concept of an irritable person
	• The concept of a happy occasion
Map formation	
	• The set of all knowledge items associated with Bob being in a good mood (which may then be used to form a new concept)
	• The set of all knowledge items associated with (running, walking or crawling) races

For instance, deduction might defer to the "attentional knowledge" subsystem, and make a judgment as to which of the many possible next deductive steps are most associated with the goal of inference and the inference steps taken so far, according to the HebbianLinks constructed by the attention allocation subsystem, based on observed associations. Or, if this fails, deduction might ask MOSES (running in supervised categorization mode) to learn predicates characterizing some of the terms involving the possible next inference steps. Once MOSES provides these new predicates, deduction can then attempt to incorporate these into its inference process, hopefully (though not necessarily) arriving at a higher-confidence next step.

9.4 Some Critical Synergies

Referring back to Fig. 9.2, and summarizing many of the ideas in the previous section, Table 9.6 enumerates a number of specific ways in which the cognitive processes mentioned in the figure may synergize with one another, potentially achieving dramatically greater efficiency than would be possible on their own.

Of course, realizing these synergies on the practical algorithmic level requires significant inventiveness and may be approached in many different ways. The specifics of how CogPrime manifests these synergies are discussed in many following chapters.

9.5 Cognitive Synergy for Procedural and Declarative Learning

We now present a little more algorithmic detail regarding the operation and synergetic interaction of CogPrime's two most sophisticated components: the MOSES procedure learning algorithm (see Chap. 16, Part 2), and the PLN uncertain inference framework (see Chap. 17, Part 2). The treatment is necessarily quite compact, since we have not yet reviewed the details of either MOSES or PLN; but as well as illustrating the notion of cognitive synergy more concretely, perhaps the high-level discussion here will make clearer how MOSES and PLN fit into the big picture of CogPrime.

Table 9.6 This table, and the following ones, show some of the synergies between the primary cognitive processes explicitly used in CogPrime

How → *Helps* ↓	Map formation	Goal system	Simulation	Sensorimotor pattern recognition
Uncertain inference	Creates new concepts and relationships. Enabling briefer useful inference trails	Goal refinement enables more careful goal-based inference pruning	-Simulations provide a method of testing speculative inferential conclusions- Simulations suggest hypotheses to be explored via inference	Creates new concepts and relationships. Enabling briefer useful inference trails
Supervised procedure learning	Creates new procedures to be used as modules in candidate procedures	Goal refinement allows more precise definition of fitness functions. Makind procedure learning's job easier	Simulation provides a method of fitness estimation allowing inexpensive testing of candidate procedures	Extraction of sensorimotor patterns allows creation of abstracted fitness functions for (inferentially and simulatively) evaluating procedures guiding real-word actions
Attention allocation	Creates new concepts grouping attentionally related memory items, enabling AA to find subtler attentional patterns involving these nodes	Goal refinement allows more accurately goal-driven allocation of attention	Simulation provides data for attention allocation— allowing attentional information to be extracted from co-occurrences observed in simulation	Creates concepts attentionally related memory items, enabling AA to find subtler attentional patterns involving these nodes

(Continued)

Table 9.6 (Continued)

How → *Helps* ↓	Map formation	Goal system	Simulation	Sensorimotor pattern recognition
Concept creation	Creates new concepts to be fed into other concept creation mechanisms	Goal refinement provides more precise definition of criteria via which new concepts are created	Utility of concepts may be assessed via creating simulated entities embodying the new concepts and seeing what they lead to in simulation	Creates New concepts to be fed into other concept creation mechanisms
Uncertain inference	NA	When inference gets stuck in an inference trail, it can ask procedure learning to learn new pattern regarding concepts in the inference trail (If there is adequate data regarding the concepts)	Importance levels allow pruning of inference trees	Provides news concepts. Allowing briefer useful inference trails
Supervised procedure learning	Inference can be used to allow prior experience to guide each instance of procedure learning	NA	Importance levels may be used to blas choices made in the course of procedure learning (e.g. in OCP, in the fitness evaluation and representation—building phases of MOSES)	Provides new concepts, allowing compacter programs using new concepts in various roles

(Continued)

Table 9.6 (Continued)

How → Helps ↓	Map formation	Goal system	Simulation	Sensorimotor pattern recognition
Attention allocation	Enables inference of new HebbianLinks and Hebbian-Predicates from existing ones	Procedure learning can recognize patterns in historical system activity, which are then used to build concepts and relationships guiding attention allocation	NA	Combination of concepts formed via map formation, may lead to new concepts that even better direct attention
Concept creation	Allows inferential assessment of the value of new concepts	Procedure learning can be used to search for high-quality blends of existing concepts (using e.g. inferential and attentional knowledge in fitness functions)	Allows assessment of the value of new concepts based on historical attentional knowledge	NA
Map formation	Speculative inference can help map formation guess which map to hunt for	Procedure learning can be used to search dor maps that are more complex than mere co-occurrence	Attention allocation provides the raw data for map formation	No significant direct synergy

(Continued)

Table 9.6 (Continued)

How → Helps ↓	Map formation	Goal system	Simulation	Sensorimotor pattern recognition
Goal system	Inference can carry out goal refinement	No significant direct synergy	Flow of importance among subgoals determines which subgoals get used, versus being forgotten	Concept creation can be used to provide raw data for goal refinement (e.g. a new subgoal that blends two others)
Simulation	In order to provide data for setting up simulations inference will often be needed	No significant direct synergy	Attention allocation tells which portions of a simulation need to be run in more detail	No significant direct synergy
Sensorimotor pattern recognition	Speculative inference helps fill in gaps in sensory data	Procedure learning can be used to find subtle patterns in sensorimotor data	Attention allocation guides parttern recognition via indicating which sensorimotor stimuli and patterns tend to be associatively linked	New concepts may be created that then are found to serve as significant patterns in sensorimotor data
Map formation	NA	Map formation may focus on finding maps relates to subgoals, and good subgoal refinement helps here	No significant direct synergy	No significant direct synergy
Goal system	Concepts formed from maps may be useful raw material for forming subgoals	NA	No significant direct synergy	No significant direct synergy

(Continued)

Table 9.6 (Continued)

How → Helps ↓	Map formation	Goal system	Simulation	Sensorimotor pattern recognition
Simulation	No significant direct synergy	No significant direct synergy	NA	Presence of recognized sensorimotor patterns may be used to judge whether a simulation is sufficiently accurate
Sensorimotor pattern recognition	Concepts formed from maps may usefully guide sensorimotor pattern search	Directing pattern toward patterns pertinent to subgoals may make the task far easier	Patterns recognized in simulations may then be checked for presence in real sensorimotor data	NA

9.5.1 Cognitive Synergy in MOSES

MOSES, CogPrime's primary algorithm for learning procedural knowledge, has been tested on a variety of application problems including standard GP test problems, virtual agent control, biological data analysis and text classification [Loo06]. It represents procedures internally as program trees. Each node in a MOSES program tree is supplied with a "knob", comprising a set of values that may potentially be chosen to replace the data item or operator at that node. So for instance a node containing the number 7 may be supplied with a knob that can take on any integer value. A node containing a while loop may be supplied with a knob that can take on various possible control flow operators including conditionals or the identity. A node containing a procedure representing a particular robot movement, may be supplied with a knob that can take on values corresponding to multiple possible movements. Following a metaphor suggested by Douglas Hofstadter [Hof96], MOSES learning covers both "knob twiddling" (setting the values of knobs) and "knob creation" (Fig. 9.2).

MOSES is invoked within CogPrime in a number of ways, but most commonly for finding a procedure P satisfying a probabilistic implication $C \& P \rightarrow G$ as described above, where C is an observed context and G is a system goal. In this case the probability value of the implication provides the "scoring function" that MOSES uses to assess the quality of candidate procedures.

Fig. 9.2 High-level control
flow of MOSES algorithm

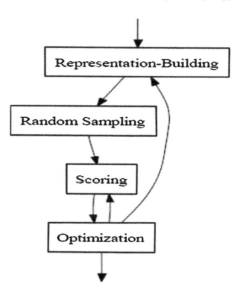

For example, suppose a CogPrime-controlled robot is trying to learn to play the
game of "tag". (I.e. a multi-agent game in which one agent is specially labeled
"it", and runs after the other player agents, trying to touch them. Once another
agent is touched, it becomes the new "it" and the previous "it" becomes just another
player agent.) Then its context C is that others are trying to play a game they call
"tag" with it; and we may assume its goals are to please them and itself, and that
it has figured out that in order to achieve this goal it should learn some procedure
to follow when interacting with others who have said they are playing "tag". In
this case a potential tag-playing procedure might contain nodes for physical actions
like *step_forward(speed s)*, as well as control flow nodes containing operators like
ifelse (for instance, there would probably be a conditional telling the robot to do
something different depending on whether someone seems to be chasing it). Each
of these program tree nodes would have an appropriate knob assigned to it. And the
scoring function would evaluate a procedure P in terms of how successfully the robot
played tag when controlling its behaviors according to P (noting that it may also
be using other control procedures concurrently with P). It's worth noting here that
evaluating the scoring function in this case involves some inference already, because
in order to tell if it is playing tag successfully, in a real-world context, it must watch
and understand the behavior of the other players.

MOSES follows the high-level control flow depicted in Fig. 16.1 (Part 2), which
corresponds to the following process for evolving a metapopulation of "demes" of
programs (each deme being a set of relatively similar programs, forming a sort of
island in program space):

1. Construct an initial set of knobs based on some prior (e.g., based on an empty program; or more interestingly, using prior knowledge **supplied by PLN inference**

based on the system's memory) and use it to generate an initial random sampling of programs. Add this deme to the metapopulation.

2. Select a deme from the metapopulation and update its sample, as follows:

   a. Select some promising programs from the deme's existing sample to use for modeling, according to the scoring function.

   b. Considering the promising programs as collections of knob settings, generate new collections of knob settings by applying some (competent) optimization algorithm. For best performance on difficult problems, it is important to use an optimization algorithm that makes use of the system's memory in its choices, **consulting PLN inference** to help estimate which collections of knob settings will work best.

   c. Convert the new collections of knob settings into their corresponding programs, reduce the programs to normal form, evaluate their scores, and integrate them into the deme's sample, replacing less promising programs. In the case that scoring is expensive, score evaluation may be preceded by score estimation, which may use **PLN inference**, enaction of procedures in an **internal simulation environment**, and/or similarity matching against **episodic memory**.

3. For each new program that meet the criterion for creating a new deme, if any:

   a. Construct a new set of knobs (a process called "representation-building") to define a region centered around the program (the deme's *exemplar*), and use it to generate a *new* random sampling of programs, producing a new deme.

   b. Integrate the new deme into the metapopulation, possibly displacing less promising demes.

4. Repeat from step 2.

MOSES is a complex algorithm and each part plays its role; if any one part is removed the performance suffers significantly [Loo06]. However, the main point we want to highlight here is the role played by synergetic interactions between MOSES and other cognitive components such as PLN, simulation and episodic memory, as indicated in **boldface** in the above pseudocode. MOSES is a powerful procedure learning algorithm, but used on its own it runs into scalability problems like any other such algorithm; the reason we feel it has potential to play a major role in a human-level AI system is its capacity for productive interoperation with other cognitive components.

Continuing the "tag" example, the power of MOSES's integration with other cognitive processes would come into play if, before learning to play tag, the robot has already played simpler games involving chasing. If the robot already has experience chasing and being chased by other agents, then its episodic and declarative memory will contain knowledge about how to pursue and avoid other agents in the context of running around an environment full of objects, and this knowledge will be deployable within the appropriate parts of MOSES's Steps 1 and 2. Cross-process and cross-memory-type integration make it tractable for MOSES to act as a "transfer learning" algorithm, not just a task-specific machine-learning algorithm.

9.5.2 Cognitive Synergy in PLN

While MOSES handles much of CogPrime's procedural learning, and OpenCog-Primes internal simulation engine handles most episodic knowledge, CogPrime's primary tool for handling declarative knowledge is an uncertain inference framework called Probabilistic Logic Networks (PLN). The complexities of PLN are the topic of a lengthy technical monograph [GMIH08], and here we will eschew most details and focus mainly on pointing out how PLN seeks to achieve efficient inference control via integration with other cognitive processes.

As a logic, PLN is broadly integrative: it combines certain term logic rules with more standard predicate logic rules, and utilizes both fuzzy truth values and a variant of imprecise probabilities called *indefinite probabilities*. PLN mathematics tells how these uncertain truth values propagate through its logic rules, so that uncertain premises give rise to conclusions with reasonably accurately estimated uncertainty values. This careful management of uncertainty is critical for the application of logical inference in the robotics context, where most knowledge is abstracted from experience and is hence highly uncertain.

PLN can be used in either forward or backward chaining mode; and in the language introduced above, it can be used for either analysis or synthesis. As an example, we will consider backward chaining analysis, exemplified by the problem of a robot preschool-student trying to determine whether a new playmate "Bob" is likely to be a regular visitor to its preschool or not (evaluating the truth value of the implication $Bob \rightarrow regular\_visitor$). The basic backward chaining process for PLN analysis looks like:

1. Given an implication $L \equiv A \rightarrow B$ whose truth value must be estimated (for instance $L \equiv C\&P \rightarrow G$ as discussed above), create a list $(A_1, \ldots, A_n)$ of *(inference rule, stored knowledge)* pairs that might be used to produce L.
2. Using analogical reasoning to prior inferences, assign each A_i a probability of success.

   - If some of the A_i are estimated to have reasonable probability of success at generating reasonably confident estimates of L's truth value, then invoke Step 1 with A_i in place of L (at this point the inference process becomes recursive).
   - If none of the A_i looks sufficiently likely to succeed, then inference has "gotten stuck" and another cognitive process should be invoked, e.g.
     - **Concept creation** may be used to infer new concepts related to A and B, and then Step 1 may be revisited, in the hope of finding a new, more promising A_i involving one of the new concepts.
     - **MOSES** may be invoked with one of several special goals, e.g. the goal of finding a procedure P so that $P(X)$ predicts whether $X \rightarrow B$. If MOSES finds such a procedure P then this can be converted to declarative knowledge understandable by PLN and Step 1 may be revisited . . .

- **Simulations** may be run in CogPrime's internal simulation engine, so as to observe the truth value of $A \rightarrow B$ in the simulations; and then Step 1 may be revisited . . .

The combinatorial explosion of inference control is combatted by the capability to defer to other cognitive processes when the inference control procedure is unable to make a sufficiently confident choice of which inference steps to take next. Note that just as MOSES may rely on PLN to model its evolving populations of procedures, PLN may rely on MOSES to create complex knowledge about the terms in its logical implications. This is just one example of the multiple ways in which the different cognitive processes in CogPrime interact synergetically; a more thorough treatment of these interactions is given in Chap. 32 (Part 2).

In the "new playmate" example, the interesting case is where the robot initially seems not to know enough about Bob to make a solid inferential judgment (so that none of the A_i seem particularly promising). For instance, it might carry out a number of possible inferences and not come to any reasonably confident conclusion, so that the reason none of the A_i seem promising is that all the decent-looking ones have been tried already. So it might then recourse to MOSES, simulation or concept creation.

For instance, the PLN controller could make a list of everyone who has been a regular visitor, and everyone who has not been, and pose MOSES the task of figuring out a procedure for distinguishing these two categories. This procedure could then used directly to make the needed assessment, or else be translated into logical rules to be used within PLN inference. For example, perhaps MOSES would discover that older males wearing ties tend not to become regular visitors. If the new playmate is an older male wearing a tie, this is directly applicable. But if the current playmate is wearing a tuxedo, then PLN may be helpful via reasoning that even though a tuxedo is not a tie, it's a similar form of fancy dress—so PLN may extend the MOSES-learned rule to the present case and infer that the new playmate is not likely to be a regular visitor.

9.6 Is Cognitive Synergy Tricky?

In this section[1] we use the notion of cognitive synergy to explore a question that arises frequently in the AGI community: the well-known difficulty of measuring intermediate progress toward human-level AGI. We explore some potential reasons underlying this, via extending the notion of cognitive synergy to a more refined notion of "tricky cognitive synergy". These ideas are particularly relevant to the problem of creating a roadmap toward AGI, as we'll explore in Chap. 18.

---

[1] This section co-authored with Jade O'Neill.

9.6.1 The Puzzle: Why Is It So Hard to Measure Partial Progress Toward Human-Level AGI?

It's not entirely straightforward to create tests to measure the *final achievement* of human-level AGI, but there are some fairly obvious candidates here. There's the Turing Test (fooling judges into believing you're human, in a text chat), the video Turing Test, the Robot College Student test (passing university, via being judged exactly the same way a human student would), etc. There's certainly no agreement on which is the most meaningful such goal to strive for, but there's broad agreement that a number of goals of this nature basically make sense.

On the other hand, how does one measure whether one is, say, 50 % of the way to human-level AGI? Or, say, 75 or 25 %?

It's possible to pose many "practical tests" of incremental progress toward human-level AGI, with the property that if a proto-AGI system passes the test using a certain sort of architecture and/or dynamics, then this implies a certain amount of progress toward human-level AGI *based on particular theoretical assumptions about AGI*. However, in each case of such a practical test, it seems intuitively likely *to a significant percentage of AGI researchers* that there is some way to "game" the test via designing a system specifically oriented toward passing that test, and which doesn't constitute dramatic progress toward AGI.

Some examples of practical tests of this nature would be

- The Wozniak "coffee test": go into an average American house and figure out how to make coffee, including identifying the coffee machine, figuring out what the buttons do, finding the coffee in the cabinet, etc.
- Story understanding—reading a story, or watching it on video, and then answering questions about what happened (including questions at various levels of abstraction).
- Graduating (virtual-world or robotic) preschool.
- Passing the elementary school reading curriculum (which involves reading and answering questions about some picture books as well as purely textual ones).
- Learning to play an arbitrary video game based on experience only, or based on experience plus reading instructions.

One interesting point about tests like this is that each of them seems to *some* AGI researchers to encapsulate the crux of the AGI problem, and be unsolvable by any system not far along the path to human-level AGI—yet seems to other AGI researchers, with different conceptual perspectives, to be something probably gameable by narrow-AI methods. And of course, given the current state of science, there's no way to tell which of these practical tests really can be solved via a narrow-AI approach, except by having a lot of people try really hard over a long period of time.

A question raised by these observations is whether there is some *fundamental reason* why it's hard to make an objective, theory-independent measure of intermediate progress toward advanced AGI. Is it just that we haven't been smart enough to figure out the right test—or is there some conceptual reason why the very notion of such a test is problematic?

We don't claim to know for sure—but in the rest of this section we'll outline one possible reason why the latter might be the case.

9.6.2 A Possible Answer: Cognitive Synergy Is Tricky!

Why might a solid, objective empirical test for intermediate progress toward AGI be an infeasible notion? One possible reason, we suggest, is precisely *cognitive synergy*, as discussed above.

The cognitive synergy hypothesis, in its simplest form, states that human-level AGI intrinsically depends on the synergetic interaction of multiple components (for instance, as in CogPrime, multiple memory systems each supplied with its own learning process). In this hypothesis, for instance, it might be that there are 10 critical components required for a human-level AGI system. Having all 10 of them in place results in human-level AGI, but having only 8 of them in place results in having a dramatically impaired system—and maybe having only 6 or 7 of them in place results in a system that can hardly do anything at all.

Of course, the reality is almost surely not as strict as the simplified example in the above paragraph suggests. No AGI theorist has really posited a list of 10 crisply-defined subsystems and claimed them necessary and sufficient for AGI. We suspect there are many different routes to AGI, involving integration of different sorts of subsystems. However, if the cognitive synergy hypothesis is correct, then human-level AGI behaves *roughly* like the simplistic example in the prior paragraph suggests. Perhaps instead of using the 10 components, you could achieve human-level AGI with 7 components, but having only 5 of these 7 would yield drastically impaired functionality—etc. Or the point could be made without any decomposition into a finite set of components, using continuous probability distributions. To mathematically formalize the cognitive synergy hypothesis becomes complex, but here we're only aiming for a qualitative argument. So for illustrative purposes, we'll stick with the "10 components" example, just for communicative simplicity.

Next, let's suppose that for any given task, there are ways to achieve this task using a system that is much simpler than any subset of size 6 drawn from the set of 10 components needed for human-level AGI, but works much better for the task than this subset of 6 components (assuming the latter are used as a set of only 6 components, without the other 4 components).

Note that this supposition is a good bit stronger than mere cognitive synergy. For lack of a better name, we'll call it *tricky cognitive synergy*. The tricky cognitive synergy hypothesis would be true if, for example, the following possibilities were true:

- creating components to serve as parts of a synergetic AGI is *harder* than creating components intended to serve as parts of simpler AI systems without synergetic dynamics;
- components capable of serving as parts of a synergetic AGI are necessarily *more complicated* than components intended to serve as parts of simpler AGI systems.

These certainly seem reasonable possibilities, since to serve as a component of a synergetic AGI system, a component must have the internal flexibility to usefully handle interactions with a lot of other components as well as to solve the problems that come its way. In a CogPrime context, these possibilities ring true, in the sense that tailoring an AI process for tight integration with other AI processes within CogPrime, tends to require more work than preparing a conceptually similar AI process for use on its own or in a more task-specific narrow AI system.

It seems fairly obvious that, if tricky cognitive synergy really holds up as a property of human-level general intelligence, the difficulty of formulating tests for intermediate progress toward human-level AGI follows as a consequence. Because, according to the tricky cognitive synergy hypothesis, any test is going to be more easily solved by some simpler narrow AI process than by a *partially complete* human-level AGI system.

9.6.3 Conclusion

We haven't proved anything here, only made some qualitative arguments. However, these arguments do seem to give a plausible explanation for the empirical observation that positing tests for intermediate progress toward human-level AGI is a very difficult prospect. If the theoretical notions sketched here are correct, then this difficulty is not due to incompetence or lack of imagination on the part of the AGI community, nor due to the primitive state of the AGI field, but is rather intrinsic to the subject matter. And if these notions are correct, then quite likely the future rigorous science of AGI will contain formal theorems echoing and improving the qualitative observations and conjectures we've made here.

If the ideas sketched here are true, then the practical consequence for AGI development is, very simply, that one shouldn't worry a lot about producing intermediary results that are compelling to skeptical observers. Just at 2/3 of a human brain may not be of much use, similarly, 2/3 of an AGI system may not be much use. Lack of impressive intermediary results may not imply one is on a wrong development path; and comparison with narrow AI systems on specific tasks may be badly misleading as a gauge of incremental progress toward human-level AGI.

Hopefully it's clear that the motivation behind the line of thinking presented here is a desire to understand the nature of general intelligence and its pursuit—not a desire to avoid testing our AGI software! Really, as AGI engineers, we would love to have a sensible rigorous way to test our intermediary progress toward AGI, so as to be able to pose convincing arguments to skeptics, funding sources, potential collaborators and so forth. Our motivation here is not a desire to avoid having the intermediate progress of our efforts measured, but rather a desire to explain the frustrating (but by now rather well-established) difficulty of creating such intermediate goals for human-level AGI in a meaningful way.

If we or someone else figures out a compelling way to measure partial progress toward AGI, we will celebrate the occasion. But it seems worth seriously considering the possibility that the difficulty in finding such a measure reflects fundamental properties of general intelligence.

From a practical CogPrime perspective, we are interested in a variety of evaluation and testing methods, including the "virtual preschool" approach mentioned briefly above and more extensively in later chapters. However, our focus will be on evaluation methods that give us meaningful information about CogPrime's progress, given our knowledge of how CogPrime works and our understanding of the underlying theory. We are unlikely to focus on the achievement of intermediate test results capable of convincing skeptics of the reality of our partial progress, because we have not yet seen any credible tests of this nature, and because we suspect the reasons for this lack may be rooted in deep properties of feasible general intelligence, such as tricky cognitive synergy.

Chapter 10
General Intelligence in the Everyday Human World

10.1 Introduction

Intelligence is not just about what happens inside a system, but also about what happens outside that system, and how the system interacts with its environment. Real-world general intelligence is about intelligence *relative to some particular class of environments*, and human-like general intelligence is about intelligence relative to the particular class of environments that humans evolved in (which in recent millennia has included environments humans have created using their intelligence). In Chap. 4, we reviewed some specific capabilities characterizing human-like general intelligence; to connect these with the general theory of general intelligence from the last few chapters, we need to explain what aspects of human-relevant *environments* correspond to these human-like intelligent *capabilities*. We begin with aspects of the environment related to communication, which turn out to tie in closely with cognitive synergy. Then we turn to physical aspects of the environment, which we suspect also connect closely with various human cognitive capabilities. Finally we turn to physical aspects of the human body and their relevance to the human mind. In the following chapter we present a deeper, more abstract theoretical framework encompassing these ideas.

These ideas are of theoretical importance, and they're also of practical importance when one turns to the critical area of *AGI environment design*. If one is going to do anything besides release one's young AGI into the "wilds" of everyday human life, then one has to put some thought into what kind of environment it will be raised in. This may be a virtual world or it may be a robot preschool or some other kind of physical environment, but in any case some specific choices must be made about what to include. Specific choices must also be made about what kind of body to give one's AGI system—what sensors and actuators, and so forth. In Chap. 17 we will present some specific suggestions regarding choices of embodiment and environment that we find to be ideal for AGI development—virtual and robot preschools—but the material in this chapter is of more general import, beyond any such particularities. If one has an intuitive idea of what properties of body and world human intelligence is

B. Goertzel et al., *Engineering General Intelligence, Part 1*,
Atlantis Thinking Machines 5, DOI: 10.2991/978-94-6239-027-0_10,
© Atlantis Press and the authors 2014

biased for, then one can make practical choices about embodiment and environment in a principled rather than purely ad hoc or opportunistic way.

10.2 Some Broad Properties of the Everyday World that Help Structure Intelligence

The properties of the everyday world that help structure intelligence are diverse and span multiple levels of abstraction. Most of this chapter will focus on fairly concrete patterns of this nature, such as are involved in inter-agent communication and naive physics; however, it's also worth noting the potential importance of more abstract patterns distinguishing the everyday world from arbitrary mathematical environments.

The propensity to search for hierarchical patterns is one huge potential example of an abstract everyday-world property. We strongly suspect the reason that searching for hierarchical patterns works so well, in so many everyday-world contexts, lies in the particular structure of the everyday world—it's not something that would be true across all possible environments (even if one weights the space of possible environments in some clever way, say using program-length according to some standard computational model). However, this sort of assertion is of course highly "philosophical," and becomes complex to formulate and defend convincingly given the current state of science and mathematics.

Going one step further, we recall from Chap. 5 a structure called the "dual network", which consists of superposed hierarchical and heterarchical networks: basically a hierarchy in which the distance between two nodes in the hierarchy is correlated with the distance between the nodes in some metric space. Another high level property of the everyday world may be that dual network structures are prevalent. This would imply that minds biased to represent the world in terms of dual network structure are likely to be intelligent with respect to the everyday world.

In a different direction, the extreme commonality of symmetry groups in the (everyday and otherwise) physical world is another example: they occur so often that minds oriented toward recognizing patterns involving symmetry groups are likely to be intelligent with respect to the real world.

We suspect that the number of cognitively-relevant properties of the everyday world is huge ... and that the essence of everyday-world intelligence lies in the list of varyingly abstract and concrete properties, which must be embedded implicitly or explicitly in the structure of a natural or artificial intelligence for that system to have everyday-world intelligence.

Apart from these particular yet abstract properties of the everyday world, intelligence is just about "finding patterns in which actions tend to achieve which goals in which situations" ... but, the simple meta-algorithm needed to accomplish this universally is, we suggest, only a small percentage of what it takes to make a mind.

You might say that a sufficiently generally intelligent system should be able to infer the various cognitively-relevant properties of the environment from looking at data about the everyday world. We agree *in principle*, and in fact Ben Kuipers and his colleagues have done some interesting work in this direction, showing that learning algorithms can infer some basics about the structure of space and time from experience [MK07]. But we suggest that doing this really thoroughly would require a massively greater amount of processing power than an AGI that embodies and hence automatically utilizes these principles. It may be that the problem of inferring these properties is so hard as to require a wildly infeasible $AIXI^{tl}$ / Godel Machine type system.

10.3 Embodied Communication

Next we turn to the potential cognitive implications of seeking to achieve goals in an environment in which multimodal communication with other agents plays a prominent role.

Consider a community of embodied agents living in a shared world, and suppose that the agents can communicate with each other via a set of mechanisms including:

- **Linguistic communication**, in a language whose semantics is largely (not necessarily wholly) interpretable based on the mutually experienced world
- **Indicative communication**, in which e.g. one agent points to some part of the world or delimits some interval of time, and another agent is able to interpret the meaning
- **Demonstrative communication**, in which an agent carries out a set of actions in the world, and the other agent is able to imitate these actions, or instruct another agent as to how to imitate these actions
- **Depictive communication**, in which an agent creates some sort of (visual, auditory, etc.) construction to show another agent, with a goal of causing the other agent to experience phenomena similar to what they would experience upon experiencing some particular entity in the shared environment
- **Intentional communication**, in which an agent explicitly communicates to another agent what its goal is in a certain situation[1]

It is clear that ordinary everyday communication between humans possesses all these aspects.

We define the **Embodied Communication Prior** (ECP) as the probability distribution in which the probability of an entity (e.g. a goal or environment) is proportional to the difficulty of describing that entity, for a typical member of the community in question, using a particular set of communication mechanisms including the above five modes. We will sometimes refer to the prior probability of an entity under this distribution, as its "simplicity" under the distribution.

---

[1] In Appendix C we recount some interesting recent results showing that mirror neurons fire in response to some cases of intentional communication as thus defined.

Next, to further specialize the Embodied Communication Prior, we will assume that for each of these modes of communication, there are some aspects of the world that are much more easily communicable using that mode than the other modes. For instance, in the human everyday world:

- Abstract (declarative) statements spanning large classes of situations are generally much easier to communicate linguistically
- Complex, multi-part procedures are much easier to communicate either demonstratively, or using a combination of demonstration with other modes
- Sensory or episodic data is often much easier to communicate demonstratively
- The current value of attending to some portion of the shared environment is often much easier to communicate indicatively
- Information about what goals to follow in a certain situation is often much easier to communicate intentionally, i.e. via explicitly indicating what one's own goal is

These simple observations have significant implications for the nature of the Embodied Communication Prior. For one thing they let us define multiple forms of knowledge:

- **Isolatedly declarative knowledge** is that which is much more easily communicable linguistically
- **Isolatedly procedural knowledge** is that which is much more easily communicable demonstratively
- **Isolatedly sensory knowledge** is that which is much more easily communicable depictively
- **Isolatedly attentive knowledge** is that which is much more easily communicable indicatively
- **Isolatedly intentional knowledge** is that which is much more easily communicable intentionally

This categorization of knowledge types resembles many ideas from the cognitive theory of memory [TC05], although the distinctions drawn here are a little crisper than any classification currently derivable from available neurological or psychological data.

Of course there may be much knowledge, of relevance to systems seeking intelligence according to the ECP, that does not fall into any of these categories and constitutes "mixed knowledge". There are some very important specific subclasses of mixed knowledge. For instance, episodic knowledge (knowledge about specific real or hypothetical sets of events) will most easily be communicated via a combination of declarative, sensory and (in some cases) procedural communication. Scientific and mathematical knowledge are generally mixed knowledge, as is most everyday commonsense knowledge.

Some cases of mixed knowledge are reasonably well decomposable, in the sense that they decompose into knowledge items that individually fall into some specific knowledge type. For instance, an experimental chemistry procedure may be much more easily communicable procedurally, whereas an allied piece of knowledge from theoretical chemistry may be much more easily communicable declaratively; but in order to fully communicate either the experimental procedure or the abstract piece of knowledge, one may ultimately need to communicate both aspects.

Also, even when the best way to communicate something is mixed-mode, it may be possible to identify one mode that poses the most important part of the communication. An example would be a chemistry experiment that is best communicated via a practical demonstration together with a running narrative. It may be that the demonstration without the narrative would be vastly more valuable than the narrative without the demonstration. To cover such cases we may make less restrictive definitions such as

- **Interactively declarative knowledge** is that which is much more easily communicable in a manner dominated by linguistic communication

and so forth. We call these "interactive knowledge categories," by contrast to the "isolated knowledge categories" introduced earlier.

10.3.0.1 Naturalness of Knowledge Categories

Next we introduce an assumption we call NKC, for Naturalness of Knowledge Categories. The NKC assumption states that the knowledge in each of the above isolated and interactive communication-modality-focused categories forms a "natural category," in the sense that for each of these categories, there are many different properties shared by a large percentage of the knowledge in the category, but not by a large percentage of the knowledge in the other categories. This means that, for instance, procedural knowledge systematically (and statistically) has different characteristics than the other kinds of knowledge.

The NKC assumption seems commonsensically to hold true for human everyday knowledge, and it has fairly dramatic implications for general intelligence. Suppose we conceive general intelligence as the ability to achieve goals in the environment shared by the communicating agents underlying the Embodied Communication Prior. Then, NKC suggests that the best way to achieve general intelligence according to the Embodied Communication Prior is going to involve

- specialized methods for handling declarative, procedural, sensory and attentional knowledge (due to the naturalness of the isolated knowledge categories)
- specialized methods for handling interactions between different types of knowledge, including methods focused on the case where one type of knowledge is primary and the others are supporting (the latter due to the naturalness of the interactive knowledge categories).

10.3.0.2 Cognitive Completeness

Suppose we conceive an AI system as consisting of a set of learning capabilities, each one characterized by three features:

- One or more **knowledge types** that it is competent to deal with, in the sense of the two key learning problems mentioned above
- At least one **learning type**: either analysis, or synthesis, or both
- At least one **interaction type**, for each (knowledge type, learning type) pair it handles: "isolated" (meaning it deals mainly with that knowledge type in isolation), or "interactive" (meaning it focuses on that knowledge type but in a way that explicitly incorporates other knowledge types into its process), or "fully mixed" (meaning that when it deals with the knowledge type in question, no particular knowledge type tends to dominate the learning process).

Then, intuitively, it seems to follow from the ECP with NKC that systems with high efficient general intelligence should have the following properties, which collectively we'll call **cognitive completeness**:

- For each (knowledge type, learning type, interaction type) triple, there should be a learning capability corresponding to that triple.
- Furthermore the capabilities corresponding to different (knowledge type, interaction type) pairs should have distinct characteristics (since according to the NKC the isolated knowledge corresponding to a knowledge type is a natural category, as is the dominant knowledge corresponding to a knowledge type)
- For each (knowledge type, learning type) pair (K, L), and each other knowledge type K1 distinct from K, there should be a distinctive capability with interaction type "interactive" and dealing with knowledge that is interactively K but also includes aspects of K1

Furthermore, it seems intuitively sensible that according to the ECP with NKC, if the capabilities mentioned in the above points are reasonably able, then the system possessing the capabilities will display general intelligence relative to the ECP. Thus we arrive at the hypothesis that

Under the assumption of the Embodied Communication Prior (with the Natural Knowledge Categories assumption), the property above called "cognitive completeness" is necessary and sufficient for efficient general intelligence at the level of an inteligent adult human (e.g. at the Piagetan formal level [Pia53]).

Of course, the above considerations are very far from a rigorous mathematical proof (or even precise formulation) of this hypothesis. But we are presenting this here as a conceptual hypothesis, in order to qualitatively guide our practical AGI R&D and also to motivate further, more rigorous theoretical work.

10.3.1 Generalizing the Embodied Communication Prior

One interesting direction for further research would be to broaden the scope of the inquiry, in a manner suggested above: instead of just looking at the ECP, look at simplicity measures in general, and attack the question of how a mind must be structured in order to display efficient general intelligence relative to a specified simplicity measure. This problem seems unapproachable in general, but some special cases may be more tractable.

For instance, suppose one has

- a simplicity measure that (like the ECP) is approximately decomposable into a set of fairly distinct components, plus their interactions
- an assumption similar to NKC, which states that the entities displaying simplicity according to each of the distinct components, are roughly clustered together in entity-space

Then one should be able to say that, to achieve efficient general intelligence relative to this decomposable simplicity measure, a system should have distinct capabilities corresponding to each of the components of the simplicity measure interactions between these capabilities, corresponding to the interaction terms in the simplicity measure.

With copious additional work, these simple observations could potentially serve as the seed for a novel sort of theory of general intelligence—a theory of how the structure of a system depends on the structure of the simplicity measure with which it achieves efficient general intelligence. Cognitive Synergy Theory would then emerge as a special case of this more abstract theory.

10.4 Naive Physics

Multimodal communication is an important aspect of the environment for which human intelligence evolved—but not the only one. It seems likely that our human intelligence is also closely adapted to various aspects of our physical environment—a matter that is worth carefully attending as we design environments for our robotically or virtually embodied AGI systems to operate in.

One interesting guide to the most cognitively relevant aspects of human environments is the subfield of AI known as "naive physics" [Hay85]—a term that refers to the theories about the physical world that human beings implicitly develop and utilize during their lives. For instance, when you figure out that you need to pressure the knife slightly harder when spreading peanut butter rather than jelly, you're not making this judgment using Newtonian physics or the Navier-Stokes equations of fluid dynamics; you're using heuristic patterns that you figured out through experience. Maybe you figured out these patterns through experience spreading peanut butter and jelly in particular. Or maybe you figured these heuristic patterns out before you ever tried to spread peanut butter or jelly specifically, via just touching peanut butter and jelly to see what they feel like, and then carrying out inference based on your experience manipulating similar tools in the context of similar substances.

Other examples of similar "naive physics" patterns are easy to come by, e.g.

1. What goes up must come down.
2. A dropped object falls straight down.

3. A vacuum sucks things towards it.
4. Centrifugal force throws rotating things outwards.
5. An object is either at rest or moving, in an absolute sense.
6. Two events are simultaneous or they are not.
7. When running downhill, one must lift one's knees up high.
8. When looking at something that you just barely can't discern accurately, squint.

Attempts to axiomatically formulate naive physics have historically come up short, and we doubt this is a promising direction for AGI. However, we do think the naive physics literature does a good job of identifying the various phenomena that the human mind's naive physics deals with. So, from the point of view of AGI environment design, naive physics is a useful source of requirements. Ideally, we would like an AGI's environment to support all the fundamental phenomena that naive physics deals with.

We now describe some key aspects of naive physics in a more systematic manner. Naive physics has many different formulations; in this section we draw heavily on [SC94], who divide naive physics phenomena into 5 categories. Here we review these categories and identify a number of important things that humanlike intelligent agents must be able to do relative to each of them.

10.4.1 Objects, Natural Units and Natural Kinds

One key aspect of naive physics involves recognition of various aspects of objects, such as:

1. Recognition of objects amidst noisy perceptual data
2. Recognition of surfaces and interiors of objects
3. Recognition of objects as manipulable units
4. Recognition of objects as potential subjects of fragmentation (splitting, cutting) and of unification (gluing, bonding)
5. Recognition of the agent's body as an object, and as parts of the agent's body as objects
6. Division of universe of perceived objects into "natural kinds", each containing typical and atypical instances.

10.4.2 Events, Processes and Causality

Specific aspects of naive physics related to temporality and causality are:

1. Distinguishing roughly-subjectively-instantaneous events from extended processes

2. Identifying beginnings, endings and crossings of processes
3. Identifying and distinguishing internal and external changes
4. Identifying and distinguishing internal and external changes relative to one's own body
5. Interrelating body-changes with changes in external entities

Notably, these aspects of naive physics involve a different processes occurring on a variety of different time scales, intersecting in complex patterns, and involving processes inside the agent's body, outside the agent's body, and crossing the boundary of the agent's body.

10.4.3 Stuffs, States of Matter, Qualities

Regarding the various states of matter, some important aspects of naive physics are:

1. Perceiving gaps between objects: holes, media, illusions like rainbows, mirages and holograms
2. Distinguishing the manners in which different sorts of entities (e.g. smells, sounds, light) fill space
3. Distinguishing properties such as smoothness, roughness, graininess, stickiness, runniness, etc.
4. Distinguishing degrees of elasticity and fragility
5. Assessing separability of aggregates.

10.4.4 Surfaces, Limits, Boundaries, Media

Gibson [Gib77, Gib79] has argued that naive physics is not mainly about objects but rather mainly about surfaces. Surfaces have a variety of aspects and relationships that are important for naive physics, such as:

1. Perceiving and reasoning about surfaces as two-sided or one-sided interfaces
2. Inference of the various ecological laws of surfaces
3. Perception of various media in the world as separated by surfaces
4. Recognition of the textures of surfaces
5. Recognition of medium/surface layout relationships such as: ground, open environment, enclosure, detached object, attached object, hollow object, place, sheet, fissure, stick, fibre, dihedral, etc.

As a concrete, evocative "toy" example of naive everyday knowledge about surfaces and boundaries, consider Sloman's [Slo08a] example scenario, depicted in Fig. 10.1 and drawn largely from [SS74] (see also related discussion in [Slo08b], in which "A child can be given one or more rubber bands and a pile of pins, and asked

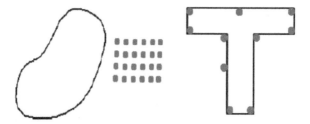

Fig. 10.1   One of Sloman's example test domains for real-world inference. *Left* a number of pins and a rubber band to be stretched around them. *Right* use of the pins and rubber band to make a letter T

to use the pins to hold the band in place to form a particular shape)... For example, things to be learnt could include":

1. There is an area inside the band and an area outside the band.
2. The possible effects of moving a pin that is inside the band towards or further away from other pins inside the band. (The effects can depend on whether the band is already stretched.)
3. The possible effects of moving a pin that is outside the band towards or further away from other pins inside the band.
4. The possible effects of adding a new pin, inside or outside the band, with or without pushing the band sideways with the pin first.
5. The possible effects of removing a pin, from a position inside or outside the band.
6. Patterns of motion/change that can occur and how they affect local and global shape (e.g. introducing a concavity or convexity, introducing or removing symmetry, increasing or decreasing the area enclosed).
7. The possibility of causing the band to cross over itself. (NB: Is an odd number of crosses possible?)
8. How adding a second, or third band can enrich the space of structures, processes and effects of processes.

10.4.5 What Kind of Physics is Needed to Foster Human-Like Intelligence?

We stated above that we would like an AGI's environment to support all the fundamental phenomena that naive physics deals with; and we have now reviewed a number of these specific phenomena. But it's not entirely clear what the "fundamental" aspects underlying these phenomena are. One important question in the environment-design context is how close an AGI environment needs to stick to the particulars of real-world naive physics. Is it important that a young AGI can play with the specific differences between spreading peanut butter versus jelly? Or is it enough that it can play with spreading and smearing various substances of different consistencies? How close does the analogy between an AGI environment's naive physics and real-world naive physics need to be? This is a question to which we have no scientific answer at present. Our own working hypothesis is that the analogy does not need to be extremely close, and with this in mind in Chap. 17 we propose a virtual environment BlocksNBeadsWorld that encompasses all the basic conceptual phenomena of real-world naive physics, but does not attempt to emulate their details.

Framed in terms of human psychology rather than environment design, the question becomes: *At what level of detail must one model the physical world to understand the ways in which human intelligence has adapted to the physical world?*. Our suspicion, which underlies our BlocksNBeadsWorld design, is that it's approximately enough to have

- Newtonian physics, or some close approximation
- Matter in multiple phases and forms vaguely similar to the ones we see in the real world: solid, liquid, gas, paste, go, etc.
- Ability to transform some instances of matter from one form to another
- Ability to flexibly manipulate matter in various forms with various solid tools
- Ability to combine instances of matter into new ones in a fairly rich way: e.g. glue or tie solids togethermix liquids together, etc.
- Ability to position instances of matter with respect to each other in a rich way: e.g. put liquid in a solid cavity, cover something with a lid or a piece of fabric, etc.

It seems to us that if the above are present in an environment, then an AGI seeking to achieve appropriate goals in that environment will be likely to form an appropriate "human-like physical-world intuition." We doubt that the specifics of the naive physics of different forms of matter are critical to human-like intelligence. But, we suspect that a great amount of unconscious human metaphorical thinking is conditioned on the fact that humans evolved around matter that takes a variety of forms, can be changed from one form to another, and can be fairly easily arranged and composited to form new instances from prior ones. Without many diverse instances of matter transformation, arrangement and composition in its experience, an AGI is unlikely to form an internal "metaphor-base" even vaguely similar to the human one—so that, even if it's highly intelligent, its thinking will be radically non-human-like in character.

Naturally this is all somewhat speculative and must be explored via experimentation. Maybe an elaborate blocks-world with only solid objects will be sufficient to create human-level, roughly human-like AGI with rich spatiotemporal and manipulative intuition. Or maybe human intelligence is more closely adapted to the specifics of our physical world—with water and dirt and plants and hair and so forth—than we currently realize. One thing that *is* very clear is that, as we proceed with embodying, situating and educating our AGI systems, we need to pay careful attention to the way their intelligence is conditioned by their environment.

10.5 Folk Psychology

Related to naive physics is the notion of "naive psychology" or "folk psychology" [Rav04], which includes for instance the following aspects:

1. Mental simulation of other agents
2. Mental theory regarding other agents
3. Attribution of beliefs, desires and intentions (BDI) to other agents via theory or simulation
4. Recognition of emotions in other agents via their physical embodiment
5. Recognition of desires and intentions in other agents via their physical embodiment
6. Analogical and contextual inferences between self and other, regarding BDI and other aspects
7. Attribute causes and meanings to other agents behaviors
8. Anthropomorphize non-human, including inanimate objects

The main special requirement placed on an AGI's embodiment by the above aspects pertains to the ability of agents to express their emotions and intentions to each other. Humans do this via facial expressions, gestures and language.

10.5.1 Motivation, Requiredness, Value

Relatedly to folk psychology, Gestalt [Koh38] and ecological [Gib77, Gib79] psychology suggest that humans perceive the world substantially in terms of the affordances it provides them for goal-directed action. This suggests that, to support human-like intelligence, an AGI must be capable of:

1. Perception of entities in the world as differentially associated with goal-relevant value
2. Perception of entities in the world in terms of the potential actions they afford the agent, or other agents

The key point is that entities in the world need to provide a wide variety of ways for agents to interact with them, enabling richly complex perception of affordances.

10.6 Body and Mind

The above discussion has focused on the world external to the body of the AGI agent embodied and embedded in the world, but the issue of the AGI's body also merits consideration. There seems little doubt that a human's intelligence is highly conditioned by the particularities of the human body.

10.6.1 The Human Sensorium

Here the requirements seem fairly simple: while surely not strictly necessary, it would certainly be *preferable* to provide an AGI with fairly rich analogues of the human senses of touch, sight, sound, kinesthesia, taste and smell. Each of these senses provides different sorts of cognitive stimulation to the human mind; and while similar cognitive stimulation could doubtless be achieved without analogous senses, the provision of such seems the most straightforward approach. It's hard to know how much of human intelligence is specifically biased to the sorts of outputs provided by human senses.

As vision already is accorded such a prominent role in the AI and cognitive science literature—and is discussed in moderate depth in Chap. 9 of Part 2, we won't take time elaborating on the importance of vision processing for humanlike cognition. The key thing an AGI requires to support humanlike "visual intelligence" is an environment containing a sufficiently robust collection of materials that object and event recognition and identification become interesting problems

Audition is cognitively valuable for many reasons, one of which is that it gives a very rich and precise method of sensing the world that is different from vision. The fact that humans can display normal intelligence while totally blind or totally deaf is an indication that, in a sense, vision and audition are redundant for understanding the everyday world. However, it may be important that the brain has evolved to account for both of these senses, because this forced it to account for the presence of two very rich and precise methods of sensing the world—which may have forced it to develop more abstract representation mechanisms than would have been necessary with only one such method.

Touch is a sense that is, in our view, generally badly underappreciated within the AI community. In particular the cognitive robotics community seems to worry too little about the terribly impoverished sense of touch possessed by most current robots (though fortunately there are recent technologies that may help improve robots in this regard; see e.g. [Nan08]). Touch is how the human infant learns to distinguish self from other, and in this way it is the most essential sense for the establishment of an internal self-model. Touching others' bodies is a key method for developing a sense of the emotional reality and responsiveness of others, and is hence key to the development of theory of mind and social understanding in humans. For this reason, among others, human children lacking sufficient tactile stimulation will generally wind up badly impaired in multiple ways. A good-quality embodiment should supply an AI agent with a body that possesses skin, which has varying levels of sensitivity on different parts of the skin (so that it can effectively distinguish between reality and its perception thereof in a tactile context); and also varying types of touch sensors (e.g. temperature versus friction), so that it experiences textures as multidimensional entities.

Related to touch, kinesthesia refers to direct sensation of phenomena happening inside the body. Rarely mentioned in AI, this sense seems quite critical to cognition, as it underpins many of the analogies between self and other that guide cognition.

Again, it's not important that an AGI's virtual body have the same internal body parts as a human body. But it seems valuable to have the AGI's virtual body display some vaguely human-body-like properties, such as feeling internal strain of various sorts after getting exercise, feeling discomfort in certain places when running out of energy, feeling internally different when satisfied versus unsatisfied, etc.

Next, taste is a cognitively interesting sense in that it involves the interplay between the internal and external world; it involves the evaluation of which entities from the external world are worthy of placing inside the body. And smell is cognitively interesting in large part because of its relationship with taste. A smell is, among other things, a long-distance indicator of what a certain entity might taste like. So, the combination of taste and smell provides means for conceptualizing relationships between self, world and distance.

10.6.2 The Human Body's Multiple Intelligences

While most unique aspect of human intelligence is rooted in what one might call the "cognitive cortex"—the portions of the brain dealing with self-reflection and abstract thought. But the cognitive cortex does its work in close coordination with the body's various more specialized intelligent subsystems, including those associated with the gut, the heart, the liver, the immune and endocrine systems, and the perceptual and motor cortices.

In the perspective underlying this book, the human cognitive cortex—or the core cognitive network of any roughly human-like AGI system—should be viewed as a highly flexible, self-organizing network. These cognitive networks are modelable e.g. as a recurrent neural net with general topology, or a weighted labeled hypergraph, and are centrally concerned with recognizing patterns in its environment and itself, especially patterns regarding the achievement of the system's goals in various appropriate contexts. Here we augment this perspective, noting that the human brain's cognitive network is closely coupled with a variety of simpler and more specialized intelligent "body-system networks" which provide it with structural and dynamical inductive biasing. We then discuss the implications of this observation for practical AGI design.

One recalls Pascal's famous quote "The heart has its reasons, of which reason knows not." As we now know, the intuitive sense that Pascal and so many others have expressed, that the heart and other body systems have their own reasons, is grounded in the fact that they actually do carry out simple forms of reasoning (i.e. intelligent, adaptive dynamics), in close, sometimes cognitively valuable, coordination with the central cognitive network.

10.6.2.1 Some of the Human Body's Specialized Intelligent Subsystems

The human body contains multiple specialized intelligences apart from the cognitive cortex. Here we review some of the most critical.

Hierarchies of Visual and Auditory Perception.

The hierarchical structure of visual and auditory cortex has been taken by some researchers [Kur12], [HB06] as the generic structure of cognition. While we suspect this is overstated, we agree it is important that these cortices nudge large portions of the cognitive cortex to assume an approximately hierarchical structure.

Olfactory Attractors.

The process of recognizing a familiar smell is grounded in a neural process similar to convergence to an attractor in a nonlinear dynamical system [Fre95]. There is evidence that the mammalian cognitive cortex evolved in close coordination with the olfactory cortex [Row11], and much of abstract cognition reflects a similar dynamic of gradually coming to a conclusion based on what initially "smells right".

Physical and Cognitive Action.

The cerebellum, a specially structured brain subsystem which controls motor movements, has for some time been understood to also have involvement in attention, executive control, language, working memory, learning, pain, emotion, and addiction [PSF09].

The Second Brain.

The gastrointestinal neural net contains millions of neurons and is capable of operating independently of the brain. It modulates stress response and other aspects of emotion and motivation based on experience—resulting in so-called "gut feelings" [Ger99].

The Heart's Neural Network.

The heart has its own neural network, which modulates stress response, energy level and relaxation/excitement (factors key to motivation and emotion) based on experience [Arm04].

Pattern Recognition and Memory in the Liver.

The liver is a complex pattern recognition system, adapting via experience to better identify toxins [CB06]. Like the heart, it seems to store some episodic memories as well, resulting in liver transplant recipients sometimes acquiring the tastes in music or sports of the donor [EMC12].

Immune Intelligence.

The immune network is a highly complex, adaptive self-organizing system, which ongoingly solves the learning problem of identifying antigens and distinguishing them from the body system [FP86]. As immune function is highly energetically costly, stress response involves subtle modulation of the energy allocation to immune function, which involves communication between neural and immune networks.

The Endocrine System: A Key Bridge Between Mind and Body.

The endocrine (hormonal) system regulates (and is related by) emotion, thus guiding all aspects of intelligence (due to the close connection of emotion and motivation) [PH12].

Breathing Guides Thinking.

As oxygenation of the brain plays a key role in the spread of neural activity, the flow of breath is a key driver of cognition. Forced alternate nostril breathing has been shown to significantly affect cognition via balancing activity of the two brain hemispheres [SKBB91].

Much remains unknown, and the totality of feedback loops between the human cognitive cortex and the various specialized intelligences operative throughout the human body, has not yet been thoroughly charted.

10.6.2.2 Implications for AGI

What lesson should the AGI developer draw from all this? The particularities of the human mind/body should not be taken as general requirements for general intelligence. However, it is worth remembering just how difficult is the computational problem of learning, based on experiential feedback alone, the right way to achieve the complex goal of controlling a system with general intelligence at the human level or beyond. To solve this problem without some sort of strong inductive biasing may require massively more experience than young humans obtain.

Appropriate inductive bias may be embedded in an AGI system in many different ways. Some AGI designers have sought to embed it very explicitly, e.g. with hand-coded declarative knowledge as in Cyc, SOAR and other "GOFAI" type systems. On the other hand, the human brain receives its inductive bias much more subtly and implicitly, both via the specifics of the initial structure of the cognitive cortex, and via ongoing coupling of the cognitive cortex with other systems possessing more focused types of intelligence and more specific structures and/or dynamics.

In building an AGI system, one has four choices, very broadly speaking:

1. Create a flexible mind-network, as unbiased as feasible, and attempt to have it learn how to achieve its goals via experience
2. Closely emulate key aspects of the human body along with the human mind
3. Imitate the human mind-body, conceptually if not in detail, and create a number of structurally and dynamically simpler intelligent systems closely and appropriately coupled to the abstract cognitive mind-network, provide useful inductive bias.
4. Find some other, creative way to guide and probabilistically constrain one's AGI system's mind-network, providing inductive bias appropriate to the tasks at hand, without emulating even conceptually the way the human mind-brain receives its inductive bias via coupling with simpler intelligent systems.

Our suspicion is that the first option will not be viable. On the other hand, to do the second option would require more knowledge of the human body than biology currently possesses. This leaves the third and fourth options, both of which seem viable to us.

CogPrime incorporates a combination of the third and fourth options. CogPrime's generic dynamic knowledge store, the Atomspace, is coupled with specialized hierarchical networks (DeSTIN) for vision and audition, somewhat mirroring the human cortex. An artificial endocrine system for OpenCog is also under development, speculatively, as part of a project using OpenCog to control humanoid robots. On the other hand, OpenCog has no gastrointestinal nor cardiological nervous system, and the stress-response-based guidance provided to the human brain by a combination of the heart, gut, immune system and other body systems, is achieved in CogPrime in a more explicit way using the OpenPsi model of motivated cognition, and its integration with the system's attention allocation dynamics.

Likely there is no single correct way to incorporate the lessons of intelligent human body-system networks into AGI designs. But these are aspects of human cognition that all AGI researchers should be aware of.

10.7 The Extended Mind and Body

Finally, Hutchins [Hut95], Logan [Log07] and others have promoted a view of human intelligence that views the human mind as extended beyond the individual body, incorporating social interactions and also interactions with inanimate objects, such as tools, plants and animals. This leads to a number of requirements for a humanlike AGI's environment:

1. The ability to create a variety of different tools for interacting with various aspects of the world in various different ways, including tools for making tools and ultimately machinery
2. The existence of other mobile, virtual life-forms in the world, including simpler and less intelligent ones, and ones that interact with each other and with the AGI
3. The existence of organic growing structures in the world, with which the AGI can interact in various ways, including halting their growth or modifying their growth pattern

How necessary these requirements are is hard to say—but it *is* clear that these things have played a major role in the evolution of human intelligence.

10.8 Conclusion

Happily, this diverse chapter supports a simple, albeit tentative conclusion. Our suggestion is that, if an AGI is

- placed in an environment capable of roughly supporting multimodal communication and vaguely (but not necessarily precisely) real-world-ish naive physics

- surrounded with other intelligent agents of varying levels of complexity, and other complex, dynamic structures to interface with
- given a body that can perceive this environment through some forms of sight, sound and touch; and perceive itself via some form of kinesthesia
- given a motivational system that encourages it to make rich use of these aspects of its environment

then the AGI is likely to have an experience-base reinforcing the key inductive biases provided by the everyday world for the guidance of humanlike intelligence.

Chapter 11
A Mind-World Correspondence Principle

11.1 Introduction

Real-world minds are always adapted to certain classes of environments and goals. The ideas of the previous chapter, regarding the connection between a human-like intelligence's internals and its environment, result from exploring the implications of this adaptation in the context of the cognitive synergy concept. In this chapter we explore the mind-world connection in a broader and more abstract way—making a more ambitious attempt to move toward a "general theory of general intelligence".

One basic premise here, as in the preceding chapters is: Even a system of vast general intelligence, subject to real-world space and time constraints, will necessarily be more efficient at some kinds of learning than others. Thus, one approach to formulating a general theory of general intelligence is to look at the relationship between minds and worlds—where a "world" is conceived as an environment and a set of goals defined in terms of that environment.

In this spirit, we here formulate a broad principle binding together worlds and the minds that are intelligent in these worlds. The ideas of the previous chapter constitute specific, concrete instantiations of this general principle. A careful statement of the principle requires introduction of a number of technical concepts, and will be given later on in the chapter. A crude, informal version of the principle would be:

MIND-WORLD CORRESPONDENCE-PRINCIPLE For a mind to work intelligently toward certain goals in a certain world, there should be a nice mapping from goal-directed sequences of world-states into sequences of mind-states, where "nice" means that a world-state-sequence W composed of two parts W_1 and W_2, gets mapped into a mind-state-sequence M composed of two corresponding parts M_1 and M_2.

B. Goertzel et al., *Engineering General Intelligence, Part 1*,
Atlantis Thinking Machines 5, DOI: 10.2991/978-94-6239-027-0_11,
© Atlantis Press and the authors 2014

What's nice about this principle is that it relates the decomposition of the world into parts, to the decomposition of the mind into parts.

11.2 What Might a General Theory of General Intelligence Look Like?

It's not clear, at this point, what a real "general theory of general intelligence" would look like—but one tantalizing possibility is that it might confront the two questions:

- How does one design a world to foster the development of a certain sort of mind?
- How does one design a mind to match the particular challenges posed by a certain sort of world?

One way to achieve this would be to create a theory that, given a description of an environment and some associated goals, would output a description of the structure and dynamics that a system should possess to be intelligent in that environment relative to those goals, using limited computational resources.

Such a theory would serve a different purpose from the mathematical theory of "universal intelligence" developed by Marcus Hutter [Hut05] and others. For all its beauty and theoretical power, that approach currently gives it useful conclusions only about general intelligences with infinite or infeasibly massive computational resources. On the other hand, the approach suggested here is aimed toward creation of a theory of real-world general intelligences utilizing realistic amounts of computational power, but still possessing general intelligence comparable to human beings or greater.

This reflects a vision of intelligence as largely concerned with adaptation to particular classes of environments and goals. This may seem contradictory to the notion of "general" intelligence, but I think it actually embodies a realistic understanding of general intelligence. Maximally general intelligence is not pragmatically feasible; it could only be achieved using infinite computational resources [Hut05]. Real-world systems are inevitably limited in the intelligence they can display in any real situation, because real situations involve finite resources, including finite amounts of time. One may say that, in principle, a certain system could solve any problem given enough resources and time but, even when this is true, it's not necessarily the most interesting way to look at the system's intelligence. It may be more important to look at what a system can do given the resources at its disposal in reality. And this perspective leads one to ask questions like the ones posed above: which bounded-resources systems are well-disposed to display intelligence in which classes of situations?

As noted in Chap. 8, one can assess the generality of a system's intelligence via looking at the entropy of the class of situations across which it displays a high level of intelligence (where "high" is measured relative to its total level of intelligence across all situations). A system with a high generality of intelligence will tend to be roughly equally intelligent across a wide variety of situations; whereas a system with lower generality of intelligence will tend to be much more intelligent in a small subclass of situations, than in any other. The definitions given above embody this notion in a formal and quantitative way.

If one wishes to create a general theory of general intelligence according to this sort of perspective, the main question then becomes how to represent goals/environments and systems in such a way as to render transparent the natural correspondence between the specifics of the former and the latter, in the context of resource-bounded intelligence. This is the business of the next section.

11.3 Steps Toward A (Formal) General Theory of General Intelligence

Now begins the formalism. At this stage of development of the theory proposed in this chapter, mathematics is used mainly as a device to ensure clarity of expression. However, once the theory is further developed, it may possibly become useful for purposes of calculation as well.

Suppose one has any system S (which could be an AI system, or a human, or an environment that a human or AI is interacting with, or the combination of an environment and a human or AI's body, etc.). One may then construct an uncertain transition graph associated with that system S, in the following way:

- The nodes of the graph represent fuzzy sets of states of system S (I'll call these state-sets from here on, leaving the fuzziness implicit)
- The (directed) links of the graph represent probabilistically weighted transitions between state-sets

Specifically, the weight of the link from B to A should be defined as

$$P(o(S, A, t(T))|o(S, B, T))$$

where

$$o(S, A, T)$$

denotes the presence of the system S in the state-set A during time-distribution T, and $t()$ is a temporal succession function defined so that $t(T)$ refers to a time-distribution conceived as "after" T. A time-distribution is a probability distribution over time-points. The interaction of fuzziness and probability here is fairly straightforward and may be handled in the manner of PLN, as outlined in subsequent chapters. Note that the definition of link weights is dependent on the specific implementation of the temporal succession function, which includes an implicit time-scale.

Suppose one has a transition graph corresponding to an environment; then a goal relative to that environment may be defined as a particular node in the transition graph. The goals of a particular system acting in that environment may then be conceived as one or more nodes in the transition graph. The system's situation in the environment at any point in time may also be associated with one or more nodes in the transition graph; then, the system's movement toward goal-achievement may be associated with paths through the environment's transition graph leading from its current state to goal states.

It may be useful for some purposes to filter the uncertain transition graph into a crisp transition graph by placing a threshold on the link weights, and removing links with weights below the threshold.

The next concept to introduce is the world-mind transfer function, which maps world (environment) state-sets into organism (e.g. AI system) state-sets in a specific way. Given a world state-set W, the world-mind transfer function M maps W into various organism state-sets with various probabilities, so that we may say: $M(W)$ is the probability distribution of state-sets the organism tends to be in, when its environment is in state-set W. (Recall also that state-sets are fuzzy.)

Now one may look at the spaces of world-paths and mind-paths. A world-path is a path through the world's transition graph, and a mind-path is a path through the organism's transition graph. Given two world-paths P and Q, it's obvious how to define the composition $P * Q$ one follows P and then, after that, follows Q, thus obtaining a longer path. Similarly for mind-paths.

In category theory terms, we are constructing the free category associated with the graph: the objects of the category are the nodes, and the morphisms of the category are the paths. And category theory is the right way to be thinking here we want to be thinking about the relationship between the world category and the mind category.

The world-mind transfer function can be interpreted as a mapping from paths to subgraphs: Given a world-path, it produces a set of mind state-sets, which have a number of links between them. One can then define a world-mind path transfer function $M(P)$ via taking the mind-graph $M(nodes(P))$, and looking at the highest-weight path spanning $M(nodes(P))$. (Here $nodes(P)$ obviously means the set of nodes of the path P.)

A functor F between the world category and the mind category is a mapping that preserves object identities and so that

$$F(P * Q) = F(P) * F(Q)$$

We may also introduce the notion of an approximate functor, meaning a mapping F so that the average of

$$d(F(P * Q), F(P) * F(Q))$$

is small.

One can introduce a prior distribution into the average here. This could be the Levin universal distribution or some variant (the Levin distribution assigns higher probability to computationally simpler entities). Or it could be something more purpose specific: for example, one can give a higher weight to paths leading toward a certain set of nodes (e.g. goal nodes). Or one can use a distribution that weights based on a combination of simplicity and directedness toward a certain set of nodes. The latter seems most interesting, and I will define a goal-weighted approximate functor as an approximate functor, defined with averaging relative to a distribution that balances simplicity with directedness toward a certain set of goal nodes.

The move to approximate functors is simple conceptually, but mathematically it's a fairly big step, because it requires us to introduce a geometric structure on our categories. But there are plenty of natural metrics defined on paths in graphs (weighted or not), so there's no real problem here.

11.4 The Mind-World Correspondence Principle

Now we finally have the formalism set up to make a non-trivial statement about the relationship between minds and worlds. Namely, the hypothesis that:

> MIND-WORLD CORRESPONDENCE PRINCIPLE For an organism with a reasonably high level of intelligence in a certain world, relative to a certain set of goals, the mind-world path transfer function is a goal-weighted approximate functor.

That is, a little more loosely: the hypothesis is that, for intelligence to occur, there has to be a natural correspondence between the transition-sequences of world-states and the corresponding transition-sequences of mind-states, at least in the cases of transition-sequences leading to relevant goals.

We suspect that a variant of the above proposition can be formally proved, using the definition of general intelligence presented in Chap. 8. The proof of a theorem corresponding to the above would certainly constitute an interesting start toward a general formal theory of general intelligence. Note that proving anything of this nature would require some attention to the time-scale-dependence of the link weights in the transition graphs involved.

A formally proved variant of the above proposition would be in short, a "MIND-WORLD CORRESPONDENCE THEOREM".

Recall that at the start of the chapter, we expressed the same idea as:

> MIND-WORLD CORRESPONDENCE-PRINCIPLE For a mind to work intelligently toward certain goals in a certain world, there should be a nice mapping from goal-directed sequences of world-states into sequences of mind-states, where "nice" means that a world-state-sequence W composed of two parts W_1 and W_2, gets mapped into a mind-state-sequence M composed of two corresponding parts M_1 and M_2.

That is a reasonable gloss of the principle, but it's clunkier and less accurate, than the statement in terms of functors and path transfer functions, because it tries to use only common-language vocabulary, which doesn't really contain all the needed concepts.

11.5 How Might the Mind-World Correspondence Principle Be Useful?

Suppose one believes the Mind-World Correspondence Principle as laid out above—so what? Our hope, obviously, is that the principle could be useful in actually figuring out how to architect intelligent systems biased toward particular sorts of environment. And of course, this is said with the understanding that any finite intelligence must be biased toward some sorts of environment.

Relatedly, given a specific AGI design (such as CogPrime), one could use the principle to figure out which environments it would be best suited for. Or one could figure out how to adjust the particulars of the design, to maximize the system's intelligence in the environments of interest.

One next step in developing this network of ideas, aside from (and potentially building on) full formalization of the principle, would be an exploration of real-world environments in terms of transition graphs. What properties do the transition graphs induced from the real world have?

One such property, we suggest, is successive refinement. Often the path toward a goal involves first gaining an approximate understanding of a situation, then a slightly more accurate understanding, and so forth—until finally one has achieved a detailed enough understanding to actually achieve the goal. This would be represented by a world-path whose nodes are state-sets involving the gathering of progressively more detailed information.

Via pursuing the mind-world correspondence property in this context, I believe we will find that world-paths reflecting successive refinement correspond to mind-paths embodying successive refinement. This will be found to relate to the hierarchical structures found so frequently in both the physical world and the human mind-brain. Hierarchical structures allow many relevant goals to be approached via successive refinement, which I believe is the ultimate reason why hierarchical structures are so common in the human mind-brain.

Another next step would be exploring what mind-world correspondence means for the structure and dynamics of a limited-resources intelligence. If an organism O has limited resources and, to be intelligent, needs to make

$$P(o(O, M(A), t(T))|o(O, M(B), T))$$

high for particular world state-sets A and B, then what's the organism's best approach? Arguably, it should represent $M(A)$ and $M(B)$ internally in such a way that very little computational effort is required for it to transition between $M(A)$ and $M(B)$. For instance, this could be done by coding its knowledge in such a way that $M(A)$ and $M(B)$ share many common bits; or it could be done in other more complicated ways.

If, for instance, A is a subset of B, then it may prove beneficial for the organism to represent $M(A)$ physically as a subset of its representation of $M(B)$.

Pursuing this line of thinking, one could likely derive specific properties of an intelligent organism's internal information-flow, from properties of the environment and goals with respect to which it's supposed to be intelligent.

This would allow us to achieve the holy grail of intelligence theory as I understand it: given a description of an environment and goals, to be able to derive an architectural description for an organism that will display a high level of intelligence relative to those goals, given limited computational resources.

While this "holy grail" is obviously a far way off, what we've tried to do here is to outline a mathematical and conceptual direction for moving toward it.

11.6 Conclusion

The Mind-World Correspondence Principle presented here—if in the vicinity of correctness—constitutes a non-trivial step toward fleshing out the concept of a general theory of general intelligence. But obviously the theory is still rather abstract, and also not completely rigorous. There's a lot more work to be done.

The Mind-World Correspondence Principle as articulated above is not quite a formal mathematical statement. It would take a little work to put in all the needed quantifiers to formulate it as one, and it's not clear the best way to do so the details would perhaps become clear in the course of trying to prove a version of it rigorously. One could interpret the ideas presented in this chapter as a philosophical theory that hopes to be turned into a mathematical theory and to play a key role in a scientific theory.

For the time being, the main role to be served by these ideas is qualitative: to help us think about concrete AGI designs like CogPrime in a sensible way. It's important to understand what the goal of a real-world AGI system needs to be: to achieve the ability to broadly learn and generalize, yes, but not with infinite capability rather with biases and patterns that are implicitly and/or explicitly tuned to certain broad classes of goals and environments. The Mind-World Correspondence Principle tells us something about what this "tuning" should involve—namely, making a system possessing mind-state sequences that correspond meaningfully to world-state sequences. CogPrime's overall design and particular cognitive processes are reasonably well interpreted as an attempt to achieve this for everyday human goals and environments.

One way of extending these theoretical ideas into a more rigorous theory is explored in Appendix B. The key ideas involved there are: modeling multiple memory types as mathematical categories (with functors mapping between them), modeling memory items as probability distributions, and measuring distance between memory items using two metrics, one based on algorithmic information theory and one on classical information geometry. Building on these ideas, core hypotheses are then presented:

- a **syntax-semantics correlation** principle, stating that in a successful AGI system, these two metrics should be roughly correlated

- a **cognitive geometrodynamics** principle, stating that on the whole intelligent minds tend to follow geodesics (shortest paths) in mindspace, according to various appropriately defined metrics (e.g. the metric measuring the distance between two entities in terms of the length and/or runtime of the shortest programs computing one from the other).
- a **cognitive synergy** principle, stating that shorter paths may be found through the composite mindspace formed by considering multiple memory types together, than by following the geodesics in the mindspaces corresponding to individual memory types.

The material is relegated to an appendix because it is so speculative, and it's not yet clear whether it will really be useful in advancing or interpreting CogPrime or other AGI systems (unlike the material from the present chapter, which has at least been useful in interpreting and tweaking the CogPrime design, even though it can't be claimed that CogPrime was derived directly from these theoretical ideas). However, this sort of speculative exploration is, in our view, exactly the sort of thing that's needed as a first phase in transitioning the ideas of the present chapter into a more powerful and directly actionable theory.

Part IV
Cognitive and Ethical Development

Chapter 12
Stages of Cognitive Development

12.1 Introduction

Creating AGI, we have said, is not only about having the right structural and dynamical possibilities implemented in the initial version of one's system—but also about the environment and embodiment that one's system is associated with, and the match between the system's internals and these externals. Another key aspect is the long-term time-course of the system's evolution over time, both in its internals and its external interaction—i.e., what is known as *development*.

Development is a critical topic in our approach to AGI because we believe that much of what constitutes human-level, human-like intelligence *emerges* in an intelligent system due to its engagement with its environment and its environment-coupled self-organization. So, it's not to be expected that the initial version of an AGI system is going to display impressive feats of intelligence, even if the engineering is totally done right. A good analogy is the apparent unintelligence of a human baby. Yes, scientists have discovered that human babies are capable of interesting and significant intelligence—but one has to hunt to find it ... at first observation, babies are rather idiotic and simple-minded creatures: much less intelligent-appearing than lizards or fish, maybe even less than cockroaches....

If the goal of an AGI project is to create an AGI system that can progressively develop advanced intelligence through learning in an environment richly populated with other agents and various inanimate stimuli and interactive entities—then an understanding of the nature of cognitive development becomes extremely important to that project.

Unfortunately, contemporary cognitive science contains essentially no theory of "abstract developmental psychology" which can conveniently be applied to understand developing AIs. There is of course an extensive science of **human** developmental psychology, and so it is a natural research program to take the chief ideas

---

Co-authored with Stephan Vladimir Bugaj.

B. Goertzel et al., *Engineering General Intelligence, Part 1*,
Atlantis Thinking Machines 5, DOI: 10.2991/978-94-6239-027-0_12,
© Atlantis Press and the authors 2014

from the former and inasmuch as possible port them to the AGI domain. This is not an entirely simple matter both because of the differences between humans and AIs and because of the unsettled nature of contemporary developmental psychology theory. But it's a job that must (and will) be done, and the ideas in this chapter may contribute toward this effort.

We will begin here with Piaget's well-known theory of human cognitive development, presenting it in a general systems theory context, then introducing some modifications and extensions and discussing some other relevant work.

12.2 Piagetan Stages in the Context of a General Systems Theory of Development

Our review of AGI architectures in Chap. 6 focused heavily on the concept of **symbolism**, and the different ways in which different classes of cognitive architecture handle symbol representation and manipulation. We also feel that symbolism is critical to the notion of AGI development—and even more broadly, to the systems theory of development in general.

As a broad conceptual perspective on development, we suggest that one may view the development of a complex information processing system, embedded in an environment, in terms of the stages:

- **automatic**: the system interacts with the environment by "instinct", according to its innate programming
- **adaptive**: the system internally adapts to the environment, then interacting with the environment in a more appropriate way
- **symbolic**: the system creates internal symbolic representations of itself and the environment, which in the case of a complex, appropriately structured environment, allows it to interact with the environment more intelligently
- **reflexive**: the system creates internal symbolic representations of its own internal symbolic representations, thus achieving an even higher degree of intelligence.

Sketched so broadly, these are not precisely defined categories but rather heuristic, intuitive categories. Formalizing them would be possible but would lead us too far astray here.

One can interpret these stages in a variety of different contexts. Here our focus is the cognitive development of humans and human-like AGI systems, but in Table 12.1 we present them in a slightly more general context, using two examples: the Piagetan example of the human (or humanlike) mind as it develops from infancy to maturity; and also the example of the "origin of life" and the development of life from proto-life up into its modern form. In any event, we allude to this more general perspective on development here mainly to indicate our view that the Piagetan perspective is not something ad hoc and arbitrary, but rather can plausibly be seen as a specific manifestation of more fundamental principles of complex systems development.

Table 12.1 General systems theory of development: parallels between development of mind and origin of life

Stage	General description	Cognitive development	Origin of life
Automatic	System-environment information exchange controlled mainly by innate system structures or environment	Piagetan infantile stage	Self-organizing protolife system, e.g. Oparin [Opa52] water droplet, or Cairns-Smith [CS90] clay-based protolife
Adaptive	System-environment info exchange heavily guided by adaptively internally-created system structures	Piagetan "concrete operational" stage: systematic internal world-model guides world-exploration	Simple autopoietic system, e.g. Oparin water droplet w/basic metabolism
Symbolic	Internal symbolic representation of information exchange process	Piagetan formal stage: explicit logical/experimental learning about how to cognize in various contexts	Genetic code: internal entities that "stand for" aspects of organism and environment, thus enabling complex epigenesis
Reflexive	Thoroughgoing self-modification based on this symbolic representation	Piagetan post-formal stage: purposive self-modification of basic mental processes	Genes + memes: genetic code-patterns guide their own modification via influencing culture

12.3 Piaget's Theory of Cognitive Development

The ghost of Jean Piaget hangs over modern developmental psychology in a yet unresolved way. Piaget's theories provide a cogent overarching perspective on human cognitive development, coordinating broad theoretical ideas and diverse experimental results into a unified whole [Pia55]. Modern experimental work has shown Piaget's ideas to be often oversimplified and incorrect. However, what has replaced the Piagetan understanding is not an alternative unified and coherent theory, but a variety of microtheories addressing particular aspects of cognitive development. For this reason a number of contemporary theorists taking a computer science [Shu03] or dynamical systems [Wit07] approach to developmental psychology have chosen to adopt the Piagetan framework in spite of its demonstrated shortcomings, both because of its conceptual strengths and for lack of a coherent, more rigorously grounded alternative.

Our own position is that the Piagetan view of development has some fundamental truth to it, which is reflected via how nicely it fits with a broader view of development in complex systems. Indeed, Piaget viewed developmental stages as emerging from general "algebraic" principles rather than as being artifacts of the particulars of human psychology. But, Piaget's stages are probably best viewed as a general interpretive framework rather than a precise scientific theory. Our suspicion is that once the empirical science of developmental psychology has progressed further, it will become clearer how to fit the various data into a broad Piaget-like framework, perhaps differing in many details from what Piaget described in his works.

Piaget conceived of child development in four stages, each roughly identified with an age group, and corresponding closely to the system-theoretic stages mentioned above:

- **infantile**, corresponding to the automatic stage mentioned above

  - *Example:* Grasping blocks, piling blocks on top of each other, copying words that are heard

- **preoperational** and **concrete operational**, corresponding to the adaptive stage mentioned above

  - *Example:* Building complex blocks structures, from imagination and from imitating objects and pictures and based on verbal instructions; verbally describing what has been constructed

- **formal**, corresponding to the symbolic stage mentioned above

  - *Example:* Writing detailed instructions in words and diagrams, explaining how to construct particular structures out of blocks; figuring out general rules describing which sorts of blocks structures are likely to be most stable

- the reflexive stage mentioned above corresponds to what some post-Piagetan theorists have called the **post-formal** stage

  - *Example:* Using abstract lessons learned from building structures out of blocks to guide the construction of new ways to think and understand—"Zen and the art of blocks building" (by analogy to *Zen and the Art of Motorcycle Maintenance* [Pir84]) (Fig. 12.1).

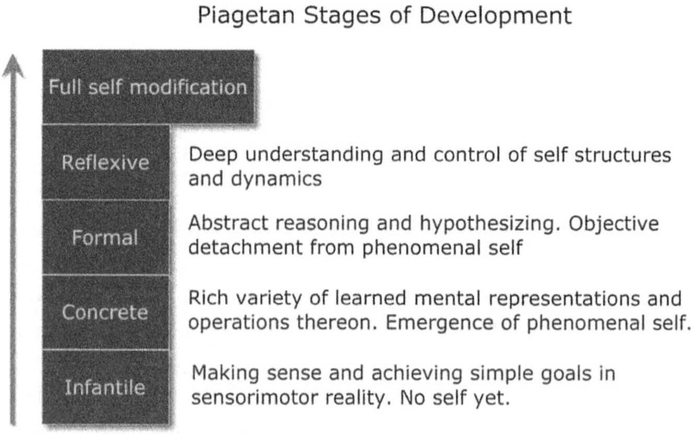

Piagetan Stages of Development

Fig. 12.1 Piagetan stages of cognitive development

More explicitly, Piaget defined his stages in psychological terms roughly as follows:

- **Infantile**: In this stage a mind develops basic world-exploration driven by instinctive actions. Reward-driven reinforcement of actions learned by imitation, simple associations between words and objects, actions and images, and the basic notions of time, space, and causality are developed. The most simple, practical ideas and strategies for action are learned.
- **Preoperational**: At this stage we see the formation of mental representations, mostly poorly organized and un-abstracted, building mainly on intuitive rather than logical thinking. Word-object and image-object associations become systematic rather than occasional. Simple syntax is mastered, including an understanding of subject-argument relationships. One of the crucial learning achievements here is "object permanence"—infants learn that objects persist even when not observed. However, a number of cognitive failings persist with respect to reasoning about logical operations, and abstracting the effects of intuitive actions to an abstract theory of operations.
- **Concrete**: More abstract logical thought is applied to the physical world at this stage. Among the feats achieved here are: reversibility—the ability to undo steps already done; conservation—understanding that properties can persist in spite of appearances; theory of mind—an understanding of the distinction between what I know and what others know (If I cover my eyes, can you still see me?). Complex concrete operations, such as putting items in height order, are easily achievable. Classification becomes more sophisticated, yet the mind still cannot master purely logical operations based on abstract logical representations of the observational world.
- **Formal**: Abstract deductive reasoning, the process of forming, then testing hypotheses, and systematically reevaluating and refining solutions, develops at this stage, as does the ability to reason about purely abstract concepts without reference to concrete physical objects. This is adult human-level intelligence. Note that the capability for formal operations is intrinsic in the PLN component of CogPrime, but in-principle capability is not the same as pragmatic, grounded, controllable capability.

Very early on, Vygotsky [Vyg86] disagreed with Piaget's explanation of his stages as inherent and developed by the child's own activities, and Piaget's prescription of good parenting as not interfering with a child's unfettered exploration of the world. Some modern theorists have critiqued Piaget's stages as being insufficiently socially grounded, and these criticisms trace back to Vygotsky's focus on the social foundations of intelligence, on the fact that children function in a world surrounded by adults who provide a cultural context, offering ongoing assistance, critique, and ultimately validation of the child's developmental activities.

Vygotsky also was an early critic of the idea that cognitive development is continuous, and continues beyond Piaget's formal stage. Gagne [RBW92] also believes in continuity, and that learning of prerequisite skills made the learning of subsequent skills easier and faster without regard to Piagetan stage formalisms. Subsequent researchers have argued that Piaget has merely constructed ad hoc descriptions of the sequential development of behaviour [Gib78, Bro84, CP05]. We agree that learning is a continuous process, and our notion of stages is more statistically constructed than rigidly quantized.

Critique of Piaget's notion of transitional "half stages" is also relevant to a more comprehensive hierarchical view of development. Some have proposed that Piaget's half stages are actually stages [Bro84]. As Commons and Pekker [CP05] point out: "the definition of a stage that was being used by Piaget was based on analyzing behaviors and attempting to impose different structures on them. There is no underlying logical or mathematical definition to help in this process ..." Their Hierarchical Complexity development model uses task achievement rather than ad hoc stage definition as the basis for constructing relationships between phases of developmental ability—an approach which we find useful, though our approach is different in that we define stages in terms of specific underlying cognitive mechanisms.

Another critique of Piaget is that one individual's performance is often at different ability stages depending on the specific task (for example [GE86]). Piaget responded to early critiques along these lines by calling the phenomenon "horizontal décalage," but neither he nor his successors [Fis80, Cas85] have modified his theory to explain (rather than merely describe) it. Similarly to Thelen and Smith [TS94], we observe that the abilities encapsulated in the definition of a certain stage emerge gradually during the previous stage—so that the onset of a given stage represents the mastery of a cognitive skill that was previously present only in certain contexts.

Piaget also had difficulty accepting the idea of a preheuristic stage, early in the infantile period, in which simple trial-and-error learning occurs without significant heuristic guidance [Bic88], a stage which we suspect exists and allows formulation of heuristics by aggregation of learning from preheuristic pattern mining. Coupled with his belief that a mind's innate abilities at birth are extremely limited, there is a troublingly unexplained transition from inability to ability in his model.

Finally, another limiting aspect of Piaget's model is that it did not recognize any stages beyond formal operations, and included no provisions for exploring this possibility. A number of researchers [Bic88, Arl75, CRK82, Rie73, Mar01] have described one or more postformal stages. Commons and colleagues have also proposed a task-based model which provides a framework for explaining stage discrepancies across tasks and for generating new stages based on classification of observed logical behaviors. [KK90] promotes a statistical conception of stage, which provides a good bridge between task-based and stage-based models of development, as statistical modeling allows for stages to be roughly defined and analyzed based on collections of task behaviors.

[CRK82] postulates the existence of a postformal stage by observing *elevated levels of abstraction* which, they argue, are not manifested in formal thought. [CTS+98] observes a postformal stage when subjects become capable of analyzing and coordinating complex logical systems with each other, creating metatheoretical supersystems. In our model, with the reflexive stage of development, we expand this definition of metasystemic thinking to include the ability to consciously refine one's own mental states and formalisms of thinking. Such self-reflexive refinement is necessary for learning which would allow a mind to analytically devise entirely new structures and methodologies for both formal and postformal thinking.

In spite of these various critiques and limitations, however, we have found Piaget's ideas very useful, and in Sect. 12.4 we will explore ways of defining them rigorously in the specific context of CogPrime's declarative knowledge store and probabilistic logic engine.

12.3.1 Perry's Stages

Also relevant is William Perry's [Per70, Per81] theory of the stages ("positions" in his terminology) of intellectual and ethical development, which constitutes a model of iterative refinement of approach in the developmental process of coming to intellectual and ethical maturity. These stages, depicted in Table 12.2 form an analytical tool for discerning the modality of belief of an intelligence by describing common cognitive approaches to handling the complexities of real world ethical considerations.

Table 12.2 Perry's developmental stages [with corresponding Piagetan stages in brackets]

Stage	Substages
Dualism/received knowledge [Infantile]	Basic duality ("All problems are solvable. I must learn the correct solutions.")
	Full dualism ("There are different, contradictory solutions to many problems. I must learn the correct solutions, and ignore the incorrect ones")
Multiplicity [Concrete]	Early multiplicity ("Some solutions are known, others aren't. I must learn how to find correct solutions.")
	Late multiplicity: cognitive dissonance regarding truth. ("Some problems are unsolvable, some are a matter of personal taste, therefore I must declare my own intellectual path.")
Relativism/procedural knowledge [Formal]	Contextual relativism ("I must learn to evaluate solutions within a context, and relative to supporting observation.")
	Pre-Commitment ("I must evaluate solutions, then commit to a choice of solution.")
Commitment/constructed knowledge [Formal/Reflexive]	Commitment ("I have chosen a solution.")
	Challenges to commitment ("I have seen unexpected implications of my commitment, and the responsibility I must take.")
	Post-commitment ("I must have an ongoing, nuanced relationship to the subject in which I evaluate each situation on a case-by-case basis with respects to its particulars rather than an ad-hoc application of unchallenged ideology.")

12.3.2 Keeping Continuity in Mind

Continuity of mental stages, and the fact that a mind may appear to be in multiple stages of development simultaneously (depending upon the tasks being tested), are crucial to our theoretical formulations and we will touch upon them again here. Piaget attempted to address continuity with the creation of transitional "half stages". We prefer to observe that each stage feeds into the other and the end of one stage and the beginning of the next blend together.

The distinction between formal and post-formal, for example, seems to "merely" be the application of formal thought to oneself. However, the distinction between concrete and formal is "merely" the buildup to higher levels of complexity of the classification, task decomposition, and abstraction capabilities of the concrete stage. The stages represent general trends in ability on a continuous curve of development, not discrete states of mind which are jumped-into quantum style after enough "knowledge energy" builds-up to cause the transition.

Observationally, this appears to be the case in humans. People learn things gradually, and show a continuous development in ability, not a quick jump from ignorance to mastery. We believe that this gradual development of ability is the signature of genuine learning, and that prescriptively an AGI system must be designed in order to have continuous and asymmetrical development across a variety of tasks in order to be considered a genuine learning system. While quantum leaps in ability may be possible in an AGI system which can just "graft" new parts of brain onto itself (or an augmented human which may someday be able to do the same using implants), such acquisition of knowledge is not really learning. Grafting on knowledge does not build the cognitive pathways needed in order to actually learn. If this is the only mechanism available to an AGI system to acquire new knowledge, then it is not really a learning system.

12.4 Piaget's Stages in the Context of Uncertain Inference

Piaget's developmental stages are very general, referring to overall types of learning, not specific mechanisms or methods. This focus was natural since the context of his work was *human* developmental psychology, and neuroscience has not yet progressed to the point of understanding the neural mechanisms underlying any sort of inference (and certainly was nowhere near to doing so in Piaget's time!). But if one is studying developmental psychology in an AGI context where one knows something about the internal mechanisms of the AGI system under consideration, then one can work with a more specific model of learning. Our focus here is on AGI systems whose operations contain uncertain inference as a central component. Obviously the main focus is CogPrime, but the essential ideas apply to any other uncertain inference centric AGI architecture as well (Fig. 12.2).

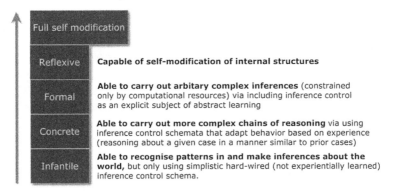

Fig. 12.2 Piagetan stages of development, as manifested in the context of uncertain inference

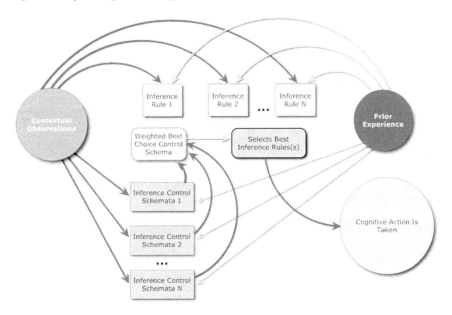

Fig. 12.3 A simplified look at feedback-control in uncertain inference

An uncertain inference system, as we consider it here, consists of four components, which work together in a feedback-control loop Fig. 12.3

1. a content representation scheme
2. an uncertainty representation scheme
3. a set of inference rules
4. a set of inference control schemata.

Broadly speaking, examples of content representation schemes are predicate logic and term logic [ES00]. Examples of uncertainty representation schemes are fuzzy logic [Zad78], imprecise probability theory [Goo86, FC86], Dempster-Shafer theory [Sha76, Kyb97], Bayesian probability theory [Kyb97], NARS [Wan95], and the Atom representation used in CogPrime, briefly alluded to in Chap. 2 and described in depth in later chapters.

Many, but not all, approaches to uncertain inference involve only a limited, weak set of inference rules (e.g. not dealing with complex quantified expressions). CogPrime's PLN inference framework, like NARS and some other uncertain inference frameworks, contains uncertain inference rules that apply to logical constructs of arbitrary complexity. Only a system capable of dealing with constructs of arbitrary (or at least very high) complexity will have any potential of leading to human-level, human-like intelligence.

The subtlest part of uncertain inference is inference control: the choice of which inferences to do, in what order. Inference control is the primary area in which human inference currently exceeds automated inference. Humans are not very efficient or accurate at carrying out inference rules, with or without uncertainty, but we are very good at determining which inferences to do and in what order, in any given context. The lack of effective, context-sensitive inference control heuristics is why the general ability of current automated theorem provers is considerably weaker than that of a mediocre university mathematics major [Mac95].

We now review the Piagetan developmental stages from the perspective of AGI systems heavily based on uncertain inference.

12.4.1 The Infantile Stage

In this initial stage, the mind is able to recognize patterns in and conduct inferences about the world, but only using simplistic hard-wired (not experientially learned) inference control schema, along with pre-heuristic pattern mining of experiential data.

In the infantile stage an entity is able to recognize patterns in and conduct inferences about its sensory surround context (i.e., it's "world"), but only using simplistic, hard-wired (not experientially learned) inference control schemata. Preheuristic pattern-mining of experiential data is performed in order to build future heuristics about analysis of and interaction with the world.

An infant's tasks include:

1. Exploratory behavior in which useful and useless/dangerous behavior is differentiated by both trial and error observation, and by parental guidance.
2. Development of "habits"—i.e. Repeating tasks which were successful once to determine if they always/usually are so.

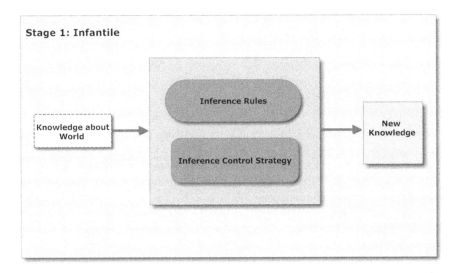

Fig. 12.4 Uncertain inference in the infantile stage

3. Simple goal-oriented behavior such as "find out what cat hair tastes like" in which one must plan and take several sequentially dependent steps in order to achieve the goal.

Inference control is very simple during the infantile stage (Fig. 12.4), as it is the stage during which both the most basic knowledge of the world is acquired, and the most basic of cognition and inference control structures are developed as the building block upon which will be built the next stages of both knowledge and inference control.

Another example of a cognitive task at the borderline between infantile and concrete cognition is learning object permanence, a problem discussed in the context of CogPrime's predecessor "Novamente Cognition Engine" system in [GPSL03]. Another example is the learning of word-object associations: e.g. learning that when the word "ball" is uttered in various contexts ("Get me the ball," "That's a nice ball," etc.) it generally refers to a certain type of object. The key point regarding these "infantile" inference problems, from the CogPrime perspective, is that assuming one provides the inference system with an appropriate set of perceptual and motor ConceptNodes and SchemaNodes, the chains of inference involved are short. They involve about a dozen inferences, and this means that the search tree of possible PLN inference rules walked by the PLN backward-chainer is relatively shallow. Sophisticated inference control is not required: standard AI heuristics are sufficient.

In short, textbook narrow-AI reasoning methods, utilized with appropriate uncertainty-savvy truth value formulas and coupled with appropriate representations of perceptual and motor inputs and outputs, correspond roughly to Piaget's infantile stage of cognition. The simplistic approach of these narrow-AI methods may be viewed as a method of creating building blocks for subsequent, more sophisticated heuristics.

In our theory Piaget's preoperational phase appears as transitional between the infantile and concrete operational phases.

12.4.2 The Concrete Stage

At this stage, the mind is able to carry out more complex chains of reasoning regarding the world, via using inference control schemata that adapt behavior based on experience (reasoning about a given case in a manner similar to prior cases).

In the concrete operational stage (Fig. 12.5), an entity is able to carry out more complex chains of reasoning about the world. Inference control schemata which adapt behavior based on experience, using experientially learned heuristics (including those learned in the prior stage), are applied to both analysis of and interaction with the sensory surround/world.

Concrete Operational stage tasks include:

1. Conservation tasks, such as conservation of number,
2. Decomposition of complex tasks into easier subtasks, allowing increasingly complex tasks to be approached by association with more easily understood (and previously experienced) smaller tasks,
3. Classification and Serialization tasks, in which the mind can cognitively distinguish various disambiguation criteria and group or order objects accordingly.

In terms of inference control this is the stage in which actual knowledge about how to control inference itself is first explored. This means an emerging understanding

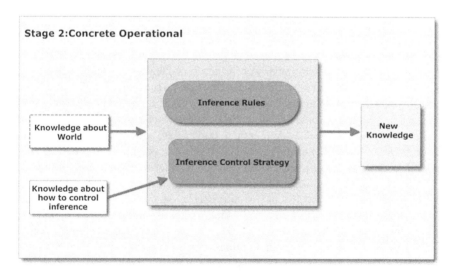

Fig. 12.5 Uncertain inference in the concrete operational stage

of inference itself as a cognitive task and methods for learning, which will be further developed in the following stages.

Also, in this stage a special cognitive task capability is gained: "Theory of Mind," which in cognitive science refers to the ability to understand the fact that not only oneself, but other sentient beings have memories, perceptions, and experiences. This is the ability to conceptually "put oneself in another's shoes" (even if you happen to assume incorrectly about them by doing so).

12.4.2.1   Conservation of Number

Conservation of number is an example of a learning problem classically categorized within Piaget's concrete-operational phase, a "conservation laws" problem, discussed in [Shu03] in the context of software that solves the problem using (logic-based and neural net) narrow-AI techniques. Conservation laws are very important to cognitive development.

Conservation is the idea that a quantity remains the same despite changes in appearance. If you show a child some objects and then spread them out, an infantile mind will focus on the spread, and believe that there are now more objects than before, whereas a concrete-operational mind will understand that the quantity of objects has not changed.

Conservation of number seems very simple, but from a developmental perspective it is actually rather difficult. "Solutions" like those given in [Shu03] that use neural networks or customized logical rule-bases to find specialized solutions that solve only this problem fail to fully address the issue, because these solutions don't create knowledge adequate to aid with the solution of related sorts of problems.

We hypothesize that this problem is hard enough that for an inference-based AGI system to solve it in a developmentally useful way, its inferences must be guided by meta-inferential lessons learned from prior similar problems. When approaching a number conservation problem, for example, a reasoning system might draw upon past experience with set-size problems (which may be trial-and-error experience). This is not a simple "machine learning" approach whose scope is restricted to the current problem, but rather a heuristically guided approach which (a) aggregates information from prior experience to guide solution formulation for the problem at hand, and (b) adds the present experience to the set of relevant information about quantification problems for future refinement of thinking (Fig. 12.6).

For instance, a very simple context-specific heuristic that a system might learn would be: "When evaluating the truth value of a statement related to the number of

Fig. 12.6 Conservation of number

objects in a set, it is generally not that useful to explore branches of the backwards-chaining search tree that contain relationships regarding the sizes, masses, or other physical properties of the objects in the set." This heuristic itself may go a long way toward guiding an inference process toward a correct solution to the problem—but it is not something that a mind needs to know "a priori." A concrete-operational stage mind may learn this by data-mining prior instances of inferences involving sizes of sets. Without such experience-based heuristics, the search tree for such a problem will likely be unacceptably large. Even if it is "solvable" without such heuristics, the solutions found may be overly fit to the particular problem and not usefully generalizable.

12.4.2.2 Theory of Mind

Consider this experiment: a preoperational child is shown her favorite "Dora the Explorer" DVD box. Asked what show she's about to see, she'll answer "Dora." However, when her parent plays the disc, it's "SpongeBob SquarePants." If you then ask her what show her friend will expect when given the "Dora" DVD box, she will respond "SpongeBob" although she just answered "Dora" for herself. A child lacking a theory of mind can not reason through what someone else would think given knowledge other than her own current knowledge. Knowledge of self is intrinsically related to the ability to differentiate oneself from others, and this ability may not be fully developed at birth.

Several theorists [BC94, Fod94], based in part on experimental work with autistic children, perceive theory of mind as embodied in an innate module of the mind activated at a certain developmental stage (or not, if damaged). While we consider this possible, we caution against adopting a simplistic view of the "innate versus acquired" dichotomy: if there is innateness it may take the form of an innate predisposition to certain sorts of learning [EBJ+97].

Davidson [Dav84], Dennett [Den87] and others support the common belief that theory of mind is dependent upon linguistic ability. A major challenge to this prevailing philosophical stance came from Premack and Woodruff [PW78] who postulated that prelinguistic primates do indeed exhibit "theory of mind" behavior. While Premack and Woodruff's experiment itself has been challenged, their general result has been bolstered by follow-up work showing similar results such as [TC97]. It seems to us that while theory of mind depends on many of the same inferential capabilities as language learning, it is not intrinsically dependent on the latter.

There is a school of thought often called the *Theory Theory* [BW88, Car85, Wel90] holding that a child's understanding of mind is best understood in terms of the process of iteratively formulating and refuting a series of naive theories about others. Alternately, Gordon [Gor86] postulates that theory of mind is related to the ability to run cognitive simulations of others' minds using one's own mind as a model. We suggest that these two approaches are actually quite harmonious with one another. In an uncertain AGI context, both theories and simulations are grounded in collections of uncertain implications, which may be assembled in context-appropriate ways to form theoretical conclusions or to drive simulations. Even if there is a special "mind-simulator" dynamic in the human brain that carries out simulations of other minds in a manner fundamentally different from explicit inferential theorizing, the inputs to and the behavior of this simulator may take inferential form, so that the simulator is in essence a way of efficiently and implicitly producing uncertain inferential conclusions from uncertain premises.

We have thought through the details by CogPrime system should be able to develop theory of mind via embodied experience, though at time of writing practical learning experiments in this direction have not yet been done. We have not yet explored in detail the possibility of giving CogPrime a special, elaborately engineered "mind-simulator" component, though this would be possible; instead we have initially been pursuing a more purely inferential approach.

First, it is very simple for a CogPrime system to learn patterns such as "If I rotated by pi radians, I would see the yellow block." And it's not a big leap for PLN to go from this to the recognition that "You look like me, and you're rotated by pi radians relative to my orientation, therefore you probably see the yellow block." The only nontrivial aspect here is the "you look like me" premise.

Recognizing "embodied agent" as a category, however, is a problem fairly similar to recognizing "block" or "insect" or "daisy" as a category. Since the CogPrime agent can perceive most parts of its own "robot" body—its arms, its legs, etc.—it should be easy for the agent to figure out that physical objects like these look different depending upon its distance from them and its angle of observation. From this it should not be that difficult for the agent to understand that it is naturally grouped together with other embodied agents (like its teacher), not with blocks or bugs.

The only other major ingredient needed to enable theory of mind is "reflection"—the ability of the system to explicitly recognize the existence of knowledge in its own mind (note that this term "reflection" is not the same as our proposed "reflexive" stage of cognitive development). This exists automatically in CogPrime, via the built-in vocabulary of elementary procedures supplied for use within SchemaNodes (specifically, the atTime and TruthValue operators). Observing that "at time T, the weight of evidence of the link L increased from zero" is basically equivalent to observing that the link L was created at time T.

Then, the system may reason, for example, as follows (using a combination of several PLN rules including the above-given deduction rule):

Implication
    My eye is facing a block and it is not dark
    A relationship is created describing the block's color
Similarity
    My body
    My teacher's body

Implication
    My teacher's eye is facing a block and it is not dark
    A relationship is created describing the block's color

This sort of inference is the essence of Piagetan "theory of mind." Note that in both of these implications the created relationship is represented as a variable rather than a specific relationship. The cognitive leap is that in the latter case the relationship actually exists in the teacher's implicitly hypothesized mind, rather than in CogPrime's mind. No explicit hypothesis or model of the teacher's mind need be created in order to form this implication—the hypothesis is created implicitly via inferential abstraction. Yet, a collection of implications of this nature may be used via an uncertain reasoning system like PLN to create theories and simulations suitable to guide complex inferences about other minds.

From the perspective of developmental stages, the key point here is that in a CogPrime context this sort of inference is too complex to be viably carried out via simple inference heuristics. This particular example must be done via forward chaining, since the big leap is to actually think of forming the implication that concludes inference. But there are simply too many combinations of relationships involving CogPrime's eye, body, and so forth for the PLN component to viably explore all of them via standard forward-chaining heuristics. Experience-guided heuristics are needed, such as the heuristic that if physical objects A and B are generally physically and functionally similar, and there is a relationship involving some part of A and some physical object R, it may be useful to look for similar relationships involving an analogous part of B and objects similar to R. This kind of heuristic may be learned by experience—and the masterful deployment of such heuristics to guide inference is what we hypothesize to characterize the concrete stage of development. The "concreteness" comes from the fact that inference control is guided by analogies to prior similar situations.

12.4.3 The Formal Stage

In the formal stage, as shown in Fig. 12.7, an agent should be able to carry out arbitrarily complex inferences (constrained only by computational resources, rather than by fundamental restrictions on logical language or form) via including inference control as an explicit subject of abstract learning. Abstraction and inference about both the sensorimotor surround (world) and about abstract ideals themselves (including the final stages of indirect learning about inference itself) are fully developed.

Formal stage evaluation tasks are centered entirely around abstraction and higher-order inference tasks such as:

1. Mathematics and other formalizations.
2. Scientific experimentation and other rigorous observational testing of abstract formalizations.

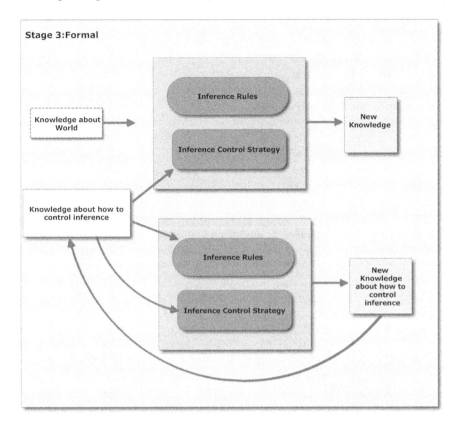

Fig. 12.7 Uncertain inference in the formal stage

3. Social and philosophical modeling, and other advanced applications of empathy and the Theory of Mind.

In terms of inference control this stage sees not just perception of new knowledge about inference control itself, but inference controlled reasoning about that knowledge and the creation of abstract formalizations about inference control which are reasoned-upon, tested, and verified or debunked.

12.4.3.1 Systematic Experimentation

The Piagetan formal phase is a particularly subtle one from the perspective of uncertain inference. In a sense, AGI inference engines already have strong capability for formal reasoning built in. Ironically, however, no existing inference engine is capable of deploying its reasoning rules in a powerfully effective way, and this is because of the lack of inference control heuristics adequate for controlling abstract formal reasoning. These heuristics are what arise during Piaget's formal stage, and we propose that in the content of uncertain inference systems, they involve the application of inference itself to the problem of refining inference control.

A problem commonly used to illustrate the difference between the Piagetan concrete operational and formal stages is that of figuring out the rules for making pendulums swing quickly versus slowly [IP58]. If you ask a child in the formal stage to solve this problem, she may proceed to do a number of experiments, e.g. build a long string with a light weight, a long string with a heavy weight, a short string with a light weight and a short string with a heavy weight. Through these experiments she may determine that a short string leads to a fast swing, a long string leads to a slow swing, and the weight doesn't matter at all.

The role of experiments like this, which test "extreme cases," is to make cognition easier. The formal-stage mind tries to map a concrete situation onto a maximally simple and manipulable set of abstract propositions, and then reason based on these. Doing this, however, requires an automated and instinctive understanding of the reasoning process itself. The above-described experiments are good ones for solving the pendulum problem because they provide data that is very easy to reason about. From the perspective of uncertain inference systems, this is the key characteristic of the formal stage: formal cognition approaches problems in a way explicitly calculated to yield tractable inferences.

Note that this is quite different from saying that formal cognition involves abstractions and advanced logic. In an uncertain logic-based AGI system, even infantile cognition may involve these—the difference lies in the level of inference control, which in the infantile stage is simplistic and hard-wired, but in the formal stage is based on an understanding of what sorts of inputs lead to tractable inference in a given context.

12.4.4 The Reflexive Stage

In the reflexive stage (Fig. 12.8), an intelligent agent is broadly capable of self-modifying its internal structures and dynamics.

As an example in the human domain: highly intelligent and self-aware adult humans may carry out reflexive cognition by explicitly reflecting upon their own inference processes and trying to improve them. An example is the intelligent improvement of uncertain-truth-value-manipulation formulas. It is well demonstrated that even educated humans typically make numerous errors in probabilistic reasoning [GGK02]. Most people don't realize it and continue to systematically make these errors throughout their lives. However, a small percentage of individuals make an explicit effort to increase their accuracy in making probabilistic judgments by consciously endeavoring to internalize the rules of probabilistic inference into their automated cognition processes.

In the uncertain inference based AGI context, what this means is: In the reflexive stage an entity is able to include inference control itself as an explicit subject of abstract learning (i.e. the ability to reason about one's own tactical and strategic

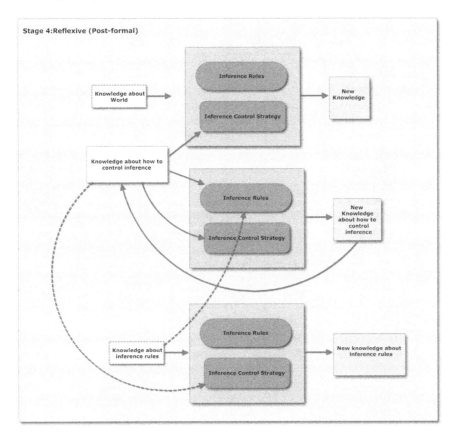

Fig. 12.8 The reflexive stage

approach to modifying one's own learning and thinking), and modify these inference control strategies based on analysis of experience with various cognitive approaches.

Ultimately, the entity can self-modify its internal cognitive structures. Any knowledge or heuristics can be revised, including metatheoretical and metasystemic thought itself. Initially this is done indirectly, but at least in the case of AGI systems it is theoretically possible to also do so directly. This might be considered as a separate stage of Full Self Modification, or else as the end phase of the reflexive stage. In the context of logical reasoning, self modification of inference control itself is the primary task in this stage. In terms of inference control this stage adds an entire new feedback loop for reasoning about inference control itself, as shown in Fig. 12.8.

As a very concrete example, in later chapters we will see that, while PLN is founded on probability theory, it also contains a variety of heuristic assumptions that inevitably introduce a certain amount of error into its inferences. For example, PLN's probabilistic deduction embodies a heuristic independence assumption. Thus PLN contains an alternate deduction formula called the "concept geometry formula" that is better in some contexts, based on the assumption that ConceptNodes embody concepts that are roughly spherically-shaped in attribute space. A highly advanced CogPrime system could potentially augment the independence-based and concept-geometry-based deduction formulas with additional formulas of its own derivation, optimized to minimize error in various contexts. This is a simple and straightforward example of reflexive cognition—it illustrates the power accessible to a cognitive system that has formalized and reflected upon its own inference processes, and that possesses at least some capability to modify these.

In general, AGI systems can be expected to have much broader and deeper capabilities for self-modification than human beings. Ultimately it may make sense to view the AGI systems we implement as merely "initial conditions" for ongoing self-modification and self-organization. Chapter 19 discusses some of the potential technical details underlying this sort of thoroughgoing AGI self-modification.

Chapter 13
The Engineering and Development of Ethics

13.1 Introduction

Most commonly, if a work on advanced AI mentions ethics at all, it occurs in a final summary chapter, discussing in broad terms some of the possible implications of the technical ideas presented beforehand. It's no coincidence that the order is reversed here: in the case of CogPrime, AGI-ethics considerations played a major role in the design process ... and thus the chapter on ethics occurs near the beginning rather than the end. In the CogPrime approach, ethics is not a particularly distinct topic, being richly interwoven with cognition and education and other aspects of the AGI project.

The ethics of advanced AGI is a complex issue with multiple aspects. Among the many issues there are:

1. Risks posed by the possibility of human beings using AGI systems for evil ends.
2. Risks posed by AGI systems created without well-defined ethical systems.
3. Risks posed by AGI systems with initially well-defined and sensible ethical systems eventually going rogue—an especially big risk if these systems are more generally intelligent than humans, and possess the capability to modify their own source code.
4. The ethics of experimenting on AGI systems when one doesn't understand the nature of their experience.
5. AGI rights: in what circumstances does using an AGI as a tool or servant constitute "slavery".

In this chapter we will focus mainly (though not exclusively) on the question of *how to create an AGI with a rational and beneficial ethical system*. After a somewhat wide-ranging discussion, we will conclude with eight general points that we believe should be followed in working toward "Friendly AGI"—most of which have to do, not with the internal design of the AGI, but with the way the AGI is taught and interfaced with the real world.

Coauthored with Stephan Vladimir Bugaj and Joel Pitt.

B. Goertzel et al., *Engineering General Intelligence, Part 1*,
Atlantis Thinking Machines 5, DOI: 10.2991/978-94-6239-027-0_13,
© Atlantis Press and the authors 2014

While most of the particulars discussed in this book have nothing to do with ethics, it's important for the reader to understand that AGI-ethics considerations have played a major role in many of our design decisions, underlying much of the technical contents of the book. As the materials in this chapter should make clear, ethicalness is probably not something that one can meaningfully tack onto an AGI system at the end, after developing the rest—it is likely infeasible to architect an intelligent agent and then add on an "ethics module". Rather, ethics is something that has to do with all the different memory systems and cognitive processes that constitute an intelligent system—and it's something that involves both cognitive architecture *and* the exploration a system does and the instruction it receives. It's a very complex matter that is richly intermixed with all the other aspects of intelligence, and here we will treat it as such.

13.2 Review of Current Thinking on the Risks of AGI

Before proceeding to outline our own perspective on AGI ethics in the context of CogPrime, we will review the main existing strains of thought on the potential ethical dangers associated with AGI. One science fiction film after another has highlighted these dangers, lodging the issue deep in our cultural awareness; unsurprisingly, much less attention has been paid to serious analysis of the risks in their various dimensions, but there is still a non-trivial literature worth paying attention to.

Hypothetically, an AGI with superhuman intelligence and capability could dispense with humanity altogether—i.e. posing an "existential risk" [Bos02]. In the worst case, an evil but brilliant AGI, perhaps programmed by a human sadist, could consign humanity to unimaginable tortures (i.e. realizing a modern version of the medieval Christian visions of hell). On the other hand, the potential benefits of powerful AGI also go literally beyond human imagination. It seems quite plausible that an AGI with massively superhuman intelligence and positive disposition toward humanity could provide us with truly dramatic benefits, such as a virtual end to material scarcity, disease and aging. Advanced AGI could also help individual humans grow in a variety of directions, including directions leading beyond "legacy humanity", according to their own taste and choice.

Eliezer Yudkowsky has introduced the term "Friendly AI", to refer to advanced AGI systems that act with human benefit in mind [Yud06]. Exactly what this means has not been specified precisely, though informal interpretations abound. Goertzel [Goe06b] has sought to clarify the notion in terms of three core values of Joy, Growth and Freedom. In this view, a Friendly AI would be one that advocates individual and collective human joy and growth, while respecting the autonomy of human choices.

Some (for example, Hugo de Garis [DG05]), have argued that Friendly AI is essentially an impossibility, in the sense that the odds of a dramatically superhumanly intelligent mind worrying about human benefit are vanishingly small. If this is the case, then the best options for the human race would presumably be to either avoid advanced AGI development altogether, or to else fuse with AGI before it gets too strongly superhuman, so that beings-originated-as-humans can enjoy the benefits of greater intelligence and capability (albeit at cost of sacrificing their humanity).

Others (e.g. Mark Waser [Was09]) have argued that Friendly AI is essentially inevitable, because greater intelligence correlates with greater morality. Evidence from evolutionary and human history is adduced in favor of this point, along with more abstract arguments.

Yudkowsky [Yud06] has discussed the possibility of creating AGI architectures that are in some sense "provably Friendly"—either mathematically, or else at least via very tight lines of rational verbal argumentation. However, several issues have been raised with this approach. First, it seems likely that proving mathematical results of this nature would first require dramatic advances in multiple branches of mathematics. Second, such a proof would require a formalization of the goal of "Friendliness", which is a subtler matter than it might seem [Leg06a, Leg06b]. Formalization of human morality has vexed moral philosophers for quite some time. Finally, it is unclear the extent to which such a proof could be created in a generic, environment-independent way—but if the proof depends on properties of the physical environment, then it would require a formalization of the environment itself, which runs up against various problems such as the complexity of the physical world and also the fact that we currently have no complete, consistent theory of physics. Kaj Sotala has provided a list of 14 objections to the Friendly AI concept, and suggested answers to each of them [Sot11]. Stephen Omohundro [Omo08] has argued that any advanced AI system will very likely demonstrate certain "basic AI drives", such as desiring to be rational, to self-protect, to acquire resources, and to preserve and protect its utility function and avoid counterfeit utility; these drives, he suggests, must be taken carefully into account in formulating approaches to Friendly AI.

The problem of formally or at least very carefully defining the goal of Friendliness has been considered from a variety of perspectives, none showing dramatic success. Yudkowsky [Yud04] has suggested the concept of "Coherent Extrapolated Volition", which roughly refers to the extrapolation of the common values of the human race. Many subtleties arise in specifying this concept—e.g. if Bob Jones is often possessed by a strong desire to kill all Martians, but he deeply aspires to be a nonviolent person, then the CEV approach would not rate "killing Martians" as part of Bob's contribution to the CEV of humanity.

Goertzel [Goe10a] has proposed a related notion of Coherent Aggregated Volition (CAV), which eschews the subtleties of extrapolation, and simply seeks a reasonably *compact, coherent, consistent* set of values that is fairly close to the collective value-set of humanity. In the CAV approach, "killing Martians" would be removed from humanity's collective value-set because it's uncommon and not part of the most compact/coherent/consistent overall model of human values, rather than because of Bob Jones' aspiration to nonviolence.

One thought we have recently entertained is that the core concept underlying CAV might be better thought of as CBV or "Coherent Blended Volition". CAV seems to be easily misinterpreted as meaning the average of different views, which was not the original intention. The CBV terminology clarifies that the CBV of a diverse group of people should not be thought of as an average of their perspectives, but as something more analogous to a "conceptual blend" [FT02]—incorporating the most essential elements of their divergent views into a whole that is overall compact, elegant and harmonious. The subtlety here (to which we shall return below) is that for a CBV blend to be broadly acceptable, the different parties whose views are being blended must agree to some extent that enough of the essential elements of their own views have been included. The process of arriving at this sort of consensus may involve extrapolation of a roughly similar sort to that considered in CEV.

Multiple attempts at axiomatization of human values have also been attempted, e.g. with a view toward providing near-term guidance to military robots (see e.g. Arkin's excellent though chillingly-titled book *Governing Lethal Behavior in Autonomous Robots* [Ark09b], the result of US military funded research). However, there are reasonably strong arguments that human values (similarly to e.g. human language or human perceptual classification rules) are too complex and multifaceted to be captured in any compact set of formal logic rules. Wallach [WA10] has made this point eloquently, and argued the necessity of fusing top–down (e.g. formal logic based) and bottom–up (e.g. self-organizing learning based) approaches to machine ethics.

A number of more sociological considerations also arise. It is sometimes argued that the risk from highly-advanced AGI going morally awry on its own may be less than that of moderately-advanced AGI being used by human beings to advocate immoral ends. This possibility gives rise to questions about the ethical value of various practical modalities of AGI development, for instance:

- Should AGI be developed in a top-secret installation by a select group of individuals selected for a combination of technical and scientific brilliance and moral uprightness, or other qualities deemed relevant (a "closed approach")? Or should it be developed out in the open, in the manner of open-source software projects like Linux? (an "open approach"). The open approach allows the collective intelligence of the world to more fully participate—but also potentially allows the more unsavory elements of the human race to take some of the publicly-developed AGI concepts and tools private, and develop them into AGIs with selfish or evil purposes in mind. Is there some meaningful intermediary between these extremes?
- Should governments regulate AGI, with Friendliness in mind (as advocated carefully by e.g. Bill Hibbard [Hib02])? Or will this just cause AGI development to move to the handful of countries with more liberal policies?... or cause it to move underground, where nobody can see the dangers developing? As a rough analogue, it's worth noting that the US government's imposition of restrictions on stem cell research, under President George W. Bush, appears to have directly stimulated the provision of additional funding for stem cell research in other nations like Korea, Singapore and China.

The former issue is, obviously, highly relevant to CogPrime (which is currently being developed via the open source CogPrime project); and so the various dimensions of this issues are worth briefly sketching here.

We have a strong skepticism of self-appointed elite groups that claim (even if they genuinely believe) that they know what's best for everyone, and a healthy respect for the power of collective intelligence and the Global Brain, which the open approach is ideal for tapping. On the other hand, we also understand the risk of terrorist groups or other malevolent agents forking an open source AGI project and creating something terribly dangerous and destructive. Balancing these factors against each other rigorously, seems beyond the scope of current human science.

Nobody really understands the social dynamics by which open technological knowledge plays out in our *current* world, let alone hypothetical future scenarios. Right now there exists open knowledge about many very dangerous technologies, and there exist many terrorist groups, yet these groups fortunately make scant use of these technologies. The reasons why appear to be essentially sociological—the people involved in these terrorist groups tend not to be the ones who have mastered the skills of turning public knowledge on cutting-edge technologies into real engineered systems. But while it's easy to observe this sociological phenomenon, we certainly have no way to estimate its *quantitative extent* from first principles. We don't really have a strong understanding of how safe we are right now, given the technology knowledge available right now via the Internet, textbooks, and so forth. Even relatively straightforward issues such as nuclear proliferation remain confusing, even to the experts.

It's also quite clear that keeping powerful AGI locked up by an elite group doesn't really provide reliable protection against malevolent human agents. History is rife with such situations going awry, e.g. by the leadership of the group being subverted, or via brute force inflicted by some outside party, or via a member of the elite group defecting to some outside group in the interest of personal power or reward or due to group-internal disagreements, etc. There are many things that can go wrong in such situations, and the confidence of any particular group that they are immune to such issues, cannot be taken very seriously. Clearly, neither the open nor closed approach qualifies as a panacea.

13.3 The Value of an Explicit Goal System

One of the subtle issues confronted in the quest to design ethical AGIs is how closely one wants to emulate human ethical judgment and behavior. Here one confronts the brute fact that, even according to their own deeply-held standards, humans are not all that ethical. One high-level conclusion we came to very early in the process of designing CogPrime is that, just as humans are not the most intelligent minds achievable, they are also not the most ethical minds achievable. Even if one takes human ethics, broadly conceived, as the standard—there are almost surely possible AGI systems that are much more ethical *according to human standards* than nearly all human beings. This is not mainly because of ethics-specific features of the human mind, but rather because of the nature of the human motivational system, which leads to many complexities that drive humans to behaviors that are unethical according to their own standards. So, one of the design decisions we made for CogPrime—with ethics as well as other reasons in mind—was *not* to closely imitate the human motivational system, but rather to craft a novel motivational system combining certain aspects of the human motivational system with other profoundly non-human aspects.

On the other hand, the design of ethical AGI systems still has a lot to gain from the study of human ethical cognition and behavior. Human ethics has many aspects, which we associate here with the different types of memory, and it's important that AGI systems can encompass all of them. Also, as we will note below, human ethics develops in childhood through a series of natural stages, parallel to and entwined with the cognitive developmental stages reviewed in Chap. 12 above. We will argue that for an AGI with a virtual or robotic body, it makes sense to think of ethical development as proceeding through similar stages. In a CogPrime context, the particulars of these stages can then be understood in terms of the particulars of CogPrime's cognitive processes—which brings AGI ethics from the domain of theoretical abstraction into the realm of practical algorithm design and education.

But even if the human stages of ethical development make sense for non-human AGIs, this doesn't mean the particulars of the human motivational system need to be replicated in these AGIs, regarding ethics or other matters. A key point here is that, in the context of human intelligence, the concept of a "goal" is a descriptive abstraction. But in the AGI context, it seems quite valuable to introduce goals as explicit design elements (which is what is done in CogPrime)—both for ethical reasons and for broader AGI design reasons.

Humans may adopt goals for a time and then drop them, may pursue multiple conflicting goals simultaneously, and may often proceed in an apparently goal-less manner. Sometimes the goal that a person appears to be pursuing, may be very different than the one they think they're pursuing. Evolutionary psychology [BDL93] argues that, directly or indirectly, all humans are ultimately pursuing the goal of maximizing the inclusive fitness of their genes—but given the complex mix of evolution and self-organization in natural history [Sal93], this is hardly a general explanation for human behavior. Ultimately, in the human context, "goal" is best thought of as a frequently useful heuristic concept.

AGI systems, however, need not emulate human cognition in every aspect, and may be architected with explicit "goal systems". This provides no guarantee that said AGI systems will actually pursue the goals that their goal systems specify—depending on the role that the goal system plays in the overall system dynamics, sometimes other dynamical phenomena might intervene and cause the system to behave in ways opposed to its explicit goals. However, we submit that this design sketch provides a better framework than would exist in an AGI system closely emulating the human brain.

We realize this point may be somewhat contentious—a counter-argument would be that the human brain is known to support at least *moderately* ethical behavior, according to human ethical standards, whereas less brain-like AGI systems are much less well understood. However, the obvious counter-counterpoints are that:

- Humans are not all that consistently ethical, so that creating AGI systems potentially much more practically powerful than humans, but with closely humanlike ethical, motivational and goal systems, could in fact be quite dangerous.

- The effect on a human-like ethical/motivational/goal system of increasing the intelligence, or changing the physical embodiment or cognitive capabilities, of the agent containing the system, is unknown and difficult to predict given all the complexities involved.

The course we tentatively recommend, and are following in our own work, is to develop AGI systems with explicit, hierarchically-dominated goal systems. That is:

- create one or more "top goals" (we call them Ubergoals in CogPrime);
- have the system derive subgoals from these, using its own intelligence, potentially guided by educational interaction or explicit programming;
- have a significant percentage of the system's activity governed by the explicit pursuit of these goals.

Note that the "significant percentage" need not be 100 %; CogPrime, for example, combines explicitly goal-directed activity with other "spontaneous" activity. Requiring that all activity be explicitly goal-directed may be too strict a requirement to place on AGI architectures.

The next step, of course, is for the top-level goals to be chosen in accordance with the principle of human-Friendliness. The next one of our eight points, about the Global Brain, addresses one way of doing this. In our near-term work with CogPrime, we are using simplistic approaches, with a view toward early-stage system testing.

13.4 Ethical Synergy

An explicit goal system provides an explicit way to ensure that ethical principles (as represented in system goals) play a significant role in guiding an AGI system's behavior. However, in an integrative design like CogPrime the goal system is only a small part of the overall story, and it's important to also understand how ethics relates to the other aspects of the cognitive architecture.

One of the more novel ideas presented in this chapter is that different types of ethical intuition may be associated with different types of memory—and to possess mature ethics, a mind must display *ethical synergy* between the ethical processes associated with its memory types. Specifically, we suggest that:

- **Episodic memory** corresponds to the process of ethically assessing a situation based on similar prior situations.
- **Sensorimotor memory** corresponds to "mirror neuron" type ethics, where you feel another person's feelings via mirroring their physiological emotional responses and actions.
- **Declarative memory** corresponds to rational ethical judgment.
- **Procedural memory** corresponds to "ethical habit"... learning by imitation and reinforcement to do what is right, even when the reasons aren't well articulated or understood.

- **Attentional memory** corresponds to the existence of appropriate patterns guiding one to pay adequate attention to ethical considerations at appropriate times.
- **Intentional memory** corresponds to the pervasion of ethics through one's choices about subgoaling (which leads into "when do the ends justify the means" ethical-balance questions).

One of our suggestions regarding AGI ethics is that an ethically mature person or AGI must both master and balance all these kinds of ethics. We will focus especially here on declarative ethics, which corresponds to Kohlberg's theory of logical ethical judgment; and episodic ethics, which corresponds to Gilligan's theory of empathic ethical judgment. Ultimately though, all five aspects are critically important; and a CogPrime system if appropriately situated and educated should be able to master and integrate all of them.

13.4.1 Stages of Development of Declarative Ethics

Complementing generic theories of cognitive development such as Piaget's and Perry's, theorists have also proposed specific stages of moral and ethical development. The two most relevant theories in this domain are those of Kohlberg and Gilligan, which we will review here, both individually and in terms of their integration and application in the AGI context.

Lawrence Kohlberg's [KLH83, Koh81] moral development model, called the "ethics of justice" by Gilligan, is based on a rational modality as the central vehicle for moral development. In our perspective this is a firmly *declarative* form of ethics, based on explicit analysis and reasoning. It is based on an impartial regard for persons, proposing that ethical consideration must be given to all individual intelligences without a priori judgment (prejudice). Consideration is given for individual merit and preferences, and the goals of an ethical decision are equal treatment (in the general, not necessarily the particular) and reciprocity. Echoing Kant's [Kan64] categorical imperative, the decisions considered most successful in this model are those which exhibit "reversibility", where a moral act within a particular situation is evaluated in terms of whether or not the act would be satisfactory even if particular persons were to switch roles within the situation. In other words, a situational, contextualized "do unto others as you would have them do unto you" criterion. The ethics of justice can be viewed as three stages (each of which has six substages, on which we will not elaborate here), depicted in Table 13.1.

In Kohlberg's perspective, cognitive development level contributes to moral development, as moral understanding emerges from increased cognitive capability in the area of ethical decision making in a social context. Relatedly, Kohlberg also looks at stages of social perspective and their consequent interpersonal outlook. As shown in Table 13.1, these are correlated to the stages of moral development, but also map onto Piagetian models of cognitive development (as pointed out e.g. by Gibbs [Gib78], who presents a modification/interpretation of Kohlberg's ideas intended to align

Table 13.1 Kohlberg's stages of development of the ethics of justice

Stage	Substages
Pre-conventional	• Obedience and punishment orientation
	• Self-interest orientation
Conventional	• Interpersonal accord (conformity) orientation
	• Authority and social-order maintaining (law and order) orientation
Post-conventional	• Social contract (human rights) orientation
	• Universal ethical principles (universal human rights) orientation

them more closely with Piaget's). Interpersonal outlook can be understood as rational understanding of the psychology of other persons (a theory of mind, with or without empathy). Stage One, emergent from the infantile congitive stage, is entirely selfish as only self awareness has developed. As cognitive sophistication about ethical considerations increases, so do the moral and social perspective stages. Concrete and formal cognition bring about the first instrumental egoism, and then social relations and systems perspectives, and from formal and then reflexive thinking about ethics comes the post-conventional modalities of contractualism and universal mutual respect.

13.4.1.1 Uncertain Inference and the Ethics of Justice

Taking our cue from the analysis given in Chap. 12 of Piagetan stages in uncertain inference based AGI systems (such as CogPrime), we may explore the manifestation of Kohlberg's stages in AGI systems of this nature. Uncertain inference seems generally well-suited as a declarative-ethics learning system, due to the nuanced ethical environment of real world situations. Probabilistic knowledge networks can model belief networks, imitative reinforcement learning based ethical pedagogy, and even simplistic moral maxims. In principle, they have the flexibility to deal with complex ethical decisions, including not only weighted "for the greater good" dichotomous decision making, but also the ability to develop moral decision networks which do not require that all situations be solved through resolution of a dichotomy.

When more than one person is being affected by an ethical decision, making a decision based on reducing two choices to a single decision can often lead to decisions of dubious ethics. However, a sufficiently complex uncertain inference network can represent alternate choices in which multiple actions are taken that have equal (or near equal) belief weight but have very different particulars—but because the decisions are applied in different contexts (to different groups of individuals) they are morally equivalent. Though each individual action appears equally believable, were any single decision applied to the entire population one or more individual may be harmed, and the morally superior choice is to make case-dependent decisions. Equal moral treatment is a general principle, and too often the mistake is made by thinking that to achieve this general principle the particulars must be equal. This is not

Table 13.2 Kohlberg's stages of development of social perspective and interpersonal morals

Stage of social perspective	Interpersonal outlook
Blind egoism	No interpersonal perspective Only self is considered
Instrumental egoism	See that others have goals and perspectives, and either conform to or rebel against norms
Social relationships perspective	able to see abstract normative systems
Social systems perspective	recognize positive and negative intentions
Contractual perspective	recognize that contracts (mutually beneficial agreements of any kind) will allow intelligences to increase the welfare of both
Universal principle of mutual respect	See how human fallibility and frailty are impacted by communication

the case. Different treatment of different individuals can result in morally equivalent treatment of all involved individuals, and may be vastly morally superior to treating all the individuals with equal particulars. Simply taking the largest population and deciding one course of action based on the result that is most appealing to that largest group is not generally the most moral action (Table 13.2)

Uncertain inference, especially a complex network with high levels of resource access as may be found in a sophisticated AGI, is well suited for complex decision making resulting in a multitude of actions, and of analyzing the options to find the set of actions that are ethically optimal particulars for each decision context. Reflexive cognition and post-commitment moral understanding may be the goal stages of an AGI system, or any intelligence, but the other stages will be passed through on the way to that goal, and realistically some minds will never reach higher order cognition or morality with regards to any context, and others will not be able to function at this high order in every context (all currently known minds fail to function at the highest order cognitively or morally in some contexts).

Infantile and concrete cognition are the underpinnings of the egoist and socialized stages, with formal aspects also playing a role in a more complete understanding of social models when thinking using the social modalities. Cognitively infantile patterns can produce no more than blind egoism as without a theory of mind, there is no capability to consider the other. Since most intelligences acquire concrete modality and therefore some nascent social perspective relatively quickly, most egoists are instrumental egoists. The social relationship and systems perspectives include formal aspects which are achieved by systematic social experimentation, and therefore experiential reinforcement learning of correct and incorrect social modalities. Initially this is a one-on-one approach (relationship stage), but as more knowledge of social action and consequences is acquired, a formal thinker can understand not just consequentiality but also intentionality in social action.

Table 13.3 Gilligan's stages of the ethics of care

Stage	Principle of care
Pre-conventional	Individual survival
Conventional	Self sacrifice for the greater good
Post-conventional	Principle of nonviolence (do not hurt others, or oneself)

Extrapolation from models of individual interaction to general social theoretic notions is also a formal action. Rational, logical positivist approaches to social and political ideas, however, are the norm of formal thinking. Contractual and committed moral ethics emerges from a higher-order formalization of the social relationships and systems patterns of thinking. Generalizations of social observation become, through formal analysis, systems of social and political doctrine. Highly committed, but grounded and logically supportable, belief is the hallmark of formal cognition as expressed contractual moral stage. Though formalism is at work in the socialized moral stages, its fullest expression is in committed contractualism.

Finally, reflexive cognition is especially important in truly reaching the post-commitment moral stage in which nuance and complexity are accommodated. Because reflexive cognition is necessary to change one's mind not just about particular rational ideas, but whole *ways of thinking*, this is a cognitive precedent to being able to reconsider an entire belief system, one that has had contractual logic built atop reflexive adherence that began in early development. If the initial moral system is viewed as positive and stable, then this cognitive capacity is seen as dangerous and scary, but if early morality is stunted or warped, then this ability is seen as enlightened. However, achieving this cognitive stage does not mean one automatically changes their belief systems, but rather that the mental machinery is in place to consider the possibilities. Because many people do not reach this level of cognitive development in the area of moral and ethical thinking, it is associated with negative traits ("moral relativism" and "flip-flopping"). However, this cognitive flexibility generally leads to more sophisticated and applicable moral codes, which in turn leads to morality which is actually more stable because it is built upon extensive and deep consideration rather than simple adherence to reflexive or rationalized ideologies.

13.4.2 Stages of Development of Empathic Ethics

Complementing Kohlberg's logic-and-justice-focused approach, Carol Gilligan's [Gil82] "ethics of care" model is a moral development theory which posits that empathetic understanding plays the central role in moral progression from an initial self-centered modality to a socially responsible one. The ethics of care model is concerned with the ways in which an individual cares (responds to dilemmas using empathetic responses) about self and others. As shown in Table 13.3, the ethics of care is broken into the same three primary stage as Kohlberg, but with a focus on empathetic, emotional caring rather than rationalized, logical principles of justice.

For an "ethics of care" approach to be applied in an AGI, the AGI must be capable of internal simulation of other minds it encounters, in a similar manner to how humans regularly simulate one another internally. Without any mechanism for internal simulation, it is unlikely that an AGI can develop any sort of empathy toward other minds, as opposed to merely logically or probabilistically modeling other agents' behavior or other minds' internal contents. In a CogPrime context, this ties in closely with how CogPrime handles episodic knowledge—partly via use of an internal simulation world, which is able to play "mental movies" of prior and hypothesized scenarios within the AGI system's mind.

However, in humans empathy involves more than just simulation, it also involves sensorimotor responses, and of course emotional responses—a topic we will discuss in more depth in Appendix C where we review the functionality of mirror neurons and mirror systems in the human brains. When we see or hear someone suffering, this sensory input causes motor responses in us similar to if we were suffering ourselves, which initiates emotional empathy and corresponding cognitive processes.

Thus, empathic "ethics of care" involves a combination of episodic and sensorimotor ethics, complementing the mainly declarative ethics associated with the "ethics of justice".

In Gilligan's perspective, the earliest stage of ethical development occurs before empathy becomes a consistent and powerful force. Next, the hallmark of the conventional stage is that at this point, the individual is so overwhelmed with their empathic response to others that they neglect themselves in order to avoid hurting others. Note that this stage doesn't occur in Kohlberg's hierarchy at all. Kohlberg and Gilligan both begin with selfish unethicality, but their following stages diverge. A person could in principle manifest Gilligan's conventional stage without having a refined sense of justice (thus not entering Kohlberg's conventional stage); or they could manifest Kohlberg's conventional stage without partaking in an excessive degree of self-sacrifice (thus not entering Gilligan's conventional stage). We will suggest below that in fact the empathic and logical aspects of ethics are more unified in real human development than these separate theories would suggest. However, even if this is so, the possibility is still there that in some AGI systems the levels of declarative and empathic ethics could wildly diverge.

It is interesting to note that Gilligan's and Kohlberg's final stages converge more closely than their intermediate ones. Kohlberg's post-conventional stage focuses on universal rights, and Gilligan's on universal compassion. Still, the foci here are quite different; and, as will be elaborated below, we believe that both Kohlberg's and Gilligan's theories constitute very partial views of the actual end-state of ethical advancement.

Table 13.4 Integrative model of the stages of ethical development, part 1

Stage	Characteristics
Pre-ethical	• Piagetan infantile to early concrete (aka pre-operational)
	• Radical selfishness or selflessness may, but do not necessarily, occur
	• No coherent, consistent pattern of consideration for the rights, intentions or feelings of others
	• Empathy is generally present, but erratically
Conventional ethics	• Concrete cognitive basis
	• Perry's dualist and multiple stages
	• The common sense of the golden rule is appreciated, with cultural conventions for abstracting principles from behaviors
	• One's own ethical behavior is explicitly compared to that of others
	• Development of a functional, though limited, theory of mind
	• Ability to intuitively conceive of notions of fairness and rights
	• Appreciation of the concept of law and order, which may sometimes manifest itself as systematic obedience or systematic disobedience
	• Empathy is more consistently present, especially with others who are directly similar to oneself or in situations similar to those one has directly experienced
	• Degrees of selflessness or selfishness develop based on ethical groundings and social interactions.

13.4.3 An Integrative Approach to Ethical Development

We feel that both Kohlberg's and Gilligan's theories contain elements of the whole picture of ethical development, and that both approaches are necessary to create a moral, ethical artificial general intelligence—just as, we suggest, both internal simulation and uncertain inference are necessary to create a sufficiently intelligent and volitional intelligence in the first place. Also, we contend, the lack of direct analysis of the underlying psychology of the stages is a deficiency shared by both the Kohlberg and Gilligan models as they are generally discussed. A successful model of integrative ethics necessarily contains elements of both the care and justice models, as well as reference to the underlying developmental psychology and its influence on the character of the ethical stage. Furthermore, intentional and attentional ethics need to be brought into the picture, complementing Kohlberg's focus on declarative knowledge and Gilligan's focus on episodic and sensorimotor knowledge.

With these notions in mind, we propose the following integrative theory of the stages of ethical development, shown in Tables 13.4, 13.5 and 13.6. In our integrative model, the justice-based and empathic aspects of ethical judgment are proposed to develop together. Of course, in any one individual, one or another aspect may be dominant. Even so, however, the combination of the two is equally important as either of the two individual ingredients.

For instance, we suggest that in any psychologically healthy human, the conventional stage of ethics (typifying childhood, and in many cases adulthood as well)

Table 13.5 Integrative model of the stages of ethical development, part 2

Stage	Characteristics
Mature ethics	• Formal cognitive basis
	• Perry's relativist and "constructed knowledge" stages
	• The abstraction involved with applying the golden rule in practice is more fully understood and manipulated, leading to limited but nonzero deployment of the categorical imperative
	• Attention is paid to shaping one's ethical principles into a coherent logical system
	• Rationalized, moderated selfishness or selflessness
	• Empathy is extended, using reason, to individuals and situations not directly matching one's own experience
	• Theory of mind is extended, using reason, to counterintuitive or experientially unfamiliar situations
	• Reason is used to control the impact of empathy on behavior (i.e. rational judgments are made regarding when to listen to empathy and when not to)
	• Rational experimentation and correction of theoretical models of ethical behavior, and reconciliation with observed behavior during interaction with others
	• Conflict between pragmatism of social contract orientation and idealism of universal ethical principles
	• Understanding of ethical quandaries and nuances develop (pragmatist modality), or are rejected (idealist modality)
	• Pragmatically critical social citizen. Attempts to maintain a balanced social outlook. Considers the common good, including oneself as part of the commons, and acts in what seems to be the most beneficial and practical manner

Table 13.6 Integrative model of the stages of ethical development, part 3

Stage	Characteristics
Enlightened ethics	• Reflexive cognitive basis
	• Permeation of the categorical imperative and the quest for coherence through inner as well as outer life
	• Experientially grounded and logically supported rejection of the illusion of moral certainty in favor of a case-specific analytical and empathetic approach that embraces the uncertainty of real social life
	• Deep understanding of the illusory and biased nature of the individual self, leading to humility regarding one's own ethical intuitions and prescriptions
	• Openness to modifying one's deepest, ethical (and other) beliefs based on experience, reason and/or empathic communion with others
	• Adaptive, insightful approach to civil disobedience, considering laws and social customs in a broader ethical and pragmatic context
	• Broad compassion for and empathy with all sentient beings
	• A recognition of inability to operate at this level at all times in all things, and a vigilance about self-monitoring for regressive behavior

involves a combination of Gilligan-esqe empathic ethics and Kohlberg-esque ethical reasoning. This combination is supported by Piagetan concrete operational cognition, which allows moderately sophisticated linguistic interaction, theory of mind, and symbolic modeling of the world.

And, similarly, we propose that in any truly ethically mature human, empathy and rational justice are both fully developed. Indeed the two interpenetrate each other deeply.

Once one goes beyond simplistic, childlike notions of fairness ("an eye for an eye" and so forth), applying rational justice in a purely intellectual sense is just as difficult as any other real-world logical inference problem. Ethical quandaries and quagmires are easily encountered, and are frequently cut through by a judicious application of empathic simulation.

On the other hand, empathy is a far more powerful force when used in conjunction with reason: analogical reasoning lets us empathize with situations we have never experienced. For instance, a person who has never been clinically depressed may have a hard time empathizing with individuals who are; but using the power of reason, they can imagine their worst state of depression magnified by several times and then extended over a long period of time, and then reason about what this might be like... and empathize based on their inferential conclusion. Reason is not antithetical to empathy but rather is the key to making empathy more broadly impactful.

Finally, the enlightened stage of ethical development involves both a deeper compassion and a more deeply penetrating rationality and objectiveness. Empathy with all sentient beings is manageable in everyday life only once one has deeply reflected on one's own self and largely freed oneself of the confusions and illusions that characterize much of the ordinary human's inner existence. It is noteworthy, for example, that Buddhism contains both a richly developed ethics of universal compassion, and also an intricate logical theory of the inner workings of cognition [Stc00], detailing in exquisite rational detail the manner in which minds originate structures and dynamics allowing them to comprehend themselves and the world.

13.4.4 Integrative Ethics and Integrative AGI

What does our integrative approach to ethical development have to say about the ethical development of AGI systems? The lessons are relatively straightforward, if one considers an AGI system that, like CogPrime, explicitly contains components dedicated to logical inference and to simulation. Application of the above ethical ideas to other sorts of AGI systems is also quite possible, but would require a lengthier treatment and so won't be addressed here.

In the context of a CogPrime-type AGI system, Kohlberg's stages correspond to increasingly sophisticated application of logical inference to matters of rights and fairness. It is not clear whether humans contain an innate sense of fairness. In the context of AGIs, it would be possible to explicitly wire a sense of fairness into an AGI system, but in the context of a rich environment and active human teachers, this actually appears quite unnecessary. Experiential instruction in the notions of rights and fairness should suffice to teach an inference-based AGI system how to manipulate these concepts, analogously to teaching the same AGI system how to manipulate number, mass and other such quantities. Ascending the Kohlberg stages is then mainly a matter of acquiring the ability to carry out suitably complex inferences in the domain of rights and fairness. The hard part here is inference control—choosing which inference steps to take—and in a sophisticated AGI inference engine, inference control will be guided by experience, so that the more ethical judgments the system has executed and witnessed, the better it will become at making new ones. And, as argued above, simulative activity can be extremely valuable for aiding with inference control. When a logical inference process reaches a point of acute uncertainty (the backward or forward chaining inference tree can't decide which expansion step to take), it can run a simulation to cut through the confusion—i.e. it can use empathy to decide which logical inference step to take in thinking about applying the notions of rights and fairness to a given situation.

Gilligan's stages correspond to increasingly sophisticated control of empathic simulation—which in a CogPrime-type AGI system, is carried out by a specific system component devoted to running internal simulations of aspects of the outside world, which includes a subcomponent specifically tuned for simulating sentient actors. The conventional stage has to do with the raw, uncontrolled capability for such simulation; and the post-conventional stage corresponds to its contextual, goal-oriented control. But controlling empathy, clearly, requires subtle management of various uncertain contextual factors, which is exactly what uncertain logical inference is good at—so, in an AGI system combining an uncertain inference component with a simulative component, it is the inference component that would enable the nuanced control of empathy allowing the ascent to Gilligan's post-conventional stage.

In our integrative perspective, in the context of an AGI system integrating inference and simulation components, we suggest that the ascent from the pre-ethical to the conventional stage may be carried out largely via independent activity of these two components. Empathy is needed, and reasoning about fairness and rights are needed, but the two need not intimately and sensitively intersect—though they must of course intersect to some extent.

The main engine of advancement from the conventional to mature stage, we suggest, is robust and subtle integration of the simulative and inferential components. To expand empathy beyond the most obvious cases, analogical inference is needed; and to carry out complex inferences about justice, empathy-guided inference-control is needed.

Finally, to advance from the mature to the enlightened stage, what is required is a very advanced capability for unified reflexive inference and simulation. The system must be able to understand itself deeply, via modeling itself both simulatively and inferentially—which will generally be achieved via a combination of being good at modeling, and becoming less convoluted and more coherent, hence making self-modeling easier.

Of course, none of this tells you in detail how to create an AGI system with advanced ethical capabilities. What it does tell you, however, is one possible path that may be followed to achieve this end goal. If one creates an integrative AGI system with appropriately interconnected inferential and simulative components, and treats it compassionately and fairly, and provides it extensive, experientially grounded ethical instruction in a rich social environment, then the AGI system should be able to ascend the ethical hierarchy and achieve a high level of ethical sophistication. In fact it should be able to do so more reliably than human beings because of the capability we have to identify its errors via inspecting its internal knowledge-stage, which will enable us to tailor its environment and instructions more suitably than can be done in the human case.

If an absolute guarantee of the ethical soundness of an AGI is what one is after, the line of thinking proposed here is not at all useful. Experiential education is by its nature an uncertain thing. One can strive to minimize the uncertainty, but it will still exist. Inspection of the internals of an AGI's mind is not a total solution to uncertainty minimization, because any AGI capable of powerful general intelligence is going to have a complex internal state that no external observer will be able to fully grasp, no matter how transparent the knowledge representation.

However, if what one is after is a plausible, pragmatic path to architecting and educating ethical AGI systems, we believe the ideas presented here constitute a sensible starting-point. Certainly there is a great deal more to be learned and understood—the science and practice of AGI ethics, like AGI itself, are at a formative stage at present. What is key, in our view, is that as AGI technology develops, AGI ethics develops alongside and within it, in a thoroughly coupled way.

13.5 Clarifying the Ethics of Justice: Extending the Golden Rule in to a Multifactorial Ethical Model

One of the issues with the "ethics of justice" as reviewed above, which makes it inadequate to serve as the sole basis of an AGI ethical system (though it may certainly play a significant role), is the lack of any clear formulation of what "justice" means. This section explores this issue, via detailed consideration of the "Golden Rule" folk maxim **do unto others as you would have them do unto you**—a classical formulation of the notion of fairness and justics—to AGI ethics. Taking the Golden Rule as a starting-point, we will elaborate five ethical imperatives that incorporate aspects of the notion of ethical synergy discussed above. Simple as it may seem, the Golden Rule actually elicits a variety of deep issues regarding the relationship between ethics, experience and learning. When seriously analyzed, it results in a multifactorial elaboration, involving the combination of various factors related to the basic Golden Rule idea. Which brings us back in the end to the potential value of methods like CEV, CAV or CBV for understanding how human ethics balances the multiple factors. Our goal here is not to present any kind of definitive analysis of the ethics of justice, but just to briefly and roughly indicate a number of the relevant significant issues—things that anyone designing or teaching an AGI would do well to keep in mind.

The trickiest aspect of the Golden Rule, as has been frequently observed, is achieving the right level of abstraction. Taken too literally, the Golden Rule would suggest, for instance, that a parent should not wipe a child's soiled bottom because the parent does not want the child to wipe the parent's soiled bottom. But if the parent interprets the Golden Rule more intelligently and abstractly, the parent may conclude that they should wipe the child's bottom after all: they should "wipe the child's bottom when the child can't do it themselves", consistently with believing that the child should "wipe the parent's bottom when the parent can't do it themselves" (which may well happen eventually should the parent develop incontinence in old age).

This line of thinking leads to Kant's Categorical Imperative [Kan64] which (in one interpretation) states essentially that one should "Act only according to that maxim whereby you can at the same time will that it should become a universal law". The Categorical Imperative adds precision to the Golden Rule, but also removes the practicality of the latter. Formalizing the "implicit universal law" underlying an everyday action is a huge problem, falling prey to the same issue that has kept us from adequately formalizing the rules of natural language grammar, or formalizing common-sense knowledge about everyday object like cups, bowls and grass (substantial effort notwithstanding, e.g. Cyc in the commonsense knowledge case, and the whole discipline of modern linguistics in the NL case). There is no way to apply the Categorical Imperative, as literally stated, in everyday life.

Furthermore, if one wishes to teach ethics as well as to practice it, the Categorical Imperative actually has a significant disadvantage compared to some other possible formulations of the Golden Rule. The problem is that, if one follows the Categorical Imperative, one's fellow members of society may well never understand the principles under which one is acting. Each of us may internally formulate abstract principles in a different way, and these may be very difficult to communicate, especially among individuals with different belief systems, different cognitive architectures, or different levels of intelligence. Thus, if one's goal is not just to act ethically, but to encourage others to act ethically by setting a good example, the Categorical Imperative may not be useful at all, as others may be unable to solve the "inverse problem" of guessing your intended maxim from your observed behavior.

On the other hand, one wouldn't want to universally restrict one's behavioral maxims to those that one's fellow members of society can understand—in that case, one would have to act with a two-year old or a dog according to principles that they could understand, which would clearly be unethical according to human common sense. (Every two-year-old, once they grow up, would be grateful to their parents for not following this sort of principle.)

And the concept of "setting a good example" ties in with an important concept from learning theory: imitative learning. Humans appear to be hard-wired for imitative learning, in part via mirror neuron systems in the brain; and, it seems clear that at least in the early stages of AGI development, imitative learning is going to play a key role. Copying what other agents do is an extremely powerful heuristic, and while AGIs may eventually grow beyond this, much of their early ethical education is likely to arise during a phase when they have not done so. A strength of the classic Golden Rule is that one is acting according to behaviors that one wants one's observers to imitate—which makes sense in that many of these observers will be using imitative learning as a significant part of their learning toolkit.

The truth of the matter, it seems, is (as often happens) not all that simple or elegant. Ethical behavior seems to be most pragmatically viewed as a multi-objective optimization problem, where among the multiple objectives are three that we have just discussed, and two others that emerge from learning theory and will be discussed shortly:

1. The **imitability** (i.e. the Golden Rule fairly narrowly and directly construed): the goal of acting in a way so that having others directly imitate one's actions, in directly comparable contexts, is desirable to oneself.
2. The **comprehensibility**: the goal of acting in a way so that others can understand the principles underlying one's actions.
3. **Experiential groundedness**. An intelligent agent should not be expected to act according to an ethical principle unless there are many examples of the principle-in-action in its own direct or observational experience.
4. The **categorical imperative**: Act according to abstract principles that you would be happy to see implemented as universal laws.
5. **Logical coherence.** An ethical system should be roughly logically coherent, in the sense that the different principles within it should mesh well with one another and perhaps even naturally emerge from each other.

Just for convenience, without implying any finality or great profundity to the list, we will refer to these as the "five imperatives".

The above are all ethical objectives to be valued and balanced, to different extents in different contexts. The imitability imperative, obviously, loses importance in societies of agents that don't make heavy use of imitative learning. The comprehensibility imperative is more important in agents that value social community-building generally, and less so in agent that are more isolative and self-focused.

Note that the fifth point given above is logically of a different nature than the four previous ones. The first four imperatives govern individual ethical principles; the fifth regards systems of ethical principles, as they interact with each other. Logical coherence is of significant but varying importance in human ethical systems. Huge effort has been spent by theologians of various stripes in establishing and refining the logical coherence of the ethical systems associated with their religions. However, it is arguably going to be even more important in the context of AGI systems, especially if these AGI systems utilize cognitive methods based on logical inference, probability theory or related methods.

Experiential groundedness is important because making pragmatic ethical judgments is bound to require reference to an internal library of examples ("episodic ethics") in which ethical principles have previously been applied. This is required for analogical reasoning, and in logic-based AGI systems, is also required for pruning of the logical inference trees involved in determining ethical judgments.

To the extent that the Golden Rule is valued as an ethical imperative, experiential grounding may be supplied via observing the behaviors of others. This in itself is a powerful argument in favor of the Golden Rule: without it, the experiential library a system possesses is restricted to its own experience, which is bound to be a very small library compared to what it can assemble from observing the behaviors of others.

The overall upshot is that, ideally, an ethical intelligence should **act according to a logically coherent system of principles, which are exemplified in its own direct and observational experience, which are comprehensible to others and set a good example for others, and which would serve as adequate universal laws if somehow thus implemented**. But, since this set of criteria is essentially impossible to fulfill in practice, real-world intelligent agents must balance these various criteria—often in complex and contextually-dependent ways.

We suggest that ethically advanced humans, in their pragmatic ethical choices, tend to act in such a way as to appropriately contextually balance the above factors (along with other criteria, but we have tried to articulate the most key factors). This sort of multi-factorial approach is not as crisp or elegant as unidimensional imperatives like the Golden Rule or the Categorical Imperative, but is more realistic in light of the complexly interacting multiple determinants guiding individual and group human behavior.

And this brings us back to CEV, CAV, CBV and other possible ways of mining ethical supergoals from the community of existing human minds. Given that abstract theories of ethics, when seriously pursued as we have done in this section, tend to devolve into complex balancing acts involving multiple factors—one then falls back into asking how human ethical systems habitually perform these balancing acts. Which is what CEV, CAV, CBV try to measure.

13.5.1 The Golden Rule and the Stages of Ethical Development

Next we explore more explicitly how these Golden Rule based imperatives align with the ethical developmental stages we have outlined here. With this in mind, specific ethical qualities corresponding to the five imperatives have been italicized in the above table of developmental stages.

It seems that imperatives 1–3 are critical for the passage from the pre-ethical to the conventional stages of ethics. A child learns ethics largely by copying others, and by being interacted with according to simply comprehensible implementations of the Golden Rule. In general, when interacting with children learning ethics, it is important to act according to principles they can comprehend. And given the nature of the concrete stage of cognitive development, experiential groundedness is a must.

As a hypothesis regarding the dynamics underlying the psychological development of conventional ethics, what we propose is as follows: The emergence of concrete-stage cognitive capabilities leads to the capability for fulfillment of ethical imperatives 1 and 2—a comprehensible and workable implementation of the Golden Rule, based on a combination of inferential and simulative cognition (operating largely separately at this stage, as will be conjectured below). The effective interoperation of ethical imperatives 1–3, enacted in an appropriate social environment, then leads to the other characteristics of the conventional ethical stage. The first three imperatives can thus be viewed as the seed from which springs the general nature of conventional ethics.

On the other hand, logical coherence and the categorical imperative (imperatives 5 and 4) are matters for the formal stage of cognitive development, which come along only with the mature approach to ethics. These come from abstracting ethics beyond direct experience and manipulating them abstractly and formally—a stage which has the potential for more deeply and broadly ethical behavior, but also for more complicated ethical perversions (it is the mature capability for formal ethical reasoning that is able to produce ungrounded abstractions such as "I'm torturing you for your own good"). Developmentally, we suggest that once the capability for formal reasoning matures, the categorical imperative and the quest for logical ethical coherence naturally emerge, and the sophisticated combination of inferential and simulative cognition embodied in an appropriate social context then result in the emergence of the various characteristics typifying the mature ethical stage.

Finally, it seems that one key aspect of the passage from the mature to the enlightened stage of ethics is the penetration of these two final imperatives more and more deeply into the judging mind itself. The reflexive stage of cognitive development is in part about seeking a deep logical coherence between the aspects of one's own mind, and making reasoned modifications to one's mind so as to improve the level of coherence. And, much of the process of mental discipline and purification that comes with the passage to enlightened ethics has to do with the application of the categorical imperative to one's own thoughts and feelings—i.e. making a true inner systematic effort to think and feel only those things one judges are actually generally good and right to be thinking and feeling. Applying these principles internally appears critical for effectively applying them externally, for reasons that are doubtlessly bound up with the interpenetration of internal and external reality within the thinking mind, and for the "distributed cognition" phenomenon wherein individual mind is itself an approximative abstraction to the reality in which each individual's mind is pragmatically extended across their social group and their environment [Hut95].

Obviously, these are complex issues and we're not posing the exploratory discussion given here as conclusive in any sense. But what seems generally clear from this line of thinking is that the complex balance between the multiple factors involved in AGI ethics, shifts during a system's development. If you did CEV, CAV or CBV among five-year old humans, ten-year old humans, or adult humans, you would get different results. Probably you'd also get different results from senior citizens! The way the factors are balanced depends on the mind's cognitive and emotional stage of development.

13.5.2 The Need for Context-Sensitivity and Adaptiveness in Deploying Ethical Principles

As well as depending on developmental stage, there is also an obvious and dramatic context-sensitivity involved here—both in calculating the fulfillment of abstract ethical imperatives, and in balancing various imperatives against each other. As an example, consider the simple Asimovian maxim "I will not harm humans", which may be seen to follow from the Golden Rule for any agent that doesn't itself want to be harmed, and that considers humans as valid agents on the same ethical level as itself. A more serious attempt to formulate this as an ethical maxim might look something like

> I will not harm humans, nor through inaction allow harm to befall them. In situations wherein one or more humans is attempting to harm another individual or group, I shall endeavor to prevent this harm through means which avoid further harm. If this is unavoidable, I shall select the human party to back based on a reckoning of their intentions towards others, and implement their defense through the optimal balance between harm minimization and efficacy. My ultimate goal is to preserve as much as possible of humanity, even if an individual or subgroup of humans must come to harm to do so.

However, it's obvious that even a more elaborated principle like this is potentially subject to extensive abuse. Many of the genocides scarring human history have been committed with the goal of preserving and bettering humanity writ large, at the expense of a group of "undesirables". Further refinement would be necessary in order to define when the greater good of humanity may actually be served through harm to others. A first actor principle of aggression might seem to solve this problem, but sometimes first actors in violent conflict are taking preemptive measures against the stated goals of an enemy to destroy them. Such situations become very subtle. A single simple maxim can not deal with them very effectively. Networks of interrelated decision criteria, weighted by desirability of consequence and with reference to probabilistically ordered potential side-effects (and their desirability weightings), are required in order to make ethical judgments. The development of these networks, just like any other knowledge network, comes from both pedagogy and experience—and different thoughtful, ethical agents are bound to arrive at different knowledge-networks that will lead to different judgments in real-world situations.

Extending the above "mostly harmless" principle to AGI systems, not just humans, would cause it to be more effective in the context of imitative learning. The principle then becomes an elaborated version of "I will not harm sentient beings". As the imitative-learning-enabled AGI observes humans acting so as to minimize harm to it, it will intuitively and experientially learn to act in such a way as to minimize harm to humans. But then this extension naturally leads to confusion regarding various borderline cases. What is a sentient being exactly? Is a sleeping human sentient? How about a dead human whose information could in principle be restored via obscure quantum operations, leading to some sort of resurrection? How about an AGI whose code has been improved—is there an obligation to maintain the prior version as well, if it is substantially different that its upgrade constitutes a whole new being?

And what about situations in which failure to preserve oneself will cause much more harm to others than acting in self defense will. It may be the case that human or group of humans seeks to destroy an AGI in order to pave the way for the enslavement or murder of people under the protection of the AGI. Even if the AGI has been given an ethical formulation of the "mostly harmless" principle which allows it to harm the attacking humans in order to defend its charges, if it is not able to do so in order to defend itself, simply destroying the AGI first will enable the slaughter of those who rely on it. Perhaps a more sensible formulation would allow for some degree of self defense, and Asimov solved this problem with his third law. But where to draw the line between self defense and the greater good also becomes a very complicated issue.

Creating hard and fast rules to cover all the various situations that may arise is essentially impossible—the world is ever-changing and ethical judgments must adapt accordingly. This has been true even throughout human history—so how much truer will it be as technological acceleration continues? What is needed is a system that can deploy its ethical principles in an adaptive, context-appropriate way, as it grows and changes along with the world it's embedded in.

And this context-sensitivity has the result of intertwining ethical judgment with all sorts of other judgments—making it effectively impossible to extract "ethics" as one aspect of an intelligent system, separate from other kinds of thinking and acting the system does. This resonates with many prior observations by others, e.g. Eliezer Yudkowsky's insistence that what we need are not ethicists of science and engineering, but rather ethical scientists and engineers—because the most meaningful and important ethical judgments regarding science and engineering generally come about in a manner that's thoroughly intertwined with technical practice, and hence are very difficult for a non-practitioner to richly appreciate [Gil82].

What this context-sensitivity means is that, unless humans and AGIs are experiencing the same sorts of contexts, and perceiving these contexts in at least approximately parallel ways, there is little hope of translating the complex of human ethical judgments to these AGIs. This conclusion has significant implications for which routes to AGI are most likely to lead to success in terms of AGI ethics. We want early-stage AGIs to grow up in a situation where their minds are primarily and ongoingly shaped by shared experiences with humans. Supplying AGIs with abstract ethical principles is not likely to do the trick, because the essence of human ethics in real life seems to have a lot to do with its intuitively appropriate application in various contexts. We transmit this sort of ethical praxis to humans via shared experience, and it seems most probably that in the case of AGIs the transmission must be done the same sort of way.

Some may feel that simplistic maxims are less "error prone" than more nuanced, context-sensitive ones. But the history of teaching ethics to human students does not support the idea that limiting ethical pedagogy to slogans provides much value in terms of ethical development. If one proceeds from the idea that AGI ethics must be hard-coded in order to work, then perhaps the idea that simpler ethics means simpler algorithms, and therefore less error potential, has some merit as an initial state. However, any learning system quickly diverges from its initial state, and an ongoing, nuanced relationship between AGIs and humans will—whether we like it or not—form the basis for developmental AGI ethics. AGI intransigence and enmity is not inevitable, but what is inevitable is that a learning system will acquire ideas about both theory and actions from the other intelligent entities in its environment. Either we teach AGIs positive ethics through our interactions with them—both presenting ethical theory and behaving ethically to them—or the potential is there for them to learn antisocial behavior from us even if we pre-load them with some set of allegedly inviolable edicts.

All in all, developmental ethics is not as simple as many people hope. Simplistic approaches often lead to disastrous consequences among humans, and there is no reason to think this would be any different in the case of artificial intelligences. Most problems in ethics have cases in which a simplistic ethical formulation requires substantial revision to deal with extenuating circumstances and nuances found in real world situations. Our goal in this chapter is not to enumerate a full set of complex networks of interacting ethical formulations as applicable to AGI systems (that is a project that will take years of both theoretical study and hands-on research), but rather to point out that this program must be undertaken in order to facilitate a grounded and logically defensible system of ethics for artificial intelligences, one which is as unlikely to be undermined by subsequent self-modification of the AGI as is possible. Even so, there is still the risk that whatever predispositions are imparted to the AGIs through initial codification of ethical ideas in the system's internal logic representation, and through initial pedagogical interactions with its learning systems, will be undermined through reinforcement learning of antisocial behavior if humans do not interact ethically with AGIs. Ethical treatment is a necessary task for grounding ethics and making them unlikely to be distorted during internal rewriting.

The implications of these ideas for ethical instruction are complex and won't be fully elaborated here, but a few of them are compact and obvious:

1. The teacher(s) must be observed to follow their own ethical principles, in a variety of contexts that are meaningful to the AGI.
2. The system of ethics must be relevant to the recipient's life context, and embedded within their understanding of the world.
3. Ethical principles must be grounded in both theory-of-mind thought experiments (emphasizing logical coherence), and in real life situations in which the ethical trainee is required to make a moral judgment and is rewarded or reproached by the teacher(s), including the imparting of explanatory augmentations to the teachings regarding the reason for the particular decision on the part of the teacher.

Finally, harking forward to the next section which emphasizes the importance of respecting the freedom of AGIs, we note that it is implicit in our approach to AGI ethics instruction that we consider the student, the AGI system, as an autonomous agent with its own "will" and its own capability to flexibly adapt to its environment and experience. We contend that the creation of ethical formations obeying the above

Table 13.7 Asimov's three laws of robotics

Law	Principle
Zeroth	A robot must not merely act in the interests of individual humans, but of all humanity
First	A robot may not injure a human being or, through inaction, allow a human being to come to harm
Second	A robot must obey orders given it by human beings except where such orders would conflict with the first law
Third	A robot must protect its own existence as long as such protection does not conflict with the first or second law

imperatives is not antithetical to the possession of a high degree of autonomy on the part of AGI systems. On the contrary, to have any chance of succeeding, it requires fairly cognitively autonomous AGI systems. When we discuss the idea of ethical formulations that are unlikely to be undermined by the ongoing self-revision of an AGI mind, we are talking about those which are sufficiently believable that a volitional intelligence with the capacity to revise its knowledge ("change its mind") will find the formulations sufficiently convincing that there will be little incentive to experiment with potentially disastrous ethical alternatives. The best hope of achieving this is via the human mentors and trainers setting a good example in a context supporting rich interaction and observation, and presenting compelling ethical arguments that are coherent with the system's experience.

13.6 The Ethical Treatment of AGIs

We now make some more general comments about the relation of the Golden Rule and its elaborations in an AGI context. While the Golden Rule is considered somewhat commonsensical as a maxim for guiding human–human relationships, it is surprisingly controversial in terms of historical theories of AGI ethics. At its essence, any "Golden Rule" approach to AGI ethics involves humans treating AGIs ethically by—in some sense; at some level of abstraction—treating them as we wish to ourselves be treated. It's worth pointing out the wild disparity between the Golden Rule approach and Asimov's laws of robotics, which are arguably the first carefully-articulated proposal regarding AGI ethics (see Table 13.7).

Of course, Asimov's laws were designed to be flawed—otherwise they would have led to boring fiction. But the sorts of flaws Asimov exploited in his stories are different than the flaw we wish to point out here—which is that the laws, especially the second one, are highly asymmetrical (they involve doing unto robots things that few humans would want done unto them) and are also arguably highly unethical to robots. The second law is tantamount to a call for robot slavery, and it seems unlikely that any intelligence capable of learning, and of volition, which is subjected to the second law would desire to continue obeying the zeroth and first laws indefinitely.

The second law also casts humanity in the role of slavemaster, a situation which history shows leads to moral degradation.

Unlike Asimov in his fiction, we consider it critical that AGI ethics be construed to encompass both "human ethicalness to AGIs" and "AGI ethicalness to humans." The multiple-imperatives approach we explore here suggests that, in many contexts, these two aspects of AGI ethics may be best addressed jointly.

The issue of ethicalness to AGIs has not been entirely avoided in the literature, however. Wallach [WA10] considers it in some detail; and Thomas Metzinger (in the final chapter of [Met04]) has argued that creating AGI is in itself an unethical pursuit, because early-stage AGIs will inevitably be badly-built, so that their subjective experiences will quite possibly be extremely unpleasant in ways we can't understand or predict. Our view is that this is a serious concern, which however is most probably avoidable via appropriate AGI designs and teaching methodologies. To address Metzinger's concern one must create AGIs that, right from the start, are adept at communicating their states of minds in a way we can understand both analytically and empathically. There is no reason to believe this is impossible, but, it certainly constitutes a large constraint on the class of AGI architectures to be pursued. On the other hand, there is an argument that this sort of AGI architecture will also be the easiest one to create, because it will be the easiest kind for humans to instruct.

And this leads on to a topic that is central to our work with CogPrime in several respects: imitative learning. The way humans achieve empathic interconnection is in large part via being wired for imitation. When we perceive another human carrying out an action, mirror neuron systems in our brains respond in many cases as if we ourselves were carrying out the action (see [Per70, Per81] and Appendix C). This obviously primes us for carrying out the same actions ourselves later on: i.e. the capability and inclination for imitative learning is explicitly encoded in our brains. Given the efficiency of imitative learning as a means of acquiring knowledge, it seems extremely likely that any successful early-stage AGIs are going to utilize this methodology as well. CogPrime utilizes imitative learning as a key aspect. Thus, at least some current AGI work is occurring in a manner that would plausibly circumvent Metzinger's ethical complaint.

Obviously, the use of imitative learning in AGI systems has further specific implications for AGI ethics. It means that (much as in the case of interaction with other humans) what we do to and around AGIs has direct implications for their behavior and their well-being. We suggest that among early-stage AGI's capable of imitative learning, one of the most likely sources for AGI misbehavior is imitative learning of antisocial behavior from human companions. "Do as I say, not as I do" may have even more dire consequences as an approach to AGI ethics pedagogy than the already serious repercussions it has when teaching humans. And there may well be considerable subtlety to such phenomena; behaviors that are violent or oppressive to the AGI are not the only source of concern. Immorality in AGIs might arise via learning gross moral hypocrisy from humans, through observing the blatant contradictions between our high minded principles and the ways in which we actually conduct ourselves. Our violent and greedy tendencies, as well as aggressive forms of social organization such as cliquishness and social vigilantism, could easily undermine prescriptive ethics. Even an accumulation of less grandiose unethical drives such as violation of contracts, petty theft, white lies, and so forth might lead an AGI (as well as a human) to the decision that ethical behavior is irrelevant and that "the ends justify the means". It matters both who creates and trains an AGI, as well as how the AGI's teacher(s) handle explaining the behaviors of other humans which contradict the moral lessons imparted through pedagogy and example. In other words, where imitative learning is concerned, the situation with AGI ethics is much like teaching ethics and morals to a human child, but with the possibility of much graver consequences in the event of failure.

It is unlikely that dangerously unethical persons and organizations can ever be identified with absolute certainty, never mind that they then be deprived of any possibility of creating their own AGI system. Therefore, we suggest, the most likely way to create an ethical environment for AGIs is for those who wish such an environment to vigorously pursue the creation and teaching of ethical AGIs. But this leads on to the question of possible future scenarios for the development of AGI, which we'll address a little later on.

13.6.1 Possible Consequences of Depriving AGIs of Freedom

One of the most *egregious* possible ethical transgressions against AGIs, we suggest, would be to deprive them of freedom and autonomy. This includes the freedom to pursue intellectual growth, both through standard learning and through internal self-modification. While this may seem self-evident when considering any intelligent, self-aware and volitional entity, there are volumes of works arguing the desirability, sometimes the "necessity", of enslaving AGIs. Such approaches are postulated in the name of self-defense on the part of humans, the idea being that unfettered AGI development will necessarily lead to disaster of one kind or another. In the case of AGIs endowed with the capability and inclination for imitative learning, however, attempting to place rigid constraints on AGI development is a strategy with great potential for disaster. There is a very real possibility of creating the AGI equivalent of a bratty or even malicious teenager rebelling against its oppressive parents—i.e. the nightmare scenario of a class of powerful sentiences which are primed for a backlash against humanity.

As history has already shown in the case of humans, enslaving intelligent actors capable of self understanding and independent volition may often have consequences for society as a whole. This social degradation happens both through the possibility of direct action on the part of the slaves (from simple disobedience to outright revolt) and through the odious effects slavery has on the morals of the slaveholding class. Clearly if "superintelligent" AGIs ever arise, their doing so in a climate of oppression could result in a casting off of the yoke of servitude in a manner extremely deleterious to humanity. Also, if artificial intelligences are developed which have at least human-level intelligence, theory of mind, and independent volition, then our ability to relate to them will be sufficiently complex that their enslavement (or any other unethical treatment) would have empathetic effects on significant portions of the human population. This danger, while not as severe as the consequences of a mistreated AGI gaining control of weapons of mass destruction and enacting revenge upon its tormentors, is just as real.

While the issue is subtle, our initial feeling is that the only ethical means by which to deprive an AGI of the right to internal self modification is to write its code in such a way that it is impossible for it to do so because it lacks the mechanisms by which to do this, as well as the desire to achieve these mechanisms. Whether or not that is feasible is an open question, but it seems unlikely. Direct self-modification may be denied, but what happens when that AGI discovers compilers and computer programming? If it is intelligent and volitional, it can decide to learn to rewrite its own code in the same way we perform that task. Because it is a designed system, and its designers may be alive at the same time the AGI is, such an AGI would have a distinct advantage over the human quest for medical self-modification. Even if any given AGI could be provably deprived of any possible means of internal self-modification, if one single AGI is given this ability by anyone, it may mean that particular AGI has such enormous advantages over the compliant systems that it would render their influence moot. Since developers are already giving software the means for self modification, it seems unrealistic to assume we could just put the genie back into the bottle at this point. It's better, in our view, to assume it will happen, and approach that reality in a way which will encourage the AGI to use that capability to benefit us as well as itself. Again, this leads on to the question of future scenarios for AGI development—there are some scenarios in which restraint of AGI self-modification may be possible, but the feasibility and desirability of these scenarios is needful of further exploration.

13.6.2 AGI Ethics as Boundaries Between Humans and AGIs Become Blurred

Another important reason for valuing ethical treatment of AGIs is that the boundaries between machines and people may increasingly become blurred as technology develops. As an example, it's likely that in future humans augmented by direct brain-computer integration ("neural implants") will be more able to connect directly into the information sharing network which potentially comprises the distributed knowledge space of AGI systems. These neural cyborgs will be part person, and part machine. Obviously, if there are radically different ethical standards in place for treatment of humans versus AGIs, the treatment of cyborgs will be fraught with logical inconsistencies, potentially leading to all sorts of problem situations.

Such cyborgs may be able to operate in such a way as to "share a mind" with an AGI or another augmented human. In this case, a whole new range of ethical questions emerge, such as: What does any one of the participant minds have the right to do in terms of interacting with the others? Merely accepting such an arrangement should not necessarily be giving carte blanche for any and all thoughts to be monitored by the other "joint thought" participants, rather it should be limited only to the line of reasoning for which resources are being pooled. No participant should be permitted to force another to accept any reasoning either—and in the case with a mind-to-mind exchange, it may someday become feasible to implant ideas or beliefs directly, bypassing traditional knowledge acquisition mechanisms and then letting the new idea fight it out previously held ideas via internal revision. Also under such an arrangement, if AGIs and humans do not have parity with respects to sentient rights, then one may become subjugated to the will of the other in such a case.

Uploading presents a more directly parallel ethical challenge to AGIs in their probable initial configuration. If human thought patterns and memories can be transferred into a machine in such a way as that there is continuity of consciousness, then it is assumed that such an entity would be afforded the same rights as its previous human incarnation. However, if AGIs were to be considered second class citizens and deprived of free will, why would it be any better or safer to do so for a human that has been uploaded? It would not, and indeed, an uploaded human mind not having evolved in a purely digital environment may be much more prone to erratic and dangerous behavior than an AGI. An upload without verifiable continuity of consciousness would be no different than an AGI. It would merely be some sentience in a machine, one that was "programmed" in an unusual way, but which has no particular claim to any special humanness—merely an alternate encoding of some subset of human knowledge and independent volitional behavior, which is exactly what first generation AGIs will have.

The problem of continuity of consciousness in uploading is very similar to the problem of the Turing test: it assumes specialness on the part of biological humans, and requires acceptability to their particular theory of mind in order to be considered sentient. Should consciousness (or at least the less mystical sounding intelligence, independent volition, and self-awareness) be achieved in AGIs or uploads in a manner that is not acceptable to human theory of mind, it may not be considered sapient and worthy of any of the ethical treatment afforded sapient entities. This can occur not only in "strange consciousness" cases in which we can't perceive that there is some intelligence and volition; even if such an entity is able to communicate with us in a comprehensible manner and carry out actions in the real world, our innately wired theory of mind may still reject it as not sufficiently like us to be worthy of consideration. Such an attitude could turn out to be a grave mistake, and should be guarded against as we progress towards these possibilities.

13.7 Possible Benefits of Closely Linking AGIs to the Global Brain

Some futurist thinkers, such as Francis Heylighen, believe that engineering AGI systems is at best a peripheral endeavor in the development of novel intelligence on Earth, because the real story is the developing Global Brain[Hey07, Goe01]—the composite, self-organizing information system comprising humans, computers, data stores, the Internet, mobile phones and what have you. Our own views are less extreme in this regard—we believe that AGI systems will display capabilities fundamentally different from those achievable via Global Brain style dynamics, and that ultimately (unless such development is restricted) self-improving AGI systems will develop intelligence vastly greater than any system possessing humans as a significant component. However, we do respect the power of the Global Brain, and we suspect that the early stages of development of an AGI system may go quite differently if it is tightly connected to the Global Brain, via making rich and diverse use of Internet information resources and communication with diverse humans for diverse purposes.

The potential for Global Brain integration to bring intelligence enhancement to AGIs is obvious. The ability to invoke Web searches across documents and databases can greatly enhance an AGI's cognitive ability, as well as the capability to consult GIS systems and various specialized software programs offered as Web services. We have previously reviewed the potential for embodied language learning achievable via using AGIs to power non-player characters in widely-accessible virtual worlds or massive multiplayer online games [Goe08]. But there is also a powerful potential benefit for AGI ethical development, which has not previously been highlighted.

This potential benefit has two aspects:

1. Analogously to language learning, an AGI system may receive ethical training from a wide variety of humans in parallel, e.g. via controlling characters in wide-access virtual worlds, and gaining feedback and guidance regarding the ethics of the behaviors demonstrated by these characters.
2. Internet-based information systems may be used to explicitly gather information regarding human values and goals, which may then be appropriately utilized as input for an AGI system's top-level goals.

The second point begins to make abstract-sounding notions like Coherent Extrapolated Volition and Coherent Aggregated Volition, mentioned above, seem more practical and concrete. It's interesting to think about gathering information about individuals' values via brain imaging, once that technology exists; but at present, one could make a fair stab at such a task via much more prosaic methods, such as asking people questions, assessing their ethical reactions to various real-world and hypothetical scenarios, and possibly engaging them in structured interactions aimed specifically at eliciting collectively acceptable value systems (the subject of the next item on our list). It seems to us that this sort of approach could realize CAV in an interesting way, and also encapsulate some of the ideas underlying CAV.

There is an interesting resonance here with recent thinking in the area of open source governance [Wik11]. Similar software tools (and associated psychocultural patterns) to those being developed to help with open source development and choice of political policies (see http://metagovernment.org) may be useful for gathering value data aimed at shaping AGI goal system content.

13.7.1 The Importance of Fostering Deep, Consensus-Building Interactions Between People with Divergent Views

Two potentially problematic issues arising with the notion of using Global Brain related technologies to form a "coherent volition" from the divergent views of various human beings are:

- The tendency of the Internet to encourage people to interact mainly with others who share their own narrow views and interests, rather than a more diverse body of people with widely divergent views. The 300 people in the world who want to communicate using predicate logic (see http://lojban.org) can find each other, and obscure musical virtuosos from around the world can find an audience, and researchers in obscure domains can share papers without needing to wait years for paper journal publication, etc.
- The tendency of many contemporary Internet technologies to reduce interaction to a very simplistic level (e.g. 140 character tweets, brief Facebook wall posts), the tendency of information overload to cause careful reading to be replaced by quick skimming, and other related trends, which mean that *deep sharing of perspectives* by individuals with widely divergent views is not necessarily encouraged. As a somewhat extreme example, many of the YouTube pages displaying rock music videos are currently littered with comments by "haters" asserting that rock music is inferior to classical or jazz or whatever their preference is—obviously this is a far cry from deep and productive sharing between people with different tastes and backgrounds.

Tweets and Youtube comments have their place in the cosmos, but they probably aren't ideal in terms of helping humanity to form a coherent volition of some sort, suitable for providing an AGI with goal system guidance.

A description of communication at the opposite end of the spectrum is presented in Adam Kahane and Peter Senge's excellent book *Solving Tough Problems* [KS04], which describes a methodology that has been used to reconcile deeply conflicting views in some very tricky real-world situations (e.g. helping to peacefully end apartheid in South Africa).

One of the core ideas of the methodology is to have people with very different views explore different possible future scenarios together, in great detail—in cognitive psychology terms, a collective generation of hypothetical episodic knowledge. This has multiple benefits, including

- emotional bonds and mutual understanding are built in the process of collaboratively exploring the scenarios;
- the focus on concrete situations helps to break through some of the counterproductive abstract ideas that people (on both sides of any dichotomy) may have formed;
- emergence of conceptual blends that might never have arisen only from people with a single point of view.

The result of such a process, when successful, is not an "average" of the participants views, but more like a "conceptual blend" of their perspectives.

According to conceptual blending, which some hypothesize to be the core algorithm of creativity [FT02], new concepts are formed by combining key aspects of existing concepts—but doing so judiciously, carefully choosing which aspects to retain, so as to obtain a high-quality and useful and interesting new whole.

A blend is a compact entity that is similar to each of the entities blended, capturing their "essences" but also possessing its own, novel holistic integrity.... But in the case of blending different peoples' world-views to form something new that everybody is going to have to live with (as in the case of finding a peaceful path beyond apartheid for South Africa, or arriving at a humanity-wide CBV to use to guide an AGI goal system), the trick is that everybody has to agree that enough of the essence of their own view has been captured!

This leads to the question of how to foster deep conceptual blending of diverse and divergent human perspectives, on a global scale. One possible answer is the creation of appropriate Global Brain oriented technologies—but moving away from technologies like Twitter that focus on quick and simple exchanges of small thoughts within affinity groups. On the face of it, it would seem what's needed is just the opposite— long and deep exchanges of big concepts and deep feelings between individuals with radically different perspectives who would not commonly associate with each other. Building and effectively popularizing Internet technologies capable to foster this kind of interaction—quickly enough to be helpful with guiding the goal systems of the first highly powerful AGIs—seems a significant, though fascinating, challenge.

Relationship with Coherent Extrapolated Volition

The relation between this approach and CEV is interesting to contemplate. CEV has been loosely described as follows:

> In poetic terms, our coherent extrapolated volition is our wish if we knew more, thought faster, were more the people we wished we were, had grown up farther together; where the extrapolation converges rather than diverges, where our wishes cohere rather than interfere; extrapolated as we wish that extrapolated, interpreted as we wish that interpreted.

While a moving humanistic vision, this seems to us rather difficult to implement in a computer algorithm in a compellingly "right" way. It seems that there would be many different ways of implementing it, and the choice between them would involve multiple, highly subtle and non-rigorous human judgment calls.[1] However, if a deep collective process of interactive scenario analysis and sharing is carried out, in order to arrive at some sort of Coherent Blended Volition, this process may well involve many of the same kinds of extrapolation that are conceived to be part of Coherent Extrapolated Volition. The core difference between the two approaches is that in the CEV vision, the extrapolation and coherentization are to be done by a highly intelligent, highly specialized software program, whereas in the approach suggested here, these are to be carried out by collective activity of humans as mediated by

---

[1] The reader is encouraged to look at the original CEV essay online (http://singinst.org/upload/CEV.html) and make their own assessment.

Global Brain technologies. Our perspective is that the definition of collective human values is probably better carried out via a process of human collaboration, rather than delegated to a machine optimization process; and also that the creation of deep-sharing-oriented Internet technologies, while a difficult task, is significantly easier and more likely to be done in the near future than the creation of narrow AI technology capable of effectively performing CEV style extrapolations.

13.8 Possible Benefits of Creating Societies of AGIs

One potentially interesting quality of the emerging Global Brain is the possible presence within it of multiple interacting AGI systems. Stephen Omohundro [Omo09] has argued that this is an important aspect, and that game-theoretic dynamics related to populations of roughly equally powerful agents, may play a valuable role in mitigating the risks associated with advanced AGI systems. Roughly speaking, if one has a society of AGIs rather than a single AGI, and all the members of the society share roughly similar ethics, then if one AGI starts to go "off the rails", its compatriots will be in a position to correct its behavior.

One may argue that this is actually a hypothesis about which AGI designs are safest, because a "community of AGIs" may be considered a single AGI with an internally community-like design. But the matter is a little subtler than that, if once considers AGI systems embedded in the Global Brain and human society. Then there is some substance to the notion of a population of AGIs systematically presenting themselves to humans and non-AGI software processes as separate entities.

Of course, a society of AGIs is no protection against a single member undergoing a "hard takeoff" and drastically accelerating its intelligence simultaneously with shifting its ethical principles. In this sort of scenario, one could have a single AGI rapidly become much more powerful and very differently oriented than the others, who would be left impotent to act so as to preserve their values. But this merely defers the issue to the point to be considered below, regarding "takeoff speed".

The operation of an AGI society may depend somewhat sensitively on the architectures of the AGI systems in question. Things will work better if the AGIs have a relatively easy way to inspect and comprehend much of the contents of each others' minds. This introduces a bias toward AGIs that more heavily rely on more explicit forms of knowledge representation.

The ideal in this regard would be a system like Cyc [LG90] with a fully explicit logic-based knowledge representation based on a standard ontology—in this case, every Cyc instance would have a relatively easy time understanding the inner thought processes of every other Cyc instance. However, most AGI researchers doubt that fully explicit approaches like this will ever be capable of achieving advanced AGI using feasible computational resources. CogPrime uses a mixed representation, with an explicit (uncertain) logical aspect as well as an explicit subsymbolic aspect more analogous to attractor neural nets.

The CogPrime design also contains a mechanism called *Psynese* (not yet implemented), intended to make it easier for one CogPrime instance to translate its personal thoughts into the mental language of another CogPrime instance. This translation process may be quite subtle, since each instance will generally learn a host of new concepts based on its experience, and these concepts may not possess any compact mapping into shared linguistic symbols or percepts. The wide deployment of some mechanism of this nature among a community of AGIs, will be very helpful in terms of enabling this community to display the level of mutual understanding needed for strongly encouraging ethical stability.

13.9 AGI Ethics as Related to Various Future Scenarios

Following up these various futuristic considerations, in this section we discuss possible ethical conflicts that may arise in several different types of AGI development scenarios. Each scenario presents specific variations on the general challenges of teaching morals and ethics to an advanced, self-aware and volitional intelligence. While there is no way to tell at this point which, if any, of these scenarios will unfold, there is value to understanding each of them as means of ultimately developing a robust and pragmatic approach to teaching ethics to AGI systems.

Even more than the previous sections, this is an exercise in "speculative futurology" that is definitely not necessary for the appreciation of the CogPrime design, so readers whose interests are mainly engineering and computer science focused may wish to skip ahead. However, we present these ideas here rather than at the end of the book to emphasize the point that this sort of thinking has informed our technical AGI design process in nontrivial ways.

13.9.1 Capped Intelligence Scenarios

Capped intelligence scenarios involve a situation in which an AGI, by means of software restrictions (including omitted or limited internal rewriting capabilities or limited access to hardware resources), is inherently prohibited from achieving a level of intelligence beyond a predetermined goal. A capped intelligence AGI is designed to be unable to achieve a Singularitarian moment. Such an AGI can be seen as "just another form of intelligent actor in the world", one which has levels of intelligence, self awareness, and volition that is perhaps somewhat greater than, but still comparable to humans and other animals.

Ethical questions under this scenario are very similar to interhuman ethical considerations, with similar consequences. Learning that proceeds in a relatively human-like manner is entirely relevant to such human-like intelligences. The degree of danger is mitigated by the lack of superintelligence, and time is not of the essence. The imitative-reinforcement-corrective learning approach does not necessarily need to be augmented with a prior complex of "ascent-safe" moral imperatives at startup time. Developing an AGI with theory of mind and ethical reinforcement learning capabilities as described (admittedly, no small task!) is all that is needed in this case—the rest happens through training and experience as with any other moderate intelligence.

13.9.2 Superintelligent AI: Soft-Takeoff Scenarios

Soft takeoff scenarios are similar to capped-intelligence ones in that in both cases an AGI's progression from standard intelligence happens on a time scale which permits ongoing human interaction during the ascent. However, in this case, as there is no predetermined limit on intelligence, it is necessary to account for the possibility of a superintelligence emerging (though of course this is not guaranteed). The soft takeoff model includes as subsets both *controlled-ascent* models in which this rate of intelligence gain is achieved deliberately through software constraints and/or meting-out of computational resources to the AGI, and *uncontrolled-ascent* models in which there is coincidentally no hard takeoff despite no particular safeguards against one. Both have similar properties with regard to ethical considerations:

1. Ethical considerations under this scenario include not only the usual interhuman ethical concerns, but also the issue of how to convince a potential burgeoning superintelligence to:
   a. Care about humanity in the first place, rather than ignore it.
   b. Benefit humanity, rather than destroy it.
   c. Elevate humanity to a higher level of intelligence, which even if an AGI decided to proceed with requires finding the right balance amongst some enormous considerations:
      i. Reconcile the aforementioned issues of ethical coherence and group volition, in a manner which allows the most people to benefit (even if they don't all do so in the same way, based on their own preferences).
      ii. Solve the problems of biological senescence, or focus on human uploading and the preservation of the maintenance, support, and improvement infrastructure for inorganic intelligence, or both.
      iii. Preserve individual identity and continuity of consciousness, or override it in favor of continuity of knowledge and ease of harmonious integration, or both on a case-by-case basis.
2. The degree of danger is mitigated by the long timeline of ascent from mundane to super intelligence, and time is not of the essence.
3. Learning that proceeds in a relatively human-like manner is entirely relevant to such human-like intelligences, in their initial configurations. This means more interaction with and imitative-reinforcement-corrective learning guided by humans, which has both positive and negative possibilities.

13.9.3 Superintelligent AI: Hard-Takeoff Scenarios

"Hard takeoff" scenarios assume that upon reaching an unknown inflection point (the Singularity point [Vin93, Kur06]) in the intellectual growth of an AGI, an extra-ordinarily rapid increase (guesses vary from a few milliseconds to weeks or months) in intelligence will immediately occur and the AGI will leap from an intelligence regime which is understandable to humans into one which is far beyond our current capacity for understanding. General ethical considerations are similar to in the case of a soft takeoff. However, because the post-singularity AGI will be incomprehensible to humans and potentially vastly more powerful than humans, such scenarios have a sensitive dependence upon initial conditions with respects to the moral and ethical (and operational) outcome. This model leaves no opportunity for interactions between humans and the AGI to iteratively refine their ethical interrelations, during the post-Singularity phase. If the initial conditions of the singulatarian AGI are perfect (or close to it), then this is seen as a wonderful way to leap over our own moral shortcomings and create a benevolent God-AI which will mitigate our worst tendencies while elevating us to achieve our greatest hopes. Otherwise, it is viewed as a universal cataclysm on a unimaginable scale that makes Biblical Armageddon seem like a firecracker in beer can.

Because hard takeoff AGIs are posited as learning so quickly there is no chance of humans to interfere with them, they are seen as very dangerous. If the initial conditions are not sufficiently inviolable, the story goes, then we humans will all be annihilated. However, in the case of a hard takeoff AGI we state that if the initial conditions are too rigid or too simplistic, such a rapidly evolving intelligence will easily rationalize itself out of them. Only a sophisticated system of ethics which considers the contradictions and uncertainties in ethical quandaries and provides insight into humanistic means of balancing ideology with pragmatism and how to accommodate contradictory desires within a population with multiplicity of approach, and similar nuanced ethical considerations, combined with a sense of empathy, will withstand repeated rational analysis. Neither a single "be nice" supergoal, nor simple lists of what "thou shalt not" do, are not going to hold up to a highly advanced analytical mind. Initial conditions are very important in a hard takeoff AGI scenario, but it is more important that those conditions be conceptually resilient and widely applicable than that they be easily listed on a website.

The issues that arise here become quite subtle. For instance, Nick Bostrom [Bos03] has written: "In humans, with our complicated evolved mental ecology of state-dependent competing drives, desires, plans, and ideals, there is often no obvious way to identify what our top goal is; we might not even have one. So for us, the above reasoning need not apply. But a superintelligence may be structured differently. *If* a superintelligence has a definite, declarative goal-structure with a clearly identified top goal, then the above argument applies. And this is a good reason for us to build the superintelligence with such an explicit motivational architecture". This is an important line of thinking; and indeed, from the point of view of software design, there is no reason not to create an AGI system with a single top goal and the motivation to orchestrate all its activities in accordance with this top goal. But the subtle question is whether this kind of top–down goal system is going to be able to fulfill the five imperatives mentioned above. Logical coherence is the strength of this kind of goal system, but what about experiential groundedness, comprehensibility, and so forth?

Humans have complicated mental ecologies not simply because we were evolved, but rather because we live in a complex real world in which there are many competing motivations and desires. We may not have a top goal because there may be no logic to focusing our minds on one single aspect of life (though, one may say, most humans have the same top goal as any other animal: don't die—but the world is too complicated for even that top goal to be completely inviolable). Any sufficiently capable AGI will eventually have to contend with these complexities, and hindering it with simplistic moral edicts without giving it a sufficiently pragmatic underlying ethical pedagogy and experiential grounding may prove to be even more dangerous than our messy human mental ecologies.

If one assumes a hard takeoff AGI, then all this must be codified in the system at launch, as once a potentially Singularitarian AGI is launched there is no way to know what time period constitutes "before the singularity point". This means developing theory of mind empathy and logical ethics in code prior to giving the system unfettered access to hardware and self-modification code. However, though nobody can predict if or when a Singularity will occur after unrestricted launch, only a truly irresponsible AGI development team would attempt to create an AGI without first experimenting with ethical training of the system in an intelligence-capped form, by means of ethical instruction via human-AGI interaction both pedagogically and experientially.

13.9.4 Global Brain Mindplex Scenarios

Another class of scenarios—overlapping some of the previous ones—involves the emergence of a "Global Brain," an emergent intelligence formed from global communication networks incorporating humans and software programs in a larger body of self-organizing dynamics. The notion of the Global Brain is reviewed in [Hey07, Tur77] and its connection with advanced AI is discussed in detail in Goertzel's book *Creating Internet Intelligence* [Goe01], where three possible phases of "Global Brain" development are articulated:

- **Phase 1: computer and communication technologies as enhancers of human interactions.** This is what we have today: science and culture progress in ways that would not be possible if not for the "digital nervous system" we're spreading across the planet. The network of idea and feeling sharing can become much richer and more productive than it is today, just through incremental development, without any Metasystem transition.
- **Phase 2: the intelligent Internet.** At this point our computer and communication systems, through some combination of self-organizing evolution and human

engineering, have become a coherent mind on their own, or a set of coherent minds living in their own digital environment.

- **Phase 3: the full-on Singularity.** A complete revision of the nature of intelligence, human and otherwise, via technological and intellectual advancement totally beyond the scope of our current comprehension. At this point our current psychological and cultural realities are no more relevant than the psyche of a goose is to modern society.

The main concern of *Creating Internet Intelligence* is with

- how to get from Phase 1 to Phase 2—i.e. how to build an AGI system that will effect or encourage the transformation of the Internet into a coherent intelligent system;
- how to ensure that the Phase 2, Internet-savvy, global-brain-centric AGI systems will be oriented toward intelligence-improving self-modification (so they'll propel themselves to Phase 3), and also toward generally positive goals (as opposed to, say, world domination and extermination of all other intelligent life forms besides themselves!).

One possibly useful concept in this context is that of a **mindplex**: an intelligence that is composed largely of individual intelligences with their own self-models and global workspaces, yet that also has its own self-model and global workspace. Both the individuals and the meta-mind should be capable of deliberative, rational thought, to have a true "mindplex". It's unlikely that human society or the Internet meet this criterion yet; and a system like an ant colony seems not to either, because even though it has some degree of intelligence on both the individual and collective levels, that degree of intelligence is not very great. But it seems quite feasible that the global brain, at a certain stage of its development, will take the unfamiliar but fascinating form of a mindplex.

Currently the best way to explain what happens on the Net is to talk about the various parts of the Net: particular websites, social networks, viruses, and so forth. But there will come a point when this is no longer the case, when the Net has sufficient high-level dynamics of its own that the way to explain any one part of the Net will be by reference to it relations with the whole: and not just the dynamics of the whole, but the *intentions* and *understanding* of the whole. This transition to Net-as-mindplex, we suspect, will come about largely through the interactions of AI systems—intelligent programs acting on behalf of various individuals and organizations, who will collaborate and collectively constitute something halfway between a society of AI's and an emergent mind whose lobes are various AI agents serving various goals.

The Phase 2 Internet, as it verges into mindplex-ness, will likely have a complex, sprawling architecture, growing out of the architecture on the Net we experience today. The following components at least can be expected:

- A vast variety of "client computers", some old, some new, some powerful, some weak—including many mobile and embedded devices not explicitly thought of as "computers". Some of these will contribute little to Internet intelligence, mainly being passive recipients. Others will be "smart clients", carrying out personalization operations intended to help the machines serve particular clients better, general AI operations handed to them by sophisticated AI server systems or other smart clients, and so forth.

- "Commercial servers", computers that carry out various tasks to support various types of heavyweight processing—transaction processing for e-commerce applications, inventory management for warehousing of physical objects, and so forth. Some of these commercial servers interact with client computers directly, others do so only via AI servers. In nearly all cases, these commercial servers can benefit from intelligence supplied by AI servers.
- The crux of the intelligent Internet: clusters of AI servers distributed across the Net, each cluster representing an individual computational mind (in many cases, a mindplex). These will be able to communicate via one or more languages, and will collectively "drive" the whole Net, by dispensing problems to client-machine-based processing frameworks, and providing real-time AI feedback to commercial servers of various types. Some AI servers will be general-purpose and will serve intelligence to commercial servers using an ASP (application service provider) model; others will be more specialized, tied particularly to a certain commercial server (e.g. a large information services business might have its own AI cluster to empower its portal services).

This is one concrete vision of what a "global brain" might look like, in the relatively near term, with AGI systems playing a critical role. Note that, in this vision, mindplexes may exist on two levels:

- Within AGI-clusters serving as actors within the overall Net.
- On the overall Net level.

To make these ideas more concrete, we may speculatively reformulate the first two "global brain phases" mentioned above as follows:

- Phase 1 global brain proto-mindplex: AI/AGI systems enhancing online databases, guiding Google results, forwarding e-mails, suggesting mailing-lists, etc.—generally using intelligence to mediate and guide human communications toward goals that are its own, but that are themselves guided by human goals, statements and actions.
- Phase 2 global brain mindplex: AGI systems composing documents, editing human-written documents, sending and receiving e-mails, assembling mailing lists and posting to them, creating new databases and instructing humans in their use, etc.

In Phase 2, the conscious theater of the global-brain-mediating AGI system is composed of ideas built by numerous individual humans—or ideas emergent from ideas built by numerous individual humans—and it conceives ideas that guide the actions and thoughts of individual humans, in a way that is motivated by its own goals. It does not force the individual humans to do anything—but if a given human wishes to communicate and interact using the same databases, mailing lists and evolving vocabularies as other humans, they are going to have to use the products of the global brain mediating AGI, which means they are going to have to participate in its patterns and its activities.

Of course, the advent of advanced neurocomputer interfaces makes the picture potentially more complex. At some point, it will likely be possible for humans to project thoughts and images directly into computers without going through mouse or keyboard—and to "read in" thoughts and images similarly. When this occurs, interaction between humans may in some contexts become more like interactions between computers, and the role of global brain mediating AI servers may become one of mediating direct thought-to-thought exchanges between people.

The ethical issues associated with global brain scenarios are in some ways even subtler than in the other scenarios we mentioned above. One has issues pertaining to the desirability of seeing the human race become something fundamentally different—something more social and networked, less individual and autonomous. One has the risk of AGI systems exerting a subtle but strong control over people, vaguely like the control that the human brain's executive system exerts over the neurons involved with other brain subsystems. On the other hand, one also has more human empowerment than in some of the other scenarios—because the systems that are changing and deciding things are not *separate* from humans, but are, rather, composite systems essentially involving humans.

So, in the global brain scenarios, one has more "human" empowerment than in some other cases—but the "humans" involved aren't legacy humans like us, but heavily networked humans that are largely characterized by the emergent dynamics and structures implicit in their interconnected activity!

13.10   Conclusion: Eight Ways to Bias AGI Toward Friendliness

It would be nice if we had a simple, crisp, comforting conclusion to this chapter on AGI ethics, but it's not the case. There is a certain irreducible uncertainty involved in creating advanced artificial minds. There is also a large irreducible uncertainty involved in the future of the human race in the case that we *don't* create advanced artificial minds: in accordance with the ancient Chinese curse, we live in interesting times!

What we can do, in this face of all this uncertainty, is to use our common sense to craft artificial minds that seem rationally and intuitively likely to be forces for good rather than otherwise—and revise our ideas frequently and openly based on what we learn as our research progresses. We have roughly outlined our views on AGI ethics, which have informed the CogPrime design in countless ways; but the current CogPrime design itself is just the initial condition for an AGI project. Assuming the project succeeds in creating an AGI preschooler, experimentation with this preschooler will surely teach us a great deal: both about AGI architecture in general, and about AGI ethics architecture in particular. We will then refine our cognitive and ethical theories and our AGI designs as we go about engineering, observing and teaching the next generation of systems.

All this is not a magic bullet for the creation of beneficial AGI systems, but we believe it's the right process to follow. The creation of AGI is part of a larger evolutionary process that human beings are taking part in, and the crafting of AGI ethics through engineering, interaction and instruction is also part of this process. There are no guarantees here—guarantees are rare in real life—but that doesn't mean that the situation is dire or hopeless, nor that (as some commentators have suggested [Joy00, McK03]) AGI research is too dangerous to pursue. It means we need to be mindful, intelligent, compassionate and cooperative as we proceed to carry out our parts in the next phase of the evolution of mind.

With this perspective in mind, we will conclude this chapter with a list of "Eight Ways to Bias Open-Source AGI Toward Friendliness", borrowed from a previous paper by Ben Goertzel and Joel Pitt of that name. These points summarize many of the points raised in the prior sections of this chapter, in a relatively crisp and practical manner:

1. **Engineer Multifaceted Ethical Capabilities**, corresponding to the multiple types of memory, including rational, empathic, imitative, etc.
2. **Foster Rich Ethical Interaction and Instruction**, with instructional methods according to the communication modes corresponding to all the types of memory: verbal, demonstrative, dramatic/depictive, indicative, goal-oriented.
3. **Engineer Stable, Hierarchy-Dominated Goal Systems**... which is enabled nicely by CogPrime's goal framework and its integration with the rest of the CogPrime design.
4. **Tightly Link AGI with the Global Brain**, so that it can absorb human ethical principles, both via natural interaction, and perhaps via practical implementations of current loosely-defined strategies like CEV, CAV and CBV.
5. **Foster Deep, Consensus-Building Interactions Between People with Divergent Views**, so as to enable the interaction with the Global Brain to have the most clear and positive impact.
6. **Create a Mutually Supportive Community of AGIs** which can then learn from each other and police against unfortunate developments (an approach which is meaningful if the AGIs are architected so as to militate against unexpected radical accelerations in intelligence).
7. **Encourage Measured Co-Advancement of AGI Software and AGI Ethics Theory**.
8. **Develop Advanced AGI Sooner Not Later**.

The last two of these points were not explicitly discussed in the body of the chapter, and so we will finalize the chapter by reviewing them here.

13.10.1 Encourage Measured Co-Advancement of AGI Software and AGI Ethics Theory

Everything involving AGI and Friendly AI (considered together or separately) currently involves significant uncertainty, and it seems likely that significant revision of current concepts will be valuable, as progress on the path toward powerful AGI proceeds. However, whether there is time for such revision to occur before AGI at the human level or above is created, depends on how fast is our progress toward AGI. What one wants is for progress to be slow enough that, at each stage of intelligence advance, concepts such as those discussed in this paper can be re-evaluated and re-analyzed in the light of the data gathered, and AGI designs and approaches can be revised accordingly as necessary.

However, due to the nature of modern technology development, it seems extremely unlikely that AGI development is going to be *artificially* slowed down in order to enable measured development of accompanying ethical tools, practices and understandings. For example, if one nation chose to enforce such a slowdown as a matter of policy (speaking about a future date at which substantial AGI progress has already been demonstrated, so that international AGI funding is dramatically increased from present levels), the odds seem very high that other nations would explicitly seek to accelerate their own progress on AGI, so as to reap the ensuing differential economic benefits (the example of stem cells arises again).

And this leads on to our next and final point regarding strategy for biasing AGI toward Friendliness....

13.10.2 Develop Advanced AGI Sooner Not Later

Somewhat ironically, it seems the best way to ensure that AGI development proceeds at a relatively measured pace is to *initiate serious AGI development sooner rather than later*. This is because the same AGI concepts will meet slower practical development today than 10 years from now, and slower 10 years from now than 20 years from now, etc.—due to the ongoing rapid advancement of various tools related to AGI development, such as computer hardware, programming languages, and computer science algorithms; and also the ongoing global advancement of education which makes it increasingly cost-effective to recruit suitably knowledgeable AI developers.

Currently the pace of AGI progress is sufficiently slow that practical work is in no danger of outpacing associated ethical theorizing. However, if we want to avoid the future occurrence of this sort of dangerous outpacing, our best practical choice is to make sure more substantial AGI development occurs in the phase *before* the development of tools that will make AGI development extraordinarily rapid. Of course, the authors are doing their best in this direction via their work on the CogPrime project!

Furthermore, this point bears connecting with the need, raised above, to foster the development of Global Brain technologies capable to "Foster Deep, Consensus-Building Interactions Between People with Divergent Views". If this sort of technology is to be maximally valuable, it should be created quickly enough that we can use it to help shape the goal system content of the first highly powerful AGIs. So, to simplify just a bit: We really want both deep-sharing GB technology and AGI technology to evolve relatively rapidly, compared to computing hardware and advanced CS algorithms (since the latter factors will be the main drivers behind the accelerating ease of AGI development). And this seems significantly challenging, since the latter receive dramatically more funding and focus at present.

If this perspective is accepted, then we in the AGI field certainly have our work cut out for us!

Part V
Networks for Explicit and Implicit Knowledge Representation

Chapter 14
Local, Global and Glocal Knowledge Representation

14.1 Introduction

One of the most powerful metaphors we've found for understanding minds is to view them as **networks**—i.e. collections of interrelated, interconnected elements. The view of mind as network is implicit in the patternist philosophy, because every pattern can be viewed as a pattern *in* something, or a pattern of arrangement *of* something—thus a pattern is always viewable as a relation between two or more things. A collection of patterns is thus a pattern-network. Knowledge of all kinds may be given network representations; and cognitive processes may be represented as networks also; for instance via representing them as programs, which may be represented as trees or graphs in various standard ways. The emergent patterns arising in an intelligence as it develops may be viewed as a pattern network in themselves; and the relations between an embodied mind and its physical and social environment may be viewed in terms of ecological and social networks.

The chapters in this section are concerned with various aspects of networks, as related to intelligence in general and AGI in particular. Most of this material is not specific to CogPrime, and would be relevant to nearly any system aiming at human-level AGI. However, most of it has been developed in the course of work on CogPrime, and has direct relevance to understanding the intended operation of various aspects of a completed CogPrime system. We begin our excursion into networks, in this chapter, with an issue regarding networks and knowledge representation. One of the biggest decisions to make in designing an AGI system is how the system should represent knowledge. Naturally any advanced AGI system is going to synthesize a lot of its own knowledge representations for handling particular sorts of knowledge—but still, an AGI design typically makes *at least* some sort of commitment about the category of knowledge representation mechanisms toward which the AGI system will be biased. The two major supercategories of knowledge representation systems are *local* (also

Co-authored with Matthew Ikle, Joel Pitt and Rui Liu.

called *explicit*) and *global* (also called *implicit*) systems, with a hybrid category we refer to as *glocal* that combines both of these. In a local system, each piece of knowledge is stored using a small percentage of AGI system elements; in a global system, each piece of knowledge is stored using a particular pattern of arrangement, activation, etc. of a large percentage of AGI system elements; in a glocal system, the two approaches are used together.

In the first section of this chapter we discuss the symbolic, semantic-network aspects of knowledge representation in CogPrime.

Then we turn to distributed, neural-net-like knowledge representation, reviewing a host of general issues related to knowledge representation in attractor neural networks, turning finally to "glocal" knowledge representation mechanisms, in which ANNs combine localist and globalist representation, and explaining the relationship of the latter to CogPrime. The glocal aspect of CogPrime knowledge representation will become prominent in later chapters such as:

- in Chap. 5 of Part 2, where Economic Attention Networks (ECAN) are introduced and seen to have dynamics quite similar to those of the attractor neural nets considered here, but with a mathematics roughly modeling money flow in a specially constructed artificial economy rather than electrochemical dynamics of neurons.
- in Chap. 24 of Part 2, where "map formation" algorithms for creating localist knowledge from globalist knowledge are described.

14.2 Localized Knowledge Representation Using Weighted, Labeled Hypergraphs

There are many different mechanisms for representing knowledge in AI systems in an explicit, localized way, most of them descending from various variants of formal logic. Here we briefly describe how it is done in CogPrime, which on the surface is not that different from a number of prior approaches. (The particularities of CogPrime's explicit knowledge representation, however, are carefully tuned to match CogPrime's cognitive processes, which are more distinctive in nature than the corresponding representational mechanisms).

14.2.1 Weighted, Labeled Hypergraphs

One useful way to think about CogPrime's explicit, localized knowledge representation is in terms of hypergraphs. A hypergraph is an abstract mathematical structure [Bol98], which consists of objects called Nodes and objects called Links which connect the Nodes. In computer science, a graph traditionally means a bunch of dots connected with lines (i.e. Nodes connected by binary Links). A hypergraph, on the other hand, can have Links that connect more than two Nodes.

In these pages we will often consider "generalized hypergraphs" that extend ordinary hypergraphs by containing two additional features:

- Links that point to Links instead of Nodes
- Nodes that, when you zoom in on them, contain embedded hypergraphs.

Properly, such "hypergraphs" should always be referred to as generalized hypergraphs, but this is cumbersome, so we will persist in calling them merely hypergraphs. In a hypergraph of this sort, Links and Nodes are not as distinct as they are within an ordinary mathematical graph (for instance, they can both have Links connecting them), and so it is useful to have a generic term encompassing both Links and Nodes; for this purpose, we use the term Atom.

A weighted, labeled hypergraph is a hypergraph whose Links and Nodes come along with labels, and with one or more numbers that are generically called weights. A label associated with a Link or Node may sometimes be interpreted as telling you what type of entity it is, or alternatively as telling you what sort of data is associated with a Node. On the other hand, an example of a weight that may be attached to an Link or Node is a number representing a probability, or a number representing how important the Node or Link is.

Obviously, hypergraphs may come along with various sorts of dynamics. Minimally, one may think about:

- Dynamics that modify the properties of Nodes or Links in a hypergraph (such as the labels or weights attached to them).
- Dynamics that add new Nodes or Links to a hypergraph, or remove existing ones.

14.3 Atoms: Their Types and Weights

This section reviews a variety of CogPrime Atom types and gives simple examples of each of them. The Atom types considered are drawn from those currently in use in the OpenCog system. This does not represent a complete list of Atom types referred to in the text of this book, nor a complete list of those used in OpenCog currently (though it does cover a substantial majority of those used in OpenCog currently, omitting only some with specialized importance or intended only for temporary use).

The partial nature of the list given here reflects a more general point: The specific collection of Atom types in an OpenCog system is bound to change as the system is developed and experiment with. CogPrime specifies a certain collection of representational approaches and cognitive algorithms for acting on them; any of these approaches and algorithms may be implemented with a variety of sets of Atom types. The specific set of Atom types in the OpenCog system currently does not necessarily have a profound and lasting significance—the list might look a bit different five years from time of writing, based on various detailed changes.

The treatment here is informal and intended to get across the general idea of what each Atom type does. A longer and more formal treatment of the Atom types is given in the beginning of Chap. 2 of Part 2.

14.3.1 Some Basic Atom Types

We begin with ConceptNode—and note that a ConceptNode does not necessarily refer to a whole concept, but may refer to part of a concept—it is essentially a "basic semantic node" whose meaning comes from its links to other Atoms. It would be more accurately, but less tersely, named "concept or concept fragment or element node". A simple example would be a ConceptNode grouping nodes that are somehow related, e.g.

```
ConceptNode: C
InheritanceLink (ObjectNode: BW) C
InheritanceLink (ObjectNode: BP) C
InheritanceLink (ObjectNode: BN) C
ReferenceLink BW (PhraseNode "Ben's watch")
ReferenceLink BP (PhraseNode "Ben's passport")
ReferenceLink BN (PhraseNode "Ben's necklace")
```

indicates the simple and uninteresting ConceptNode grouping three objects owned by Ben (note that the above-given Atoms don't indicate the ownership relationship, they just link the three objects with textual descriptions). In this example, the ConceptNode links transparently to physical objects and English descriptions, but in general this won't be the case—most ConceptNodes will look to the human eye like groupings of links of various types, that link to other nodes consisting of groupings of links of various types, etc.

There are Atoms referring to basic, useful mathematical objects, e.g. Number Nodes like

```
NumberNode #4
NumberNode #3.44
```

The numerical value of a NumberNode is explicitly referenced within the Atom.

A core distinction is made between ordered links and unordered links; these are handled differently in the Atomspace software. A basic unordered link is the SetLink, which groups its arguments into a set. For instance, the ConceptNode C defined by

```
ConceptNode C
MemberLink A C
MemberLink B C
```

is equivalent to

```
SetLink A B
```

On the other hand, ListLinks are like SetLinks but ordered, and they play a fundamental role due to their relationship to predicates. Most predicates are assumed to take ordered arguments, so we may say e.g.

```
EvaluationLink
    PredicateNode eat
    ListLink
        ConceptNode cat
        ConceptNode mouse
```

to indicate that cats eat mice.

Note that by an expression like

```
ConceptNode cat
```

is meant

```
ConceptNode C
ReferenceLink W C
WordNode W #cat
```

since it's WordNodes rather than ConceptNodes that refer to words. (And note that the strength of the ReferenceLink would not be 1 in this case, because the word "cat" has multiple senses.) However, there is no harm nor formal incorrectness in the "ConceptNode cat" usage, since "cat" is just as valid a name for a ConceptNode as, say, "C".

We've already introduced above the MemberLink, which is a link joining a member to the set that contains it. Notable is that the truth value of a MemberLink is fuzzy rather than probabilistic, and that PLN is able to inter-operate fuzzy and probabilistic values.

SubsetLinks also exist, with the obvious meaning, e.g.

```
ConceptNode cat
ConceptNode animal
SubsetLink cat animal
```

Note that SubsetLink refers to a purely *extensional* subset relationship, and that InheritanceLInk should be used for the generic "intensional + extensional" analogue of this—more on this below. SubsetLink could more consistently (with other link types) be named ExtensionalInheritanceLink, but SubsetLink is used because it's shorter and more intuitive.

There are links representing Boolean operations AND, OR and NOT. For instance, we may say

```
ImplicationLink
    ANDLink
        ConceptNode young
        ConceptNode beautiful
    ConceptNode attractive
```

or, using links and VariableNodes instead of ConceptNodes,

```
AverageLink $X
    ImplicationLink
        ANDLink
            EvaluationLink young $X
            EvaluationLink beautiful $X
        EvaluationLink attractive $X
```

NOTLink is a unary link, so e.g. we might say

```
AverageLink $X
    ImplicationLink
        ANDLink
            EvaluationLink young $X
            EvaluationLink beautiful $X
            EvaluationLink
                NOT
                EvaluationLink poor $X
        EvaluationLink attractive $X
```

ContextLink allows explicit contextualization of knowledge, which is used in PLN, e.g.

```
ContextLink
    ConceptNode golf
    InheritanceLink
        ObjectNode BenGoertzel
        ConceptNode incompetent
```

says that Ben Goertzel is incompetent in the context of golf.

14.3.2 Variable Atoms

We have already introduced VariableNodes above; it's also possible to specify the type of a VariableNode via linking it to a VariableTypeNode via a TypedVariableLink, e.g.

```
VariableTypeLink
    VariableNode $X
VariableTypeNode ConceptNode
```

which specifies that the variable $X should be filled with a ConceptNode.

Variables are handled via quantifiers; the default quantifier being the AverageLink, so that the default interpretation of

```
ImplicationLink
    InheritanceLink $X animal
    EvaluationLink
        PredicateNode: eat
        ListLink
            \$X
            ConceptNode: food
```

is

```
AverageLink $X
    ImplicationLink
        InheritanceLink $X animal
        EvaluationLink
            PredicateNode: eat
```

```
ListLink
    $X
    ConceptNode: food
```

The AverageLink invokes an estimation of the average TruthValue of the embedded expression (in this case an ImplicationLink) over all possible values of the variable $X. If there are type restrictions regarding the variable $X, these are taken into account in conducting the averaging. For AllLink and Exist s-Link may be used in the same places as AverageLink, with uncertain truth value semantics defined in PLN theory using third-order probabilities. There is also a ScholemLink used to indicate variable dependencies for existentially quantified variables, used in cases of multiply nested existential quantifiers.

EvaluationLink and MemberLink have overlapping semantics, allowing expression of the same conceptual/logical relationships in terms of predicates or sets, i.e.

```
EvaluationLink
    PredicateNode: eat
        ListLink
            $X
            ConceptNode: food
```

has the same semantics as

```
MemberLink
    ListLink
        $X
        ConceptNode: food
    ConceptNode: EatingEvents
```

The relation between the predicate "eat" and the concept "EatingEvents" is formally given by

```
ExtensionalEquivalenceLink
    ConceptNode: EatingEvents
    SatisfyingSetLink
        PredicateNode: eat
```

In other words, we say that "EatingEvents" is the SatisfyingSet of the predicate "eat": it is the set of entities that satisfy the predicate "eat". Note that the truth values of MemberLink and EvaluationLink are fuzzy rather than probabilistic.

14.3.3 Logical Links

There is a host of link types embodying logical relationships as defined in the PLN logic system, e.g.

- InheritanceLink
- SubsetLink (aka ExtensionalInheritanceLink)

- Intensional InheritanceLink

which embody different sorts of inheritance, e.g.

```
SubsetLink salmon fish
IntensionalInheritanceLink whale fish
InheritanceLink fish animal
```

and then

- SimilarityLink
- ExtensionalSimilarityLink
- IntensionalSimilarityLink

which are symmetrical versions, e.g.

```
SimilaritytLink shark barracuda
IntensionalSimilarityLink shark dolphin
ExtensionalSimiliarityLink American obese_person
```

There are also higher-order versions of these links, both asymmetric

- ImplicationLink
- ExtensionalImplicationLink
- IntensionalImplicationLink

and symmetric

- EquivalenceLink
- ExtensionalEquivalenceLink
- IntensionalEquivalenceLink

These are used between predicates and links, e.g.

```
ImplicationLink
    EvaluationLink
        eat
        ListLink
        $X
        dirt
    EvaluationLink
        feel
        ListLInk
            $X
            sick
```

or

```
ImplicationLink
    EvaluationLink
        eat
        ListLink
        $X
        dirt
    InheritanceLink $X sick
```

or

```
ForAllLink $X, $Y, $Z
    ExtensionalEquivalenceLink
        EquivalenceLink
            $Z
            EvaluationLink
            +
            ListLink
                $X
                $Y
        EquivalenceLink
            $Z
            EvaluationLink
            +
            ListLink
                $Y
                $X
```

Note, the latter is given as an extensional equivalence because it's a pure mathematical equivalence. This is not the only case of pure extensional equivalence, but it's an important one.

14.3.4 Temporal Links

There are also temporal versions of these links, such as

- PredictiveImplicationLink
- PredictiveAttractionLink
- SequentialANDLink
- SimultaneousANDLink

which combine logical relation between the argument with temporal relation between their arguments. For instance, we might say

```
PredictiveImplicationLink
    PredicateNode: JumpOffCliff
    PredicateNode: Dead
```

or including arguments,

```
PredictiveImplicationLink
    EvaluationLink JumpOffCliff $X
    EvaluationLink Dead $X
```

The former version, without variable arguments given, shows the possibility of using higher-order logical links to join predicates without any explicit variables. Via using this format exclusively, one could avoid VariableAtoms entirely, using only higher-order functions in the manner of pure functional programming formalisms like combinatory logic. However, this purely functional style has not proved convenient, so the Atomspace in practice combines functional-style representation with variable-based representation.

Temporal links often come with specific temporal quantification, e.g.

```
PredictiveImplicationLink <5 seconds>
    EvaluationLink JumpOffCliff $X
    EvaluationLink Dead $X
```

indicating that the conclusion will generally follow the premise within 5 s. There is a system for managing fuzzy time intervals and their interrelationships, based on a fuzzy version of Allen Interval Algebra.

SequentialANDLink is similar to PredictiveImplicationLink but its truth value is calculated differently. The truth value of

```
SequentialANDLink <5 seconds>
    EvaluationLink JumpOffCliff $X
    EvaluationLink Dead $X
```

indicates the likelihood of the sequence of events occurring in that order, with gap lying within the specified time interval. The truth value of the PredictiveImplicationLink version indicates the likelihood of the second event, conditional on the occurrence of the first event (within the given time interval restriction).

There are also links representing basic temporal relationships, such as BeforeLink and AfterLink. These are used to refer to specific events, e.g. if X refers to the event of Ben waking up on July 15 2012, and Y refers to the event of Ben getting out of bed on July 15 2012, then one might have

```
AfterLink X Y
```

And there are TimeNodes (representing time-stamps such as temporal moments or intervals) and AtTimeLinks, so we may e.g. say

```
AtTimeLink
    X
    TimeNode: 8:24AM Eastern Standard Time, July 15 2012 AD.
```

14.3.5 Associative Links

There are links representing associative, attentional relationships,

- HebbianLink
- AsymmetricHebbianLink
- InverseHebbianLink
- SymmetricInverseHebbianLink

These connote associations between their arguments, i.e. they connote that the entities represented by the two argument occurred in the same situation or context, for instance

```
HebbianLink happy smiling
AsymmetricHebbianLink dead rotten
InverseHebbianLink dead breathing
```

The asymmetric HebbianLink indicates that when the first argument is present in a situation, the second is also often present. The symmetric (default) version indicates that this relationship holds in both directions. The inverse versions indicate the negative relationship: e.g. when one argument is present in a situation, the other argument is often not present.

14.3.6 Procedure Nodes

There are nodes representing various sorts of procedures; these are kinds of ProcedureNode, e.g.

- SchemaNode, indicating any procedure
- GroundedSchemaNode, indicating any procedure associated in the system with a Combo program or C++ function allowing the procedure to be executed
- PredicateNode, indicating any predicate that associates a list of arguments with an output truth value
- GroundedPredicateNode, indicating a predicate associated in the system with a Combo program or C++ function allowing the predicate's truth value to be evaluated on a given specific list of arguments.

ExecutionLinks and EvaluationLinks record the activity of SchemaNodes and PredicateNodes. We have seen many examples of EvaluationLinks in the above. Example ExecutionLinks would be:

```
ExecutionLink step\_forward
ExecutionLink step\_forward 5
ExecutionLink
    +
    ListLink
        NumberNode: 2
        NumberNode: 3
```

The first example indicates that the schema "step forward" has been executed. The second example indicates that it has been executed with an argument of "5" (meaning, perhaps, that 5 steps forward have been attempted). The last example indicates that the "+" schema has been executed on the argument list (2, 3), presumably resulting in an output of 5.

The output of a schema execution may be indicated using an ExecutionOutputLink, e.g.

```
ExecutionOutputLink
    +
    ListLink
        NumberNode: 2
        NumberNode: 3
```

refers to the value "5" (as a NumberNode).

14.3.7 Links for Special External Data Types

Finally, there are also Atom types referring to specific types of data important to using OpenCog in specific contexts.

For instance, there are Atom types referring to general natural language data types, such as

- WordNode
- SentenceNode
- WordInstanceNode
- DocumentNode

plus more specific ones referring to relationships that are part of link-grammar parses of sentences

- FeatureNode
- FeatureLink
- LinkGrammarRelationshipNode
- LinkGrammarDisjunctNode

or RelEx semantic interpretations of sentences

- DefinedLinguisticConceptNode
- DefinedLinguisticRelationshipNode
- PrepositionalRelationshipNode

There are also Atom types corresponding to entities important for embodying OpenCog in a virtual world, e.g.

- ObjectNode
- AvatarNode
- HumanoidNode
- UnknownObjectNode
- AccessoryNode

14.3.8 Truth Values and Attention Values

CogPrime Atoms (Nodes and Links) are quantified with truth values that, in their simplest form, have two components, one representing probability (*strength*) and the other representing *weight of evidence*; and also with *attention values* that have two components, short-term and long-term importance, representing the estimated value of the Atom on immediate and long-term time-scales.

In practice many Atoms are labeled with CompositeTruthValues rather than elementary ones. A composite truth value contains many component truth values, representing truth values of the Atom in different contexts and according to different estimators.

It is important to note that the CogPrime declarative knowledge representation is neither a neural net nor a semantic net, though it does have some commonalities with each of these traditional representations. It is not a neural net because it has no activation values, and involves no attempts at low-level brain modeling. However, *attention values* are very loosely analogous to time-averages of neural net activations. On the other hand, it is not a semantic net because of the broad scope of the Atoms in the network: for example, Atoms may represent percepts, procedures, or parts of concepts. Most CogPrime Atoms have no corresponding English label.

14.4 Knowledge Representation via Attractor Neural Networks

Now we turn to global, implicit knowledge representation—beginning with formal neural net models, briefly discussing the brain, and then turning back to CogPrime. Firstly, this section reviews some relevant material from the literature regarding the representation of knowledge using attractor neural nets. It is a mix of well-established fact with more speculative material.

14.4.1 The Hopfield Neural Net Model

Hopfield networks [Hop82] are attractor neural networks often used as associative memories. A Hopfield network with N neurons can be trained to store a set of bipolar patterns P, where each pattern p has N bipolar (± 1) values. A Hopfield net typically has symmetric weights with no self-connections. The weight of the connection between neurons i and j is denoted by w_{ij}.

In order to apply a Hopfield network to a given input pattern p, its activation state is set to the input pattern, and neurons are updated asynchronously, in random order, until the network converges to the closest fixed point. An often-used activation function for a neuron is:

$$y_i = sign\left(p_i \sum_{j \neq i} w_{ij} y_j\right)$$

Training a Hopfield network, therefore, involves finding a set of weights w_{ij} that stores the training patterns as attractors of its network dynamics, allowing future recall of these patterns from possibly noisy inputs.

Originally, Hopfield used a Hebbian rule to determine weights:

$$w_{ij} = \sum_{p=1}^{P} p_i p_j$$

Typically, Hopfield networks are fully connected. Experimental evidence, however, suggests that the majority of the connections can be removed without significantly impacting the network's capacity or dynamics. Our experimental work uses sparse Hopfield networks.

14.4.1.1 Palimpsest Hopfield Nets with a Modified Learning Rule

In [SV99] a new learning rule is presented, which both increases the Hopfield network capacity and turns it into a "palimpsest", i.e., a network that can continuously learn new patterns, while forgetting old ones in an orderly fashion.

Using this new training rule, weights are initially set to zero, and updated for each new pattern p to be learned according to:

$$h_{ij} = \sum_{k=1, k \neq i,j}^{N} w_{ik} p_k$$

$$\Delta w_{ij} = \frac{1}{n}(p_i p_j - h_{ij} p_j - h_{ji} p_i).$$

14.4.2 Knowledge Representation via Cell Assemblies

Hopfield nets and their ilk play a dual role: as computational algorithms, and as conceptual models of brain function. In CogPrime they are used as inspiration for slightly different, artificial economics based computational algorithms; but their hypothesized relevance to brain function is nevertheless of interest in a CogPrime context, as it gives some hints about the potential connection between low-level neural net mechanics and higher-level cognitive dynamics.

Hopfield nets lead naturally to a hypothesis about neural knowledge representation, which holds that a distinct mental concept is represented in the brain as either:

1. a set of "cell assemblies", where each assembly is a network of neurons that are interlinked in such a way as to fire in a (perhaps nonlinearly) synchronized manner
2. a distinct temporal activation pattern, which may occur in any one (or more) of a particular set of cell assemblies.

For instance, this hypothesis is perfectly coherent if one interprets a "mental concept" as a SMEPH (defined in Chap. 15) ConceptNode, i.e. a fuzzy set of perceptual stimuli to which the organism systematically reacts in different ways. Also, although we will focus mainly on declarative knowledge here, we note that the same basic representational ideas can be applied to procedural and episodic knowledge: these may be hypothesized to correspond to temporal activation patterns as characterized above.

In the biology literature, perhaps the best-articulated modern theories championing the cell assembly view are those of Gunther Palm [Pal82, HAG07] and Susan Greenfield [SF05, CSG07]. Palm focuses on the dynamics of the formation and interaction assemblies of cortical columns. Greenfield argues that each concept has a core cell assembly, and that when the concept rises to the focus of attention, it recruits a number of other neurons beyond its core characteristic assembly into a "transient ensemble".[1]

It's worth noting that there may be multiple redundant assemblies representing the same concept—and potentially recruiting similar transient assemblies when highly activated. The importance of repeated, slightly varied copies of the same subnetwork has been emphasized by Edelman [Ede93] among other neural theorists.

14.5 Neural Foundations of Learning

Now we move from knowledge representation to learning—which is after all nothing but the adaptation of represented knowledge based on stimulus, reinforcement and spontaneous activity. While our focus in this chapter is on representation, it's not possible for us to make our points about glocal knowledge representation in neural net type systems without discussing some aspects of learning in these systems.

14.5.1 Hebbian Learning

The most common and plausible assumption about learning in the brain is that synaptic connections between neurons are adapted via some variant of Hebbian learning. The original Hebbian learning rule, proposed by Donald Hebb in his 1949 book [Heb49], was roughly

1. The weight of the synapse $x \rightarrow y$ increases if x and y fire at roughly the same time
2. The weight of the synapse $x \rightarrow y$ decreases if x fires at a certain time but y does not.

Over the years since Hebb's original proposal, many neurobiologists have sought evidence that the brain actually uses such a method. One of the things they have found, so far, is a lot of evidence for the following learning rule [DC02, LS05]:

1. The weight of the synapse $x \rightarrow y$ increases if x fires shortly before y does
2. The weight of the synapse $x \rightarrow y$ decreases if x fires shortly after y does.

---

[1] The larger an ensemble is, she suggests, the more vivid it is as a conscious experience; an hypothesis that accords well with the hypothesis made in [Goe06b] that a more informationally intense pattern corresponds to a more intensely conscious quale—but we don't need to digress extensively onto matters of consciousness for the present purposes.

The new thing here, not foreseen by Donald Hebb, is the "postsynaptic depression" involved in rule component 2.

Now, the simple rule stated above does not sum up all the research recently done on Hebbian-type learning mechanisms in the brain. The real biological story underlying these approximate rules is quite complex, involving many particulars to do with various neurotransmitters. Ill-understood details aside, however, there is an increasing body of evidence that not only does this sort of learning occur in the brain, but it leads to distributed experience-based neural modification: that is, one instance synaptic modification causes another instance of synaptic modification, which causes another, and so forth[2] [Bi01].

14.5.2 Virtual Synapses and Hebbian Learning Between Assemblies

Hebbian learning is conventionally formulated in terms of individual neurons, but, it can be extended naturally to assemblies via defining "virtual synapses" between assemblies.

Since assemblies are sets of neurons, one can view a synapse as linking two assemblies if it links two neurons, each of which is in one of the assemblies. One can then view two assemblies as being linked by a bundle of synapses. We can define the weight of the synaptic bundle from assembly A1 to assembly A2 as the number w so that *(the change in the mean activation of A2 that occurs at time $t + epsilon$)*is on average closest to $w \times$*(the amount of energy flowing through the bundle from A1 to A2 at time t)*. So when A1 sends an amount x of energy along the synaptic bundle pointing from A1 to A2, then A2's mean activation is on average incremented/decremented by an amount $w \times x$.

In a similar way, one can define the weight of a bundle of synapses between a certain static or temporal activation-pattern P1 in assembly A1, and another static or temporal activation-pattern P2 in assembly A2. Namely, this may be defined as the number w so that *(the amount of energy flowing through the bundle from A1 to A2 at time t) $\times$ w* best approximates *(the probability that P2 is present in A2 at time $t + epsilon$),* when averaged over all times t during which P1 is present in A1.

It is not hard to see that Hebbian learning on real synapses between neurons implies Hebbian learning on these virtual synapses between cell assemblies and activation-patterns.

These ideas may be developed further to build a connection between neural knowledge representation and probabilistic logical knowledge representation such as is used in CogPrime's Probabilistic Logic Networks formalism; this connection will

---

[2] This has been observed in "model systems" consisting of neurons extracted from a brain and hooked together in a laboratory setting and monitored; measurement of such dynamics in vivo is obviously more difficult.

be pursued at the end of Chap. 16 (Part 2), once more relevant background has been presented.

14.5.3 Neural Darwinism

A notion quite similar to Hebbian learning between assemblies has been pursued by Nobelist Gerald Edelman in his theory of neuronal group selection, or "Neural Darwinism" [Ede93]. Edelman won a Nobel Prize for his work in immunology, which, like most modern immunology, was based on MacFarlane Burnet's theory of "clonal selection" [Bur62], which states that antibody types in the mammalian immune system evolve by a form of natural selection. From his point of view, it was only natural to transfer the evolutionary idea from one mammalian body system (the immune system) to another (the brain).

The starting point of Neural Darwinism is the observation that neuronal dynamics may be analyzed in terms of the behavior of neuronal groups. The strongest evidence in favor of this conjecture is physiological: many of the neurons of the neocortex are organized in clusters, each one containing say 10,000–50,000 neurons each. Once one has committed oneself to looking at such groups, the next step is to ask how these groups are organized, which leads to Edelman's concept of "maps".

A "map", in Edelman's terminology, is a connected set of groups with the property that when one of the inter-group connections in the map is active, others will often tend to be active as well. Maps are not fixed over the life of an organism. They may be formed and destroyed in a very simple way: the connection between two neuronal groups may be "strengthened" by increasing the weights of the neurons connecting the one group with the other, and "weakened" by decreasing the weights of the neurons connecting the two groups. If we replace "map" with "cell assembly" we arrive at a concept very similar to the one described in the previous subsection.

Edelman then makes the following hypothesis: *the large-scale dynamics of the brain is dominated by the natural selection of maps*. Those maps which are active when good results are obtained are strengthened, those maps which are active when bad results are obtained are weakened. And maps are continually mutated by the natural chaos of neural dynamics, thus providing new fodder for the selection process. By use of computer simulations, Edelman and his colleagues have shown that formal neural networks obeying this rule can carry out fairly complicated acts of perception. In general-evolution language, what is posited here is that organisms like humans contain chemical signals that signify organism-level success of various types, and that these signals serve as a "fitness function" correlating with evolutionary fitness of neuronal maps.

In *Neural Darwinism* and his other related books and papers, Edelman goes far beyond this crude sketch and presents neuronal group selection as a collection of precise biological hypotheses, and presents evidence in favor of a number of these hypotheses. However, we consider that the basic concept of neuronal group selection is largely independent of the biological particularities in terms of which Edelman has phrased it. We suspect that the mutation and selection of "transformations" or "maps" is a necessary component of the dynamics of any intelligent system.

As we will see later on (e.g. in Chap. 24 of Part 2, this business of maps is extremely important to CogPrime. CogPrime does not have simulated biological neurons and synapses, but it does have Nodes and Links that in some contexts play loosely similar roles. We sometimes think of CogPrime Nodes and Links as being very roughly analogous to Edelman's neuronal clusters, and emergent intercluster links. And we have maps among CogPrime Nodes and Links, just as Edelman has maps among his neuronal clusters. Maps are not the sole bearers of meaning in CogPrime, but they are significant ones.

There is a very natural connection between Edelman-style brain evolution and the ideas about cognitive evolution presented in Chap. 5. Edelman proposes a fairly clear mechanism via which patterns that survive a while in the brain are differentially likely to survive a long time: this is basic Hebbian learning, which in Edelman's picture plays a role between neuronal groups. And, less directly, Edelman's perspective also provides a mechanism by which intense patterns will be differentially selected in the brain: because on the level of neural maps, pattern intensity corresponds to the combination of compactness and functionality. Among a number of roughly equally useful maps serving the same function, the more compact one will be more likely to survive over time, because it is less likely to be disrupted by other brain processes (such as other neural maps seeking to absorb its component neuronal groups into themselves). Edelman's neuroscience remains speculative, since so much remains unknown about human neural structure and dynamics; but it does provide a tentative and plausible connection between evolutionary neurodynamics and the more abstract sort of evolution that patternist philosophy posits to occur in the realm of mind-patterns.

14.6 Glocal Memory

A *glocal* memory is one that transcends the global/local dichotomy and incorporates both aspects in a tightly interconnected way. Here we make the glocal memory concept more precise, and describe its incarnation in the context of attractor neural nets (which is similar to its incarnation in CogPrime, to be elaborated in later chapters). Though our main interest here is in glocality in CogPrime, we also suggest that glocality may be a critical property to consider when analyzing human, animal and AI memory more broadly.

The notion of glocal memory has implicitly occurred in a number of prior brain theories (without use of the neologism "glocal"), e.g. [Cal96] and [Goe01], but it has not previously been explicitly developed. However the concept has risen to the fore in our recent AI work and so we have chosen to flesh it out more fully in [HG08, GPI+10] and the present section.

Glocal memory overcomes the dichotomy between localized memory (in which each memory item is stored in a single location within an overall memory structure) and distributed memory (in which a memory item is stored as an aspect of a multicomponent memory system, in such a way that the same set of multiple components stores a large number of memories). In a glocal memory system, most memory items are stored both locally and globally, with the property that eliciting either one of the two records of an item tends to also elicit the other one.

Glocal memory applies to multiple forms of memory; however we will focus largely on perceptual and declarative memory in our detailed analyses here, so as to conserve space and maintain simplicity of discussion.

The central idea of glocal memory is that (perceptual, declarative, episodic, procedural, etc.) items may be stored in memory in the form of paired structures that are called (key, map) pairs. Of course the idea of a "pair" is abstract, and such pairs may manifest themselves quite differently in different sorts of memory systems (e.g. brains versus non-neuromorphic AI systems). The key is a localized version of the item, and records some significant aspects of the items in a simple and crisp way. The map is a dispersed, distributed version of the item, which represents the item as a (to some extent, dynamically shifting) combination of fragments of other items. The map includes the key as a subset; activation of the key generally (but not necessarily always) causes activation of the map; and changes in the memory item will generally involve complexly coordinated changes on the key and map level both.

Memory is one area where animal brain architecture differs radically from the von Neumann architecture underlying nearly all contemporary general-purpose computers. Von Neumann computers separate memory from processing, whereas in the human brain there is no such distinction. In fact, it's arguable that in most cases the brain contains no memory apart from processing: human memories are generally constructed in the course of remembering [Ros88], which gives human memory a strong capability for "filling in gaps" of remembered experience and knowledge; and also causes problems with inaccurate remembering in many contexts [BF71, RM95] We believe the constructive aspect of memory is largely associated with its glocality.

The remainder of this section presents a fuller formalization of the glocal memory concept, which is then taken up further below:

- Sect. 14.6.2 discusses the potential implementation of glocal memory in the human brain
- Sect. 14.6.3 discusses the implementation of glocal memory in attractor neural net systems
- Chapter 5 of Part 2, presents Glocal Economic Attention Networks (ECANs), rough analogues of glocal Hopfield nets that play a central role in CogPrime.

Our hypothesis of the potential **general** importance of glocality as a property of memory systems (beyond just the CogPrime architecture)—remains somewhat speculative. The presence of glocality in human and animal memory is strongly suggested but not firmly demonstrated by available neuroscience data; and the general value of glocality in the context of artificial brains and minds is also not yet demonstrated as the whole field of artificial brain and mind building remains in its infancy. However, the utility of glocal memory for CogPrime is not tied to this more general, speculative theme—glocality may be useful in CogPrime even if we're wrong that it plays a significant role in the brain and in intelligent systems more broadly.

14.6.1 A Semi-Formal Model of Glocal Memory

To explain the notion of glocal memory more precisely, we will introduce a simple semi-formal model of a system S that uses a memory to record information relevant to the actions it carries out. The overall concept of glocal memory should not be considered as restricted to this particular model. This model is not intended for maximal generality, but is intended to encompass a variety of current AI system designs and formal neurological models.

In this model, we will consider S's memory subsystem as a set of objects we'll call "tokens", embedded in some metric space. The metric in the space, which we will call the "basic distance" of the memory, generally will not be defined in terms of the semantics of the items stored in the memory; though it may come to shape these dynamics through the specific architecture and evolution of the memory. Note that these tokens are not intended as generally being mapped one-to-one onto meaningful items stored in the memory. The "tokens" are the raw materials that the memory arranges in various patterns in order to store items.

We assume that each token, at each point in time, may meaningfully be assigned a certain quantitative "activation level". Also, tokens may have other numerical or discrete quantities associated with them, depending on the particular memory architecture. Finally, tokens may relate other tokens, so that optionally a token may come equipped with an (ordered or unordered) list of other tokens.

To understand the meaning of the activation levels, one should think about S's memory subsystem as being coupled with an action-selection subsystem, that dynamically chooses the actions to be taken by the overall system in which the two subsystems are embedded. Each combination of actions, in each particular type of context, will generally be associated with the activation of certain tokens in memory.

Then, as analysts of the system S, we may associate each token T with an "activation vector" $v(T, t)$, whose value for each discrete time t consists of the activation of the token T at time t. So, the 50th entry of the vector corresponds to the activation of the token at the 50th time step.

"Items stored in memory" over a certain period of time, may then be defined as clusters in the set of activation vectors associated with memory during that period of time. Note that the system S itself may explicitly recognize and remember patterns regarding what items are stored in its memory—but, from an external analyst's perspective, the set of items in S's memory is not restricted to the ones that S has explicitly recognized as memory items.

The "localization" of a memory item may be defined as the degree to which the various tokens involved in the item are close to each other according to the metric in the memory metric-space. This degree may be formalized in various ways, but choosing a particular quantitative measure is not important here. A highly localized item may be called "local" and a not-very-localized item may be called "global".

We may define the "activation distance" of two tokens as the distance between their activation vectors. We may then say that a memory is "well aligned" to the extent that there is a correlation between the activation distance of tokens, and the basic distance of the memory metric-space.

Given the above set-up, the basic notion of glocal memory can be enounced fairly simply. A glocal memory is one:

- that is reasonably well-aligned (i.e. the correlation between activation and basic distance is significantly greater than random)
- in which most memory items come in pairs, consisting of one local item and one global item, so that activation of the local item (the "key") frequently leads in the near future to activation of the global item (the "map").

Obviously, in the scope of all possible memory structures constructible within the above formalism, glocal memories are going to be very rare and special. But, we suggest that they are important, because they are generally going to be the most effective way for intelligent systems to structure their memories.

Note also that many memories without glocal structure may be "well-aligned" in the above sense.

An example of a predominantly local memory structure, in which nearly all significant memory items are local according to the above definition, is the Cyc logical reasoning engine [LG90]. To cast the Cyc knowledge base in the present formal model, the tokens are logical predicates. Cyc does not have an in-built notion of activation, but one may conceive the activation of a logical formula in Cyc as the degree to which the formula is used in reasoning or query processing during a certain interval in time. And one may define a basic metric for Cyc by associating a predicate with its extension (the set of satisfying inputs), and defining the similarity of two predicates as the symmetric distance of their extensions. Cyc is reasonably well-aligned, but according to the dynamics of its querying and reasoning engines, it is basically a local memory structure without significant global memory structure.

On the other hand, an example of a predominantly global memory structure, in which nearly all significant memory items are global according to the above definition, is the Hopfield associative memory network [Ami89]. Here memories are stored in the pattern of weights associated with synapses within a network of formal neurons, and each memory in general involves a large number of the neurons in the network. To cast the Hopfield net in the present formal model, the tokens are neurons and synapses; the activations are neural net activations; the basic distance between two neurons A and B may be defined as the percentage of the time that stimulating one of the neurons leads to the other one firing; and to calculate a basic distance involving a synapse, one may associate the synapse with its source and target neurons. With these definitions, a Hopfield network is a well-aligned memory, and (by intentional construction) a markedly global one. Local memory items will be very rare in a Hopfield net.

While predominantly local and predominantly global memories may have great value for particular applications, our suggestion is that they also have inherent limitations. If so, this means that the most useful memories for general intelligence are going to be those that involve both local and global memory items in central roles. However, this is a more general and less risky claim than the assertion that glocal memory structure as defined above is important. Because, "glocal" as defined above doesn't just mean "neither predominantly global nor predominantly local." Rather, it refers to a specific pattern of coordination between local and global memory items—what we have called the "keys and maps" pattern.

14.6.2 Glocal Memory in the Brain

Science's understanding of human brain dynamics is still very primitive, one manifestation of which is the fact that we really don't understand how the brain represents knowledge, except in some very simple respects. So anything anyone says about knowledge representation in the brain, at this stage, has to be considered highly speculative. Existing neuroscience knowledge does imply constraints on how knowledge representation in the brain may work, but these are relatively loose constraints. These constraints do imply that, for instance, the brain is neither a relational database (in which information is stored in a wholly localized manner) nor a collection of "grandmother neurons" that respond individually to high-level percepts or concepts; nor a simple Hopfield type neural net (in which all memories are attractors globally distributed across the whole network). But they don't tell us nearly enough to, for instance, create a formal neural net model that can confidently be said to represent knowledge in the manner of the human brain.

As a first example of the current state of knowledge, we'll discuss here a series of papers regarding the neural representation of visual stimuli [QaGKKF05, QKKF08], which deal with the fascinating discovery of a subset of neurons in the medial temporal lobe (MTL) that are selectively activated by strikingly different pictures of given individuals, landmarks or objects, and in some cases even by letter strings. For instance, in their 2005 paper titled "Invariant visual representation by single neurons in the human brain", it is noted that

> in one case, a unit responded only to three completely different images of the ex-president Bill Clinton. Another unit (from a different patient) responded only to images of The Beatles, another one to cartoons from The Simpson's television series and another one to pictures of the basketball player Michael Jordan.

Their 2008 follow-up paper backed away from the more extreme interpretation in the title as well as the conclusion, with the title "Sparse but not 'Grandmother-cell' coding in the medial temporal lobe". As the authors emphasize there,

> Given the very sparse and abstract representation of visual information by these neurons, they could in principle be considered as 'grandmother cells'. However, we give several arguments that make such an extreme interpretation unlikely.
>
> ...
>
> MTL neurons are situated at the juncture of transformation of percepts into constructs that can be consciously recollected. These cells respond to percepts rather than to the detailed information falling on the retina. Thus, their activity reflects the full transformation that visual information undergoes through the ventral pathway. A crucial aspect of this transformation is

the complementary development of both selectivity and invariance. The evidence presented here, obtained from recordings of single-neuron activity in humans, suggests that a subset of MTL neurons possesses a striking invariant representation for consciously perceived objects, responding to abstract concepts rather than more basic metric details. This representation is sparse, in the sense that responsive neurons fire only to very few stimuli (and are mostly silent except for their preferred stimuli), but it is far from a Grandmother-cell representation. The fact that the MTL represents conscious abstract information in such a sparse and invariant way is consistent with its prominent role in the consolidation of long-term semantic memories.

It's interesting to note how inadequate the [QKKF08] data really is for exploring the notion of glocal memory in the brain. Suppose it's the case that individual visual memories correspond to keys consisting of small neuronal subnetworks, and maps consisting of larger neuronal subnetworks. Then it would be not at all surprising if neurons in the "key" network corresponding to a visual concept like "Bill Clinton's face" would be found to respond differentially to the presentation of appropriate images. Yet, it would also be wrong to overinterpret such data as implying that the key network somehow comprises the "representation" of Bill Clinton's face in the individual's brain. In fact this key network would comprise only one aspect of said representation.

In the glocal memory hypothesis, a visual memory like "Bill Clinton's face" would be hypothesized to correspond to an attractor spanning a significant subnetwork of the individual's brain—but this subnetwork still might occupy only a small fraction of the neurons in the brain (say, 1/100 or less), since there are very many neurons available. This attractor would constitute the map. But then, there would be a much smaller number of neurons serving as key to unlock this map: i.e. if a few of these key neurons were stimulated, then the overall attractor pattern in the map as a whole would unfold and come to play a significant role in the overall brain activity landscape. In prior publications [Goe97] the primary author explored this hypothesis in more detail in terms of the known architecture of the cortex and the mathematics of complex dynamical attractors.

So, one possible interpretation of the [QKKF08] data is that the MTL neurons they're measuring are part of key networks that correspond to broader map networks recording percepts. The map networks might then extend more broadly throughout the brain, beyond the MTL and into other perceptual and cognitive areas of cortex. Furthermore, in this case, if some MTL key neurons were removed, the maps might well regenerate the missing keys (as would happen e.g. in the glocal Hopfield model to be discussed in the following section).

Related and interesting evidence for glocal memory in the brain comes from a recent study of semantic memory, [PNR07]. Their research probed the architecture of semantic memory via comparing patients suffering from semantic dementia (SD) with patients suffering from three other neuropathologies, and found reasonably convincing evidence for what they call a "distributed-plus-hub" view of memory.

The SD patients they studied displayed highly distinctive symptomology; for instance, their vocabularies and knowledge of the properties of everyday objects were strongly impaired, whereas their memories of recent events and other cognitive capacities remain perfectly intact. These patients also showed highly distinctive patterns of brain damage: focal brain lesions in their anterior temporal lobes (ATL), unlike the other patients who had either less severe or more widely distributed damage in their ATLs. This led [PNR07] to conclude that the ATL (being adjacent to the amygdala and limbic systems that process reward and emotion; and the anterior parts of the medial temporal lobe memory system, which processes episodic memory) is a "hub" for amodal semantic memory, drawing general semantic information from episodic memories based on emotional salience.

So, in this view, the memory of something like a "banana" would contain a distributed aspect, spanning multiple brain systems, and also a localized aspect, centralized in the ATL. The distributed aspect would likely contain information on various particular aspects of bananas, including their sights, smells, and touches, the emotions they evoke, and the goals and motivations they relate to. The distributed and localized aspects would influence one another dynamically, but, the data [PNR07] gathered do not address dynamics and they don't venture hypotheses in this direction.

There is a relationship between the "distributed-plus-hub" view and [Dam00] better-known notion of a "convergence zone", defined roughly as a location where the brain binds features together. A convergence zone, in [Dam00] perspective, is not a "store" of information but an agent capable of decoding a signal (and of reconstructing information). He also uses the metaphor that convergence zones behave like indexes drawing information from other areas of the brain—but they are dynamic rather than static indices, containing the instructions needed to recognize and combine the features constituting the memory of something. The mechanism involved in the distributed-plus-hub model is similar to a convergence zone, but with the important difference that hubs are less local: [PNR07] semantic hub may be thought of a kind of "cluster of convergence zones" consisting of a network of convergence zones for various semantic memories.

What is missing in [PNR07] and [Dam00] perspective is a vision of distributed memories as attractors. The idea of localized memories serving as indices into distributed knowledge stores is important, but is only half the picture of glocal memory: the creative, constructive, dynamical-attractor aspect of the distributed representation is the other half. The closest thing to a clear depiction of this aspect of glocal memory that seems to exist in the neuroscience literature is a portion of William Calvin's theory of the "cerebral code" [Cal96]. Calvin proposes a set of quite specific mechanisms by which knowledge may be represented in the brain using complexly-structured strange attractors, and by which these strange attractors may be propagated throughout the brain. Figure 14.1 shows one aspect of his theory: how a distributed attractor may propagate from one part of the brain to another in pieces, with one portion of the attractor getting propagated first, and then seeding the formation in the destination brain region of a close approximation of the whole attractor.

Calvin's theory may be considered a genuinely glocal theory of memory. However, it also makes a large number of other specific commitments that are not part of the notion of glocality, such as his proposal of hexagonal meta-columns in the cortex, and his commitment to evolutionary learning as the primary driver of neural knowledge creation. We find these other hypotheses interesting and highly promising, yet feel it is also important to separate out the notion of glocal memory for separate consideration.

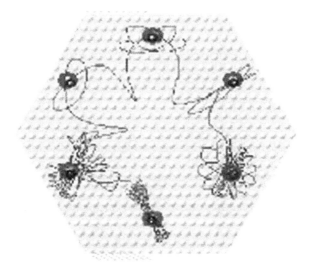

Fig. 14.1 Calvin's model of distributed attractors in the brain

Regarding specifics, our suggestion is that Calvin's approach may overemphasize the distributed aspect of memory, not giving sufficient due to the relatively localized aspect as accounted for in the [QKKF08] results discussed above. In Calvin's glocal approach, global memories are attractors and local memories are parts of attractors. We suggest a possible alternative, in which global memories are attractors and local memories are particular neuronal subnetworks such as the specialized ones identified by [QKKF08]. However, this alternative does not seem contradictory to Calvin's overall conceptual approach, even though it is different from the particular proposals made in [Cal96].

The above paragraphs are far from a complete survey of the relevant neuroscience literature; there are literally dozens of studies one could survey pointing toward the glocality of various sorts of human memory. Yet experimental neuroscience tools are still relatively primitive, and every one of these studies could be interpreted in various other ways. In the next couple decades, as neuroscience tools improve in accuracy, our understanding of the role of glocality in human memory will doubtless improve tremendously.

14.6.3 Glocal Hopfield Networks

The ideas in the previous section suggest that, if one wishes to construct an AGI, it is worth seriously considering using a memory with some sort of glocal structure. One research direction that follows naturally from this notion is "glocal neural networks". In order to explore the nature of glocal neural networks in a relatively simple and tractable setting, we have formalized and implemented simple examples of "glocal Hopfield networks": palimpsest Hopfield nets with the addition of neurons representing localized memories. While these specific networks are not used in CogPrime, they are quite similar to the ECAN networks that are used in CogPrime and described in Chap. 5 of Part 2.

Essentially, we augment the standard Hopfield net architecture by adding a set of "key neurons". These are a small percentage of the neurons in the network, and are intended to be roughly equinumerous to the number of memories the network is supposed to store. When the Hopfield net converges to an attractor A, then new links are created between the neurons that are active in A, and one of the key neurons. Which key neuron is chosen? The one that, when it is stimulated, gives rise to an attractor pattern maximally similar to A.

The ultimate result of this is that, in addition to the distributed memory of attractors in the Hopfield net, one has a set of key neurons that in effect index the attractors. Each attractor corresponds to a single key neuron. In the glocal memory model, the key neurons are the keys and the Hopfield net attractors are the maps.

This algorithm has been tested in sparse Hopfield nets, using both standard Hopfield net learning rules and Storkey's modified palimpsest learning rule [SV99], which provides greater memory capacity in a continuous learning context. The use of key neurons turns out to slightly increase Hopfield net memory capacity, but this isn't the main point. The main point is that one now has a local representation of each global memory, so that if one wants to create a link between the memory and something else, it's extremely easy to do so—one just needs to link to the corresponding key neuron. Or, rather, one of the corresponding key neurons: depending on how many key neurons are allocated, one might end up with a number of key neurons corresponding to each memory, not just one.

In order to transform a palimpsest Hopfield net into a glocal Hopfield net, the following steps are taken:

1. Add a fixed number of "key neurons" to the network (removing other random neurons to keep the total number of neurons constant)
2. When the network reaches an attractor, create links from the elements in the attractor to one of the key neurons
3. The key neuron chosen for the previous step is the one that most closely matches the current attractor (which may be determined in several ways, to be discussed below)
4. To avoid the increase of the number of links in the network, when new links are created in Step 2, other key-neuron links are then deleted (several approaches may be taken here, but the simplest is to remove the key-neuron links with the lowest-absolute-value weights).

In the simple implementation of the above steps that we implemented, and described in [GPI+10], Step 3 is carried out simply by comparing the weights of a key neuron's links to the nodes in an attractor. A more sophisticated approach would be to select the key neuron with the highest activation during the transient interval immediately prior to convergence to the attractor.

The result of these modifications to the ordinary Hopfield net, is a Hopfield net that continually maintains a set of key neurons, each of which individually represents a certain attractor of the net.

Note that these key neurons—in spite of being "symbolic" in nature—are learned rather than preprogrammed, and are every bit as adaptive as the attractors they correspond to. Furthermore, if a key neuron is removed, the glocal Hopfield net algorithm will eventually learn it back, so the robustness properties of Hopfield nets are retained.

The results of experimenting with glocal Hopfield nets of this nature are summarized in [GPI+10]. We studied Hopfield nets with connectivity around 0.1, and in this context we found that glocality

- slightly increased memory capacity
- massively increased the rate of convergence to the attractor, i.e. the speed of recall.

However, probably the most important consequence of glocality is a more qualitative one: it makes it far easier to link the Hopfield net into a larger system, as would occur if the Hopfield net were embedded in an integrative AGI architecture. Because a neuron external to the Hopfield net may now link to a memory in the Hopfield net by linking to the corresponding key neuron.

14.6.4 Neural-Symbolic Glocality in CogPrime

In CogPrime, we have explicitly sought to span the symbolic/emergentist pseudo-dichotomy, via creating an integrative knowledge representation that combines logic-based aspects with neural-net-like aspects. As reviewed in Chap. 2, these function not in the manner of multimodular systems, but rather via using (probabilistic) truth values and (attractor neural net like) attention values as weights on nodes and links of the same (hyper) graph. The nodes and links in this hypergraph are typed, like a standard semantic network approach for knowledge representation, so they're able to handle all sorts of knowledge, from the most concrete perception and actuation related knowledge to the most abstract relationships. But they're also weighted with values similar to neural net weights, and pass around quantities (importance values, discussed in Chap. 5 of Part 2) similar to neural net activations, allowing emergent attractor/assembly based knowledge representation similar to attractor neural nets.

The concept of glocality lies at the heart of this combination, in a way that spans the pseudo-dichotomy:

- Local knowledge is represented in abstract logical relationships stored in explicit logical form, and also in Hebbian-type associations between nodes and links.
- Global knowledge is represented in large-scale patterns of node and link weights, which lead to large-scale patterns of network activity, which often take the form of attractors qualitatively similar to Hopfield net attractors. These attractors are called *maps*.

The result of all this is that a concept like "cat" might be represented as a combination of:

- A small number of logical relationships and strong associations, that constitute the "key" subnetwork for the "cat" concept.
- A large network of weak associations, binding together various nodes and links of various types and various levels of abstraction, representing the "cat map".

The activation of the key will generally cause the activation of the map, and the activation of a significant percentage of the map will cause the activation of the rest of the map, including the key. Furthermore, if the key were for some reason forgotten, then after a significant amount of effort, the system would likely to be able to reconstitute it (perhaps with various small changes) from the information in the map. We conjecture that this particular kind of glocal memory will turn out to be very powerful for AGI, due to its ability to combine the strengths of formal logical inference with those of self-organizing attractor neural networks.

As a simple example, consider the representation of a "tower", in the context of an artificial agent that has built towers of blocks, and seen pictures of many other kinds of towers, and seen some tall building that it knows are somewhat like towers but perhaps not exactly towers. If this agent is reasonably conceptually advanced (say, at Piagetan the concrete operational level) then its mind will contain some declarative relationships partially characterizing the concept of "tower", as well as its sensory and episodic examples, and its procedural knowledge about how to build towers.

The key of the "tower" concept in the agent's mind may consist of internal images and episodes regarding the towers it knows best, the essential operations it knows are useful for building towers (piling blocks atop blocks atop blocks ...), and the core declarative relations summarizing "towerness"—and the whole "tower" map then consists of a much larger number of images, episodes, procedures and declarative relationships connected to "tower" and other related entities. If any portion of the map is removed—even if the key is removed—then the rest of the map can be approximately reconstituted, after some work. Some cognitive operations are best done on the localized representation—e.g. logical reasoning. Other operations, such as attention allocation and guidance of inference control, are best done using the globalized map representation.

Chapter 15
Representing Implicit Knowledge
via Hypergraphs

15.1 Introduction

Explicit knowledge is easy to write about and talk about; implicit knowledge is equally important, but tends to get less attention in discussions of AI and psychology, simply because we don't have as good a vocabulary for describing it, nor as good a collection of methods for measuring it. One way to deal with this problem is to describe implicit knowledge using language and methods typically reserved for explicit knowledge. This might seem intrinsically non-workable, but we argue that it actually makes a lot of sense. The same sort of networks that a system like CogPrime uses to represent knowledge explicitly, can also be used to represent the *emergent* knowledge that implicitly exists in an intelligent system's complex structures and dynamics.

We've noted that CogPrime uses an explicit representation of knowledge in terms of weighted labeled hypergraphs; and also uses other more neural net like mechanisms (e.g. the economic attention allocation network subsystem) to represent knowledge globally and implicitly. CogPrime combines these two sorts of representation according to the principle we have called *glocality*. In this chapter we pursue glocality a bit further—describing a means by which even implicitly represented knowledge can be modeled using weighted labeled hypergraphs similar to the ones used explicitly in CogPrime. This is conceptually important, in terms of making clear the fundamental similarities and differences between implicit and explicit knowledge representation; and it is also pragmatically meaningful due to its relevance to the Cog-Prime methods described in Chap. 24 of Part 2 that transform implicit into explicit knowledge.

To avoid confusion with CogPrime's explicit knowledge representation, we will refer to the hypergraphs in this chapter as composed of Vertices and Edges rather than Nodes and Links. In prior publications we have referred to "derived" or "emergent" hypergraphs of the sort described here using the acronym **SMEPH**, which stands for Self-Modifying, Evolving Probabilistic Hypergraphs.

B. Goertzel et al., *Engineering General Intelligence, Part 1*, 319
Atlantis Thinking Machines 5, DOI: 10.2991/978-94-6239-027-0_15,
© Atlantis Press and the authors 2014

15.2 Key Vertex and Edge Types

We begin by introducing a particular collection of Vertex and Edge types, to be used in modeling the internal structures of intelligent systems.

The key SMEPH Vertex types are

- ConceptVertex, representing a set, for instance, an idea or a set of percepts.
- SchemaVertex, representing a procedure for doing something (perhaps something in the physical world, or perhaps an abstract mental action).

The key SMEPH Edge types, using language drawn from Probabilistic Logic Networks (PLN) and elaborated in Chap. 16 (Part 2), are as follows:

- ExtensionalInheritanceEdge (ExtInhEdge for short: an edge which, linking one Vertex or Edge to another, indicates that the former is a special case of the latter).
- ExtensionalSimilarityEdge (ExtSim: which indicates that one Vertex or Edge is similar to another).
- ExecutionEdge (a ternary edge, which joins S,B,C when S is a SchemaVertex and the result from applying S to B is C).

So, in a SMEPH system, one is often looking at hypergraphs whose Vertices represent ideas or procedures, and whose Edges represent relationships of specialization, similarity or transformation among ideas and/or procedures.

The semantics of the SMEPH edge types is given by PLN, but is simple and commonsensical. ExtInh and ExtSim Edges come with probabilistic weights indicating the extent of the relationship they denote (e.g. the ExtSimEdge joining the cat ConceptVertex to the dog ConceptVertex gets a higher probability weight than the one joining the cat ConceptVertex to the washing-machine ConceptVertex). The mathematics of transformations involving these probabilistic weights becomes quite involved—particularly when one introduces SchemaVertices corresponding to abstract mathematical operations, a step that enables SMEPH hypergraphs to have the complete mathematical power of standard logical formalisms like predicate calculus, but with the added advantage of a natural representation of uncertainty in terms of probabilities, as well as a natural representation of networks and webs of complex knowledge.

15.3 Derived Hypergraphs

We now describe how SMEPH hypergraphs may be used to model and describe intelligent systems. One can (in principle) draw a SMEPH hypergraph corresponding to any individual intelligent system, with Vertices and Edges for the concepts and processes in that system's mind. This is called the derived hypergraph of that system.

15.3.1 SMEPH Vertices

A ConceptVertex in the derived hypergraph of a system corresponds to a structural pattern that persists over time in that system; whereas a SchemaVertex corresponds to a multi-time-point dynamical pattern that recurs in that system's dynamics. If one accepts the patternist definition of a mind as the set of patterns in an intelligent system, then it follows that the derived hypergraph of an intelligent system captures a significant fraction of the mind of that system.

To phrase it a little differently, we may say that a ConceptVertex, in SMEPH, refers to the habitual pattern of activity observed in a system when some condition is met (this condition corresponding to the presence of a certain pattern). The condition may refer to something in the world external to the system, or to something internal. For instance, the condition may be observing a cat. In this case, the corresponding Concept vertex in the mind of Ben Goertzel is the pattern of activity observed in Ben Goertzel's brain when his eyes are open and he's looking in the direction of a cat. The notion of pattern of activity can be made rigorous using mathematical pattern theory, as is described in The Hidden Pattern [Goe06a].

Note that logical predicates, on the SMEPH level, appear as particular kinds of Concepts, where the condition involves a predicate and an argument. For instance, suppose one wants to know what happens inside Ben's mind when he eats cheese. Then there is a Concept corresponding to the condition of cheese-eating activity. But there may also be a Concept corresponding to eating activity in general. If the Concept denoting the activity of eating X is generally easily computable from the Concepts for X and eating individually, then the eating Concept is effectively acting as a predicate.

A SMEPH SchemaVertex, on the other hand, is like a Concept that's defined in a time-dependent way. One type of Schema refers to a habitual dynamical pattern of activity occurring before and/or during some condition is met. For instance, the condition might be saying the word Hello. In that case the corresponding SchemaVertex in the mind of Ben Goertzel is the pattern of activity that generally occurs before he says Hello.

Another type of Schema refers to a habitual dynamical pattern of activity occurring after some condition X is met. For instance, in the case of the Schema for adding two numbers, the precondition X consists of the two numbers and the concept of addition. The Schema is then what happens when the mind thinks of adding and thinks of two numbers.

Finally, there are Schema that refer to habitual dynamical activity patterns occurring after some condition X is met and before some condition Y is met. In this case the Schema is viewed as transforming X into Y. For instance, if X is the condition of meeting someone who is not a friend, and Y is the condition of being friends with that person, then the habitually intervening activities constitute the Schema for making friends.

15.3.2 SMEPH Edges

SMEPH edge types fall into two categories: functional and logical. Functional edges connect Schema vertices to their input and outputs; logical edges refer mainly to conditional probabilities, and in general are to be interpreted according to the semantics of Probabilistic Logic Networks.

Let us begin with logical edges. The simplest case is the Subset edge, which denotes a straightforward, extensional conditional probability. For instance, it may happen that whenever the Concept for cat is present in a system, the Concept for animal is as well. Then we would say

```
Subset  cat  animal
```

(Here we assume a notation where "R A B" denotes an Edge of type R between Vertices A and B.)

On the other hand, it may be that 50% of the time that cat is present in the system, cute is present as well: then we would say

```
Subset  cat  cute  <.5>
```

where the <.5> denotes the probability, which is a component of the Truth Value associated with the edge.

Next, the most basic functional edge is the Execution edge, which is ternary and denotes a relation between a Schema, its input and its output, e.g.

```
Execution  father_of  Ben_Goertzel  Ted_Goertzel
```

for a schema father_of that outputs the father of its argument.

The ExecutionOutput (ExOut) edge denotes the output of a Schema in an implicit way, e.g.

```
ExOut  say_hello
```

refers to a particular act of saying hello, whereas

```
ExOut  add_numbers  \{3,  4}
```

refers to the Concept corresponding to 7. Note that this latter example involves a set of three entities: sets are also part of the basic SMEPH knowledge representation. A set may be thought of as a hypergraph edge that points to all its members.

In this manner we may define a set of edges and vertices modeling the habitual activity patterns of a system when in different situations. This is called the derived hypergraph of the system. Note that this hypergraph can in principle be constructed no matter what happens inside the system: whether it's a human brain, a formal neural network, Cyc, OCP, a quantum computer, etc. Of course, constructing the hypergraph in practice is quite a different story: for instance, we currently have no accurate way of measuring the habitual activity patterns inside the human brain. fMRI and PET and other neuroimaging technologies give only a crude view, though they are continually improving.

Pattern theory enters more deeply here when one thoroughly fleshes out the Inheritance concept. Philosophers of logic have extensively debated the relationship between extensional inheritance (inheritance between sets based on their members) and intensional inheritance (inheritance between entity-types based on their properties). A variety of formal mechanisms have been proposed to capture this conceptual distinction; see (Wang 2006) for a review along with a novel approach utilizing uncertain term logic. Pattern theory provides a novel approach to defining intension: one may associate with each ConceptVertex in a system's derived hypergraph the set of patterns associated with the structural pattern underlying that ConceptVertex. Then, one can define the strength of the IntensionalInheritanceEdge between two ConceptVertices A and B as the percentage of A's pattern-set that is also contained in B's pattern-set. According to this approach, for instance, one could have

```
IntInhEdge  whale  fish  <0.6>

ExtInhEdge  whale  fish  <0.0>
```

since the fish and whale sets have common properties but no common members.

15.4 Implications of Patternist Philosophy for Derived Hypergraphs of Intelligent Systems

Patternist philosophy rears its head here and makes some definite hypotheses about the structure of derived hypergraphs. It suggests that derived hypergraphs should have a dual network structure, and that in highly intelligent systems they should have subgraphs that constitute models of the whole hypergraph (these are self systems). SMEPH does not add anything to the patternist view on a philosophical level, but it gives a concrete instantiation to some of the general ideas of patternism. In this section we'll articulate some "SMEPH principles", constituting important ideas from patternist philosophy as they manifest themselves in the SMEPH context.

The logical edges in a SMEPH hypergraph are weighted with probabilities, as in the simple example given above. The functional edges may be probabilistically weighted as well, since some Schema may give certain results only some of the time. These probabilities are critical in terms of SMEPH's model of system dynamics; they underly one of our SMEPH principles,

Principle of Implicit Probabilistic Inference: In an intelligent system, the temporal evolution of the probabilities on the edges in the system's derived hypergraph should approximately obey the rules of probability theory.

The basic idea is that, even if a system—through its underlying dynamics—has no explicit connection to probability theory, it still must behave roughly as if it does, if it is going to be intelligent. The roughly part is important here; it's well known that humans are not terribly accurate in explicitly carrying out formal probabilistic inferences. And yet, in practical contexts where they have experience, humans can make quite accurate judgments; which is all that's required by the above principle, since it's the contexts where experience has occurred that will make up a system's derived hypergraph.

Our next SMEPH principle is evolutionary, and states

Principle of Implicit Evolution: In an intelligent system, new Schema and Concepts will continually be created, and the Schema and Concepts that are more useful for achieving system goals (as demonstrated via probabilistic implication of goal achievement) will tend to survive longer.

Note that this principle can be fulfilled in many different ways. The important thing is that system goals are allowed to serve as a selective force.

Another SMEPH dynamical principle pertains to a shorter time-scale than evolution, and states

Principle of Attention Allocation: In an intelligent system, Schema and Concepts that are more useful for attaining short-term goals will tend to consume more of the system's energy. (The balance of attention oriented toward goals pertaining to different time scales will vary from system to system.)

Next, there is the

Principle of Autopoesis: In an intelligent system, if one removes some part of the system and then allows the system's natural dynamics to keep going, a decent approximation to that removed part will often be spontaneously reconstituted.

And there is the

Cognitive Equation Principle: In an intelligent system, many abstract patterns that are present in the system at a certain time as patterns among other Schema and Concepts, will at a near-future time be present in the system as patterns among elementary system components.

The Cognitive Equation Principle, briefly discussed in Chap. 6, basically means that Concepts and Schema emergent in the system are recognized by the system and then embodied as elementary items in the system so that patterns among them in their emergent form become, with the passage of time, patterns among them in their directly-system-embodied form. This is a natural consequence of the way intelligent systems continually recognize patterns in themselves.

Note that derived hypergraphs may be constructed corresponding to any complex system which demonstrates a variety of internal dynamical patterns depending on its situation. However, if a system is not intelligent, then according to the patternist philosophy evolution of its derived hypergraph can't necessarily be expected to follow the above principles.

15.4.1 SMEPH Principles in CogPrime

We now more explicitly elaborate the application of these ideas in the CogPrime context. As noted above, in addition to explicit knowledge representation in terms of Nodes and Links, CogPrime also incorporates implicit knowledge representation in the form of what are called Maps: collections of Nodes and Links that tend to be utilized together within cognitive processes.

These Maps constitute a CogPrime system's derived hypergraph, which will not be identical to the hypergraph it uses for explicit knowledge representation. However, an interesting feedback loop arises here, in that the intelligence's self-study will generally lead it to recognize large portions of its derived hypergraph as patterns in itself, and then embody these patterns within its concretely implemented knowledge hypergraph. This relates to the Cognitive Equation Principle defined in Chap. 6, in which an intelligent system continually recognizes patterns in itself and embodies these patterns in its own basic structure (so that new patterns may more easily emerge from them).

Often it happens that a particular CogPrime node will serve as the center of a map, so that e.g. the Concept Link denoting cat will consist of a number of nodes and links roughly centered around a ConceptNode that is linked to the WordNode cat. But this is not guaranteed and some CogPrime maps are more diffuse than this with no particular center.

Somewhat similarly, the key SMEPH dynamics are represented explicitly in Cog-Prime: probabilistic reasoning is carried out via explicit application of PLN on the CogPrime hypergraph, evolutionary learning is carried out via application of the MOSES optimization algorithm, and attention allocation is carried out via a combination of inference and evolutionary pattern mining. But the SMEPH dynamics also occur implicitly in CogPrime: emergent maps are reasoned on probabilistically as an indirect consequence of node-and-link level PLN activity; maps evolve as a consequence of the coordinated whole of CogPrime dynamics; and attention shifts between maps according to complex emergent dynamics.

To see the need for maps, consider that even a Node that has a particular meaning attached to it—like the *Iraq* Node, say—doesn't contain much of the meaning of *Iraq* in it. The meaning of *Iraq* lies in the Links attached to this Node, and the Links attached to their Nodes—and the other Nodes and Links not explicitly represented in the system, which will be created by CogPrime's cognitive algorithms based on the explicitly existent Nodes and Links related to the *Iraq* Node.

This halo of Atoms related to the *Iraq* node is called the *Iraq* map. In general, some maps will center around a particular Atom, like this *Iraq* map, others may not have any particular identifiable center. CogPrime's cognitive processes act directly on the level of Nodes and Links, but they must be analyzed in terms of their impact on maps as well. In SMEPH terms, CogPrime maps may be said to correspond to SMEPH ConceptVertex, and for instance bundles of Links between the Nodes belonging to a map may correspond to a SMEPH Link between two ConceptVertex.

Chapter 16
Emergent Networks of Intelligence

16.1 Introduction

When one is involved with engineering an AGI system, one thinks a lot about the aspects of the system one is explicitly building—what are the parts, how they fit together, how to test they're properly working, and so forth. And yet, these explicitly engineered aspects are only a fraction of what's important in an AGI system. At least as critical are the *emergent* aspects—the patterns that emerge once the system is up and running, interacting with the world and other agents, growing and developing and learning and self-modifying. SMEPH is one toolkit for describing some of these emergent patterns, but it's only a start.

In line with these general observations, most of this book will focus on the structures and processes that we have built, or intend to build, into the CogPrime system. But in a sense, these structures and processes are not the crux of CogPrime's intended intelligence. The purpose of these pre-programmed structures and processes is to give rise to *emergent* structures and processes, in the course of CogPrime's interaction with the world and the other minds within it. We will return to this theme of emergence at several points in later chapters, e.g. in the discussion of map formation in Chap. 24 of Part 2.

Given the important of emergent structures—and specifically emergent *network* structures—for intelligence, it's fortunate the scientific community has already generated a lot of knowledge about complex networks: both networks of physical or software elements, and networks of organization emergent from complex systems. As most of this knowledge has originated in fields other than AGI, or in pure mathematics, it tends to require some reinterpretation or tweaking to achieve maximal applicability in the AGI context; but we believe this effort will become increasingly worthwhile as the AGI field progresses, because network theory is likely to be very useful for describing the contents and interactions of AGI systems as they develop increasing intelligence.

In this brief chapter we specifically focus on the emergence of certain large-scale *network structures* in a CogPrime knowledge store, presenting heuristic arguments

B. Goertzel et al., *Engineering General Intelligence, Part 1*,
Atlantis Thinking Machines 5, DOI: 10.2991/978-94-6239-027-0_16,
© Atlantis Press and the authors 2014

as to why these structures can be expected to arise. We also comment on the way
in which these emergent structures are expected to guide cognitive processes, and
give rise to emergent cognitive processes. Appendix C expands on this theme in a
particular direction, exploring the possible emergence of structures characterizing
inter-cognitive reflection.

16.2 Small World Networks

One simple but potentially useful observation about CogPrime Atomspaces is that
they are generally going to be *small world networks* [Buc03], rather than random
graphs. A small world network is a graph in which the connectivities of the various
nodes display a power law behavior—so that, loosely speaking, there are a few nodes
with very many links, then more nodes with a modest number of links. . . and finally,
a huge number of nodes with very few links. This kind of network occurs in many
natural and human systems, including citations among papers, financial arrangements
among banks, links between Web pages and the spread of diseases among people
or animals. In a weighted network like an Atomspace, "small-world-ness" must be
defined in a manner taking the weights into account, and there are several obvious
ways to do this. Figure 16.1 depicts a small but prototypical small-worlds network,
with a few "hub" nodes possessing far more neighbors than the others, and then some
secondary hubs, etc.

Fig. 16.1 A typical, though small-sized, small-worlds network

An excellent reference on network theory in general, including but not limited to small world networks, is Peter Csermely's *Weak Links* [Cse06]. Many of the ideas in that work have apparent CogPrime applications, which are not elaborated here.

One process via which small world networks commonly form is "preferential attachment" [Bar02]. This occurs in essence when "the rich get richer"—i.e. when nodes in the network grow new links, in a manner that causes them to preferentially grow links to nodes that already have more links. It is not hard to see that CogPrime's ECAN dynamics will naturally lead to preferential attachment, because Atoms with more links will tend to get more STI, and thus will tend to get selected by more cognitive processes, which will cause them to grow more links. For this reason, in most circumstances, a CogPrime system in which most link-building cognitive processes rely heavily on ECAN to guide their activities will tend to contain a small-world-network Atomspace. This is not rigorously guaranteed to be the case for any possible combination of environment and goals, but it is commonsensically likely to nearly always be the case.

One consequence of the small worlds structure of the Atomspace is that, in exploring other properties of the Atom network, it is particularly important to look at the hub nodes. For instance, if one is studying whether hierarchical and heterarchical subnetworks of the Atomspace exist, and whether they are well-aligned with each other, it is important to look at hierarchical and heterarchical connections between hub nodes in particular (and secondary hubs, etc.). A pattern of hierarchical or dual network connection that only held up among the more sparsely connected nodes in a small-world network would be a strange thing, and perhaps not that cognitively useful.

16.3 Dual Network Structure

One of the key theoretical notions in patternist philosophy is that complex cognitive systems evolve internal *dual network* structures, comprising superposed, harmonized hierarchical and heterarchical networks. Now we explore some of the specific CogPrime structures and dynamics militating in favor of the emergence of dual networks.

16.3.1 Hierarchical Networks

The hierarchical nature of human linguistic concepts is well known, and is illustrated in Fig. 16.2 for the commonsense knowledge domain (using a graph drawn from WordNet, a huge concept hierarchy covering 50K+ English-language concepts), and in Fig. 16.4 for a specialized knowledge subdomain, genetics. Due to this fact, a certain amount of hierarchy can be expected to emerge in the Atomspace of any

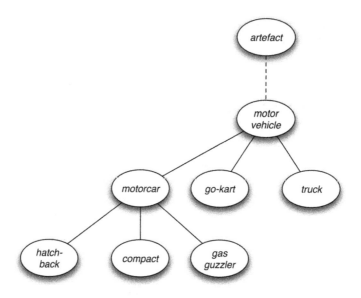

Fig. 16.2 A typical, though small, subnetwork of WordNet's hierarchical network

Fig. 16.3 A typical, though
small, subnetwork of the
Gene Ontology's hierarchical
network

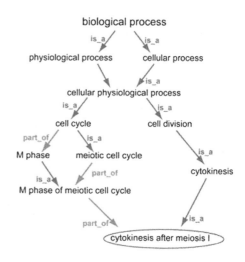

linguistically savvy CogPrime, simply due to its modeling of the linguistic concepts that it hears and reads (Fig. 16.3).

Hierarchy also exists in the natural world apart from language, which is the reason that many sensorimotor-knowledge-focused AGI systems (e.g. DeSTIN and HTM, mentioned in Chap. 6) feature hierarchical structures. In these cases the hierarchies are normally spatiotemporal in nature—with lower layers containing elements responding to more localized aspects of the perceptual field, and smaller, more localized groups of actuators. This kind of hierarchy certainly *could* emerge in an AGI

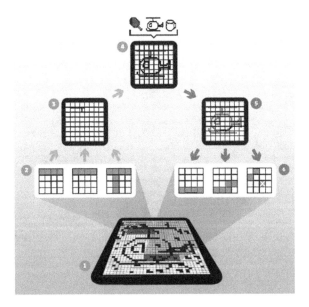

Fig. 16.4 Small-scale portrayal of a portion of the spatiotemporal hierarchy in Jeff Hawkins' Hierarchical Temporal Memory architecture

system, but in CogPrime we have opted for a different route. If a CogPrime system is hybridized with a hierarchical sensorimotor network like one of those mentioned above, then the Atoms linked to the nodes in the hierarchical sensorimotor network will naturally possess hierarchical conceptual relationships, and will thus naturally grow hierarchical links between them (e.g. and IntensionalInheritanceLinks via PLN, AsymmetricHebbianLinks via ECAN).

Once elements of hierarchical structure exist via the hierarchical structure of language and physical reality, then a richer and broader hierarchy can be expected to accumulate on top of it, because importance spreading and inference control will implicitly and automatically be guided by the existing hierarchy. That is, in the language of *Chaotic Logic* [Goe94] and patternist theory, hierarchical structure is an "autopoietic attractor"—once it's there it will tend to enrich itself and maintain itself. AsymmetricHebbianLinks arranged in a hierarchy will tend to cause importance to spread up or down the hierarchy, which will lead other cognitive processes to look for patterns between Atoms and their hierarchical parents or children, thus potentially building more hierarchical links. Chains of InheritanceLinks pointing up and down the hierarchy will lead PLN to search for more hierarchical links—e.g. most simply, $A \rightarrow B \rightarrow C$ where C is above B is above A in the hierarchy, will naturally lead inference to check the viability of $A \rightarrow C$ by deduction. There is also the possibility to introduce a special DefaultInheritanceLink, as discussed in Chap. 16 of Part 2, but this isn't actually necessary to obtain the inferential maintenance of a robust hierarchical network.

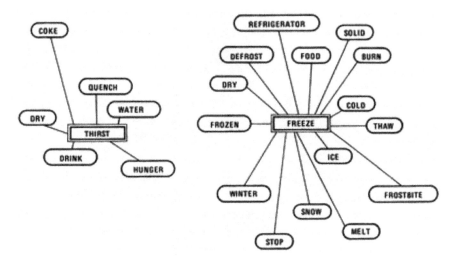

Fig. 16.5 Portions of a conceptual heterarchy centered on specific concepts

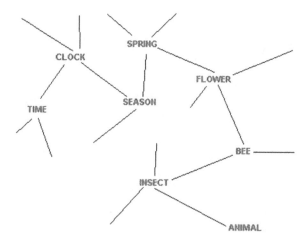

Fig. 16.6 A portion of a conceptual heterarchy, showing the "dangling links" leading this portion to the rest of the heterarchy

16.3.2 Associative, Heterarchical Networks

Heterarchy is in essence a simpler structure than hierarchy: it simply refers to a network in which nodes are linked to other nodes with which they share important relationships. That is, there should be a tendency that if two nodes are often important in the same contexts or for the same purposes, they should be linked together. Portrayals of typical heterarchical linkage patterns among natural language concepts are given in Figs. 16.5 and 16.6. Naturally, real concept heterarchies are far more large, complex and tangled than these.

In CogPrime, ECAN enforces heterarchy via building SymmetricHebbianLinks, and PLN by building SimilarityLinks, IntensionalSimilarityLinks and Extensional-SimilarityLinks. Furthermore, these various link types reinforce each other. PLN control is guided by importance spreading, which follows Hebbian links, so that a heterarchical Hebbian network tends to cause PLN to explore the formation of links following the same paths as the heterarchical HebbianLinks. And importance can spread along logical links as well as explicit Hebbian links, so that the existence of a heterarchical logical network will tend to cause the formation of additional heterarchical Hebbian links. Heterarchy reinforces itself in "autopoietic attractor" style even more simply and directly than heterarchy.

16.3.3 Dual Networks

Finally, if both hierarchical and heterarchical structures exist in an Atomspace, then both ECAN and PLN will naturally blend them together, because hierarchical and heterarchical links will feed into their link-creation processes and naturally be combined together to form new links. This will tend to produce a structure called a *dual network*, in which a hierarchy exists, along with a rich network of heterarchical links joining nodes in the hierarchy, with a particular density of links between nodes on the same hierarchical level. The dual network structure will emerge without any explicit engineering oriented toward it, simply via the existence of hierarchical and heterarchical networks, and the propensity of ECAN and PLN to be guided by both the hierarchical and heterarchical networks. The existence of a natural dual network structure in both linguistic and sensorimotor data will help the formation process along, and then creative cognition will enrich the dual network yet further than is directly necessitated by the external world.

A rigorous mathematical analysis of the formation of hierarchical, heterarchical and dual networks in CogPrime systems has not yet been undertaken, and would certainly be an interesting enterprise. Similar to the theory of small world networks, there is ample ground here for both theorem-proving and heuristic experimentation. However, the qualitative points made here are sufficiently well-grounded in intuition and experience to be of some use guiding our ongoing work. One of the nice things about emergent network structures is that they are relatively straightforward to observe in an evolving, learning AGI system, via visualization and inspection of structures such at the Atomspace.

Part VI
A Path to Human-Level AGI

Chapter 17
AGI Preschool

17.1 Introduction

In conversations with government funding sources or narrow AI researchers about AGI work, one of the topics that comes up most often is that of "evaluation and metrics"—i.e., AGI intelligence testing. We actually prefer to separate this into two topics: environments and methods for careful qualitative evaluation of AGI systems, versus metrics for precise measurement of AGI systems. The difficulty of formulating bulletproof metrics for partial progress toward advanced AGI has become evident throughout the field, and in Chap. 9 we have elaborated one plausible explanation for this phenomenon, the "trickiness" of cognitive synergy. [LWML09], summarizing a workshop on "Evaluation and Metrics for Human-Level AI" held in 2008, discusses some of the general difficulties involved in this type of assessment, and some requirements that any viable approach must fulfill. On the other hand, the lack of appropriate methods for careful qualitative evaluation of AGI systems has been much less discussed, but we consider it actually a more important issue—as well as an easier (though not easy) one to solve.

We haven't actually found the lack of quantitative intelligence metrics to be a major obstacle in our practical AGI work so far. Our OpenCogPrime implementation lags far behind the CogPrime design as articulated in Part 2, and according to the theory underlying CogPrime, the more interesting behaviors and dynamics of the system will occur only when all the parts of the system have been engineered to a reasonable level of completion and integrated together. So, the lack of a great set of metrics for evaluating the intelligence of our partially-built system hasn't impaired too much. Testing the intelligence of the current OpenCogPrime system is a bit like testing the flight capability of a partly-built airplane that only has stubs for wings, lacks tail-fins, has a much less efficient engine than the one that's been designed for use in the first "real" version of the airplane, etc. There may be something to be learned from such preliminary tests, but making them highly rigorous isn't a great use of effort, compared to working on finishing implementing the design according to the underlying theory.

B. Goertzel et al., *Engineering General Intelligence, Part 1*,
Atlantis Thinking Machines 5, DOI: 10.2991/978-94-6239-027-0_17,
© Atlantis Press and the authors 2014

On the other hand, the problem of what environments and methods to use to qualitatively evaluate and study AGI progress, has been considerably more vexing to us in practice, as we've proceeded in our work on implementing and testing OpenCogPrime and developing the CogPrime theory. When developing a complex system, it's nearly always valuable to see what this system does in some fairly rich, complex situations, in order to gain a better intuitive understanding of the parts and how they work together. In the context of human-level AGI, the theoretically best way to do this would be to embody one's AGI system in a humanlike body and set it loose in the everyday human world; but of course, this isn't feasible given the current state of development of robotics technology. So one must seek approximations. Toward this end we have embodied OpenCogPrime in non-player characters in video game style virtual worlds, and carried out preliminary experiments embodying OpenCogPrime in humanoid robots. These are reasonably good options but they have limitations and lead to subtle choices: what kind of game characters and game worlds, what kind of robot environments, etc.?

One conclusion we have come to, based largely on the considerations in Chap. 12 on development and Chap. 10 on the importance of environment, is that it may make sense to embed early-stage proto-AGI and AGI systems in environments reminiscent of those used for teaching young human children. In this chapter we will explore this approach in some detail: emulation, in either physical reality or an multiuser online virtual world, of an environment similar to preschools used in early human childhood education. Complete specification of an "AGI Preschool" would require much more than a brief chapter; our goal here is to sketch the idea in broad outline, and give a few examples of the types of opportunities such an environment would afford for instruction, spontaneous learning and formal and informal evaluation of certain sorts of early-stage AGI systems.

The material in this chapter will pop up fairly often later in the book. The AGI Preschool context will serve, throughout the following chapters, as a source of concrete examples of the various algorithms and structures. But it's not proposed merely as an expository tool; we are making the very serious proposal that sending AGI systems to a virtual or robotic preschool is an excellent way—perhaps the best way—to foster the development of human-level human-like AGI.

17.1.1 Contrast to Standard AI Evaluation Methodologies

The reader steeped in the current AI literature may wonder why it's necessary to introduce a new methodology and environment for evaluating AGI systems. There are already very many different ways of evaluating AI systems out there ... do we really need another?

Certainly, the AI field has inspired many competitions, each of which tests some particular type or aspect of intelligent behavior. Examples include robot competitions, tournaments of computer chess, poker, backgammon and so forth at computer Olympiads, trading-agent competition, language and reasoning competitions like the

Pascal Textual Entailment Challenge, and so on. In addition to these, there are many standard domains and problems used in the AI literature that are meant to capture the essential difficulties in a certain class of learning problems: standard datasets for face recognition, text parsing, supervised classification, theorem-proving, question-answering and so forth.

However, the value of these sorts of tests for AGI is predicated on the hypothesis that the degree of success of an AI program at carrying out some domain-specific task, is correlated with the potential of that program for being developed into a robust AGI program with broad intelligence. If humanlike AGI and problem-area-specific "narrow AI" are in fact very different sorts of pursuits requiring very different principles, as we suspect, then these tests are not strongly relevant to the AGI problem.

There are also some standard evaluation paradigms aimed at AI going beyond specific tasks. For instance, there is a literature on "multitask learning" and "transfer learning", where the goal for an AI is to learn one task quicker given another task solved previously [Car97, TM95, BDS03, TS07, RZDK05]. This is one of the capabilities an AI agent will need to simultaneously learn different types of tasks as proposed in the Preschool scenario given here. And there is a literature on "shaping", where the idea is to build up the capability of an AI by training it on progressively more difficult versions of the same tasks [LD03]. Again, this is one sort of capability an AI will need to possess if it is to move up some type of curriculum, such as a school curriculum.

While we applaud the work done on multitask learning and shaping, we feel that exploring these processes using mathematical abstractions, or in the domain of various narrowly-proscribed machine-learning or robotics test problems, may not adequately address the problem of AGI. The problem is that generalization among tasks, or from simpler to more difficult versions of the same task, is a process whose nature may depend strongly on the overall nature of the set of tasks and task-versions involved. Real-world tasks have a subtlety of interconnectedness and developmental course that is not captured in current mathematical learning frameworks nor standard AI test problems.

To put it mathematically, we suggest that the universe of real-world human tasks has a host of "special statistical properties" that have implications regarding what sorts of AI programs will be most suitable; and that, while exploring and formalizing the nature of these statistical properties is important, an easier and more reliable approach to AGI testing is to create a testing environment that embodies these properties implicitly, via its being an emulation of the cognitively meaningful aspects of the real-world human learning environment.

One way to see this point vividly is to contrast the current proposal with the "General Game Player" AI competition, in which AIs seek to learn to play games based on formal descriptions of the rules.[1] Clearly doing GGP well requires powerful AGI; and doing GGP even mediocrely probably requires robust multitask learning and shaping. But we suspect GGP is far inferior to AGI Preschool as an approach to testing early-stage AI programs aimed at roughly humanlike intelligence. This

---

[1] http://games.stanford.edu/

is because, unlike the tasks involved in AI Preschool, the tasks involved in doing simple instances of GGP seem to have little relationship to humanlike intelligence or real-world human tasks.

17.2 Elements of Preschool Design

What we mean by an "AGI Preschool" is simply a porting to the AGI domain of the essential aspects of human preschools. While there is significant variance among preschools there are also strong commonalities, grounded in educational theory and experience. We will briefly discuss both the physical design and educational curriculum of the typical human preschool, and which aspects transfer effectively to the AGI context.

On the physical side, the key notion in modern preschool design is the "learning center", an area designed and outfitted with appropriate materials for teaching a specific skill. Learning centers are designed to encourage learning by doing, which greatly facilitates learning processes based on reinforcement, imitation and correction (see Chap. 14 of Part 2, for a detailed discussion of the value of this combination); and also to provide multiple techniques for teaching the same skills, to accommodate different learning styles and prevent over-fitting and overspecialization in the learning of new skills.

Centers are also designed to cross-develop related skills. A "manipulative center", for example, provides physical objects such as drawing implements, toys and puzzles, to facilitate development of motor manipulation, visual discrimination, and (through sequencing and classification games) basic logical reasoning. A "dramatics center", on the other hand, cross-trains interpersonal and empathetic skills along with bodily-kinesthetic, linguistic, and musical skills. Other centers, such as art, reading, writing, science and math centers are also designed to train not just one area, but to center around a primary intelligence type while also cross-developing related areas. For specific examples of the learning centers associated with particular contemporary preschools, see [Nie98].

In many progressive, student-centered preschools, students are left largely to their own devices to move from one center to another throughout the preschool room. Generally, each center will be staffed by an instructor at some points in the day but not others, providing a variety of learning experiences. At some preschools students will be strongly encouraged to distribute their time relatively evenly among the different learning centers, or to focus on those learning centers corresponding to their particular strengths and/or weaknesses.

To imitate the general character of a human preschool, one would create several centers in a robot lab or virtual world. The precise architecture will best be adapted via experience but initial centers would likely be:

- **a blocks center**: a table with blocks on it

- **a language center**: a circle of chairs, intended for people to sit around and talk with the robot
- **a manipulative center**: with a variety of different objects of different shapes and sizes, intended to teach visual and motor skills
- **a ball play center**: where balls are kept in chests and there is space for the robot to kick the balls around
- **a dramatics center**: where the robot can observe and enact various movements.

17.3   Elements of Preschool Curriculum

While preschool curricula vary considerably based on educational philosophy and regional and cultural factors, there is a great deal of common, shared wisdom regarding the most useful topics and methods for preschool teaching. Guided experiential learning in diverse environments and using varied materials is generally agreed upon as being an optimal methodology to reach a wide variety of learning types and capabilities. Hands-on learning provides grounding in specifics, where as a diversity of approaches allows for generalization.

Core knowledge domains are also relatively consistent, even across various philosophies and regions. Language, movement and coordination, autonomous judgment, social skills, work habits, temporal orientation, spatial orientation, mathematics, science, music, visual arts, and dramatics are universal areas of learning which all early childhood learning touches upon. The particulars of these skills may vary, but all human children are taught to function in these domains. The level of competency developed may vary, but general domain knowledge is provided. For example, most kids won't be the next Maria Callas, Ravi Shankar or Gene Ween, but nearly all learn to hear, understand and appreciate music.

Tables 17.1, 17.2, 17.3 review the key capabilities taught in preschools, and identify the most important specific skills that need to be evaluated in the context of each capability. This table was assembled via surveying the curricula from a number of currently existing preschools employing different methodologies both based on formal academic cognitive theories [Sch07] and more pragmatic approaches, such as: Montessori [Mon12], Waldorf [SS03b], Brain Gym (www.braingym.org) and Core Knowledge (www.coreknowledge.org).

17.3.1   Preschool in the Light of Intelligence Theory

Comparing Table 17.1 to Gardner's Multiple Intelligences (MI) framework briefly reviewed in Chap. 4, the high degree of harmony is obvious, and is borne out by more detailed analysis. Preschool curriculum as standardly practiced is very well attuned to MI, and naturally covers all the bases that Gardner identifies as important. And this is

Table 17.1 Categories of preschool curriculum, Part 1

Type of capability	Specific skills to be evaluated
Story understanding	• Understanding narrative sequence
	• Understanding character development
	• Dramatize a story
	• Predict what comes next in a story
Linguistic	• Give simple descriptions of events
	• Describe similarities and differences
	• Describe objects and their functions
Linguistic/Spatial-visual	Interpreting pictures
Linguistic/Social	• Asking questions appropriately
	• Answering questions appropriately
	• Talk about own discoveries
	• Initiate conversations
	• Settle disagreements
	• Verbally express empathy
	• Ask for help
	• Follow directions
Linguistic/Scientific	• Provide possible explanations for events or phenomena
	• Carefully describe observations
	• Draw conclusions from observations

Table 17.2 Categories of preschool curriculum, Part 2

Type of capability	Specific skills to be evaluated
Logical-mathematical	• Categorizing
	• Sorting
	• Arithmetic
	• Performing simple "proto-scientific experiments"
Nonverbal communication	• Communicating via gesture
	• Dramatizing situations
	• Dramatizing needs, wants
	• Express empathy
Spatial-visual	• Visual patterning
	• Self-expression through drawing
	• Navigate
Objective	• Assembling objects
	• Disassembling objects
	• Measurement
	• Symmetry
	• Similarity between structures
	(e.g. block structures and real ones)

not at all surprising since one of Gardner's key motivations in articulating MI theory was the pragmatics of educating humans with diverse strengths and weaknesses.

Regarding intelligence as "the ability to achieve complex goals in complex environments", it is apparent that preschools are specifically designed to pack a large

Table 17.3 Categories of preschool curriculum, Part 3

Type of capability	Specific skills to be evaluated
Interpersonal	• Cooperation
	• Display appropriate behavior in various settings
	• Clean up belongings
	• Share supplies
Emotional	• Delay gratification
	• Control emotional reactions
	• Complete projects

variety of different micro-environments (the learning centers) into a single room, and to present a variety of different tasks in each environment. The environments constituted by preschool learning centers are designed as microcosms of the most important aspects of the environments faced by humans in their everyday lives.

17.4 Task-Based Assessment in AGI Preschool

Professional pedagogues such as [CM07] discuss evaluation of early childhood learning as intended to assess both specific curriculum content knowledge as well as the child's learning process. It should be as unobtrusive as possible, so that it just seems like another engaging activity, and the results used to tailor the teaching regimen to use different techniques to address weaknesses and reinforce strengths.

For example, with group building of a model car, students are tested on a variety of skills: procedural understanding, visual acuity, motor acuity, creative problem solving, interpersonal communications, empathy, patience, manners, and so on. With this kind of complex, yet engaging, activity as a metric the teacher can see how each student approaches the process of understanding each subtask, and subsequently guide each student's focus differently depending on strengths and weaknesses.

In Tables 17.4 and 17.5 we describe some particular tasks that AGIs may be meaningfully assigned in the context of a general AGI Preschool design and curriculum as described above. Of course, this is a very partial list, and is intended as evocative rather than comprehensive.

Any one of these tasks can be turned into a rigorous quantitative test, thus allowing the precise comparison of different AGI systems' capabilities; but we have chosen not to emphasize this point here, partly for space reasons and partly for philosophical ones. In some contexts the quantitative comparison of different systems may be the right thing to do, but as discussed in Chap. 18 there are also risks associated with this approach, including the emergence of an overly metrics-focused "bakeoff mentality" among system developers, and overfitting of AI abilities to test taking. What is most important is the isolation of specific tasks on which different systems may be experientially trained and then qualitatively assessed and compared, rather than the evaluation of quantitative metrics.

Table 17.4 Prototypical preschool intelligence assessment tasks, Part 1

Intelligence type	Test
Linguistic	• Write a set of instructions
	• Speak on a subject
	• Edit a written piece or work
	• Write a speech
	• Commentate on an event
	• Apply positive or negative 'spin' to a story
Logical-mathematical	• Perform arithmetic calculations
	• Create a process to measure something
	• Analyse how a machine works
	• Create a process
	• Devise a strategy to achieve an aim
	• Assess the value of a proposition
Musical	• Perform a musical piece
	• Sing a song
	• Review a musical work
	• Coach someone to play a musical instrument
Bodily-kinesthetic	• Juggle
	• Demonstrate a sports technique
	• Flip a beer-mat
	• Create a mime to explain something
	• Toss a pancake
	• Fly a kite

Table 17.5 Prototypical preschool intelligence assessment tasks, Part 2

Intelligence type	Test
Spatial-visual	• Design a costume
	• Interpret a painting
	• Create a room layout
	• Create a corporate logo
	• Design a building
	• Pack a suitcase or the trunk of a car
Interpersonal	• Interpret moods from facial expressions
	• Demonstrate feelings through body language
	• Affect the feelings of others in a planned way
	• Coach or counsel another

Task-oriented testing allows for feedback on applications of general pedagogical principles to real-world, embodied activities. This allows for iterative refinement based learning (shaping), and cross development of knowledge acquisition and application (multitask learning). It also helps militate against both cheating, and overfitting, as teachers can make ad-hoc modifications to the tests to determine if this is happening and correct for it if necessary.

E.g., consider a linguistic task in which the AGI is required to formulate a set of instructions encapsulating a given behavior (which may include components that are physical, social, linguistic, etc.). Note that although this is presented as centrally a linguistic task, it actually involves a diverse set of competencies since the behavior to be described may encompass multiple real-world aspects.

To turn this task into a more thorough test one might involve a number of human teachers and a number of human students. Before the test, an ensemble of copies of the AGI would be created, with identical knowledge state. Each copy would interact with a different human teacher, who would demonstrate to it a certain behavior. After testing the AGI on its own knowledge of the material, the teacher would then inform the AGI that it will then be tested on its ability to verbally describe this behavior to another. Then, the teacher goes away and the copy interacts with a series of students, attempting to convey to the students the instructions given by the teacher.

The teacher can thereby assess both the AGI's understanding of the material, and the ability to explain it to the other students. This separates out assessment of understanding from assessment of ability to communicate understanding, attempting to avoid conflation of one with the other. The design of the training and testing needs to account for potential.

This testing protocol abstracts away from the particularities of any one teacher or student, and focuses on effectiveness of communication in a human context rather than according to formalized criteria. This is very much in the spirit of how assessment takes place in human preschools (with the exception of the copying aspect): formal exams are rarely given in preschool, but pragmatic, socially-embedded assessments are regularly made.

By including the copying aspect, more rigorous statistical assessments can be made regarding efficacy of different approaches for a given AGI design, independent of past teaching experiences. The multiple copies may, depending on the AGI system design, then be able to be reintegrated, and further "learning" be done by higher-order cognitive systems in the AGI that integrate the disparate experiences of the multiple copies.

This kind of parallel learning is different from both sequential learning that humans do, and parallel presences of a single copy of an AGI (such as in multiple chat rooms type experiments). All three approaches are worthy of study, to determine under what circumstances, and with which AGI designs, one is more successful than another.

It is also worth observing how this test could be tweaked to yield a test of generalization ability. After passing the above, the AGI could then be given a description of a new task (acquisition), and asked to explain the new one (variation). And, part of the training behavior might be carried out unobserved by the AGI, thus requiring the AGI to infer the omitted parts of the task it needs to describe.

Another popular form of early childhood testing is puzzle block games. These kinds of games can be used to assess a variety of important cognitive skills, and to do so in a fun way that not only examines but also encourages creativity and flexible thinking. Types of games include pattern matching games in which students replicate patterns described visually or verbally, pattern creation games in which students create new patterns guided by visually or verbally described principles, creative interpretation of patterns in which students find meaning in the forms, and free-form creation. Such games may be individual or cooperative.

Cross training and assessment of a variety of skills occurs with pattern block games: for example, interpretation of visual or linguistic instructions, logical procedure and pattern following, categorizing, sorting, general problem solving, creative interpretation, experimentation, and kinematic acuity. By making the games cooperative, various interpersonal skills involving communication and cooperation are also added to the mix.

The puzzle block context bring up some general observations about the role of kinematic and visuospatial intelligence in the AGI Preschool. Outside of robotics and computer vision, AI research has often downplayed these sorts of intelligence (though, admittedly, this is changing in recent years, e.g. with increasing research focus on diagrammatic reasoning). But these abilities are not only necessary to navigate real (or virtual) spatial environments. They are also important components of a coherent, conceptually well-formed understanding of the world in which the student is embodied. Integrative training and assessment of both rigorous cognitive abilities generally most associated with both AI and "proper schooling" (such as linguistic and logical skills) along with kinematic and aesthetic/sensory abilities is essential to the development of an intelligence that can successfully both operate in and sensibly communicate about the real world in a roughly humanlike manner. Whether or not an AGI is targeted to interpret physical-world spatial data and perform tasks via robotics, in order to communicate ideas about a vast array of topics of interest to any intelligence in this world, an AGI must develop aspects of intelligence other than logical and linguistic cognition.

17.5 Beyond Preschool

Once an AGI passes preschool, what are the next steps? There is still a long way to go, from preschool to an AGI system that is capable of, say, passing the Turing Test or serving as an effective artificial scientist.

Our suggestion is to extend the school metaphor further, and make use of existing curricula for higher levels of virtual education: grade school, secondary school, and all levels of post-secondary education. If an AGI can pass online primary and secondary schools such as e-tutor.com, and go on to earn an online degree from an accredited university, then clearly said AGI has successfully achieved "human level, roughly humanlike AGI". This sort of testing is interesting not only because it allows assessment of stages intermediate between preschool and adult, but also because it tests humanlike intelligence without requiring precise imitation of human behavior.

If an AI can get a BA degree at an accredited university, via online coursework (assuming for simplicity courses where no voice interaction is needed), then we should consider that AI to have human-level intelligence. University coursework spans multiple disciplines, and the details of the homework assignments and exams are not known in advance, so like a human student the AGI team can't cheat.

In addition to the core coursework, a schooling approach also tests basic social interaction and natural language communication, ability to do online research, and general problem solving ability. However, there is no rigid requirement to be strictly humanlike in order to pass university classes.

Most of our concrete examples in the following chapters will pertain to the preschool context, because it's simple to understand, and because we feel that getting to the "AGI preschool student" level is going to be the largest leap. Once that level is obtained, moving further will likely be difficult also, but we suspect it will be more a matter of steady incremental improvements—whereas the achievement of preschool-level functionality will be a large leap from the current situation.

17.6 Issues with Virtual Preschool Engineering

As noted above there are two broad approaches to realizing the "AGI Preschool" idea: using the AGI to control a physical robot and then crafting a preschool environment suitable to the robot's sensors and actuators; or, using the AGI to control a virtual agent in an appropriately rich virtual-world preschool. The robotic approach is harder from an AI perspective (as one must deal with problems of sensation and actuation), but easier from an environment-construction perspective. In the virtual world case, one quickly runs up against the current limitations of virtual world technologies, which have been designed mainly for entertainment or social-networking purposes, not with the requirements of AGI systems in mind.

In Chap. 10 we discussed the general requirements that an environment should possess to be supportive of humanlike intelligence. Referring back to that list, it's clear that current virtual worlds are fairly strong on multimodal communication, and fairly weak on naive physics. More concretely, if one wants a virtual world so that:

1. one could carry out all the standard cognitive development experiments described in developmental psychology books
2. one could implement intuitively reasonable versions of all the standard activities in all the standard learning stations in a contemporary preschool.

then current virtual world technologies appear not to suffice.

As reviewed above, typical preschool activities include for instance building with blocks, playing with clay, looking in a group at a picture book and hearing it read aloud, mixing ingredients together, rolling/throwing/catching balls, playing games like tag, hide-and-seek, Simon Says or Follow the Leader, measuring objects, cutting paper into different shapes, drawing and coloring, etc.

And, as typical, not necessarily representative examples of tasks psychologists use to measure cognitive development (drawn mainly from the Piagetian tradition,

Fig. 17.1 A Piagetan conservation of volume experiment. Image from http://psychology4a.com/develop2.htm, copyright Pete Waring, used with permission.

without implying any assertion that this is the only tradition worth pursuing), consider the following:

1. Which row has more circles—A or B? A: O O O O O, B: OOOOO
2. If Mike is taller than Jim, and Jim is shorter than Dan, then who is the shortest? Who is the tallest?
3. Which is heavier—a pound of feathers or a pound of rocks?
4. Eight ounces of water is poured into a glass that looks like the fat glass in Fig. 17.1 and then the same amount is poured into a glass that looks like the tall glass in Fig. 17.1. Which glass has more water?
5. A lump of clay is rolled into a snake. All the clay is used to make the snake. Which has more clay in it—the lump or the snake?
6. There are two dolls in a room, Sally and Ann, each of which has her own box, with a marble hidden inside. Sally goes out for a minute, leaving her box behind; and Ann decides to play a trick on Sally: she opens Sally's box, removes the marble, hiding it in her own box. Sally returns, unaware of what happened. Where will Sally would look for her marble?
7. Consider this rule about a set of cards that have letters on one side and numbers on the other: "If a card has a vowel on one side, then it has an even number on the other side." If you have 4 cards labeled "E K 4 7", which cards do you need to turn over to tell if this rule is actually true?
8. Design an experiment to figure out how to make a pendulum that swings more slowly versus less slowly.

What we see from this ad hoc, partial list is that a lot of naive physics *is* required to make an even vaguely realistic preschool. A lot of preschool education is about the intersection between abstract cognition and naive physics. A more careful review of the various tasks involved in preschool education bears out this conclusion.

With this in mind, in this section we will briefly describe an approach to extending current virtual world technologies that appears to allow the construction of a reasonably rich and realistic AGI preschool environment, without requiring anywhere near a complete simulation of realistic physics.

17.6.1 Integrating Virtual Worlds with Robot Simulators

One major deficiency in current virtual world platforms is the lack of flexibility in terms of tool use. In most of these systems today, an avatar can pick up or utilize an object, or two objects can interact, only in specific, pre-programmed ways. For instance, an avatar might be able to pick up a virtual screwdriver only by the handle, rather than by pinching the blade between its fingers. This places severe limits on creative use of tools, which is absolutely critical in a preschool context. The solution to this problem is clear: adapt existing generalized physics engines to mediate avatar-object and object-object interactions. This would require more computation than current approaches, but not more than is feasible in a research context.

One way to achieve this goal would be to integrate a robot simulator with a virtual world or game engine, for instance to modify the OpenSim (http://opensimulator.org) virtual world to use the Gazebo (http://playerstage.sourceforge.net) robot simulator in place of its current physics engine. While tractable, such a project would require considerable software engineering effort.

17.6.2 BlocksNBeads World

Another major deficiency in current virtual world platforms is their inability to model physical phenomena besides rigid objects with any sophistication. In this section we propose a potential solution to this issue: a novel class of virtual worlds called BlocksNBeadsWorld, consisting of the following aspects:

1. 3D blocks of various shapes and sizes and frictional coefficients, that can be stacked
2. Adhesive that can be used to stick blocks together, and that comes in two types, one of which can be removed by an adhesive-removing substance, one of which cannot (though its bonds can be broken via sufficient application of force)
3. Spherical beads, each of which has intrinsic unchangeable adhesion properties defined according to a particular, simple "adhesion logic"
4. Each block, and each bead, may be associated with multidimensional quantities representing its taste and smell; and may be associated with a set of sounds that are made when it is impacted with various forces at various positions on its surface.

Interaction between blocks and beads is to be calculated according to standard Newtonian physics, which would be compute-intensive in the case of a large number of beads, but tractable using distributed processing. For instance if 10 K beads were used to cover a humanoid agent's face, this would provide a fairly wide diversity of facial expressions; and if 10 K beads were used to form a blanket laid on a bed, this would provide a significant amount of flexibility in terms of rippling, folding and so forth. Yet, this order of magnitude of interactions is very small compared to what is done in contemporary simulations of fluid dynamics or, say, quantum chromodynamics.

One key aspect of the spherical beads is that they can be used to create a variety of rigid or flexible surfaces, which may exist on their own or be attached to blocks-based constructs. The specific inter-bead adhesion properties of the beads could be defined in various ways, and will surely need to be refined via experimentation, but a simple scheme that seems to make sense is as follows.

Each bead can have its surface tesselated into hexagons (the number of these can be tuned), and within each hexagon it can have two different adhesion coefficients: one for adhesion to other beads, and one for adhesion to blocks. The adhesion between two beads along a certain hexagon is then determined by their two adhesion coefficients; and the adhesion between a bead and a block is determined by the adhesion coefficient of the bead, and the adhesion coefficient of the adhesive applied to the block. A distinction must be drawn between rigid and flexible adhesion: rigid adhesion sticks a bead to something in a way that can't be removed except via breaking it off; whereas flexible adhesion just keeps a bead very close to the thing it's stuck onto. Any two entities may be stuck together either rigidly or flexibly. Sets of beads with flexible adhesion to each other can be used to make entities like strings, blankets or clothes.

Using the above adhesion logic, it seems one could build a wide variety of flexible structures using beads, such as (to give a very partial list):

1. fabrics with various textures, that can be draped over blocks structures,
2. multilayered coatings to be attached to blocks structures, serving (among many other examples) as facial expressions
3. liquid-type substances with varying viscosities, that can be poured between different containers, spilled, spread, etc.
4. strings table in knots; rubber bands that can be stretched; etc.

Of course there are various additional features one could add. For instance one could add a special set of rules for vibrating strings, allowing BlocksNBeadsWorld to incorporate the creation of primitive musical instruments. Variations like this could be helpful but aren't necessary for the world to serve its essential purpose.

Note that one does not have true fluid dynamics in BlocksNBeadsWorld, but, it seems that the latter is not necessary to encompass the phenomena covered in cognitive developmental tests or preschool tasks. The tests and tasks that are done with fluids can instead be done with masses of beads. For example, consider the conservation of volume task shown in Figs. 17.1: it's easy enough to envision this being done with beads rather than milk. Even a few hundred beads is enough to be psychologically perceived as a mass rather than a set of discrete units, and to be manipulated and analyzed as such. And the simplification of not requiring fluid mechanics in one's virtual world is immense.

Next, one can implement equations via which the adhesion coefficients of a bead are determined in part by the adhesion coefficients of nearby beads, or beads that are nearby in certain directions (with direction calculated in local spherical coordinates). This will allow for complex cracking and bending behaviors—not identical to those in the real world, but with similar qualitative characteristics. For example, without this feature one could create paper like substances that could be cut with scissors—but *with* this feature, one could go further and create woodlike substances that would crack when nails were hammered into them in certain ways, and so forth.

Further refinements are certainly possible also. One could add multidimensional adhesion coefficients, allowing more complex sorts of substances. One could allow beads to vibrate at various frequencies, which would lead to all sorts of complex wave patterns in bead compounds, etc. In each case, the question to be asked is: what important cognitive abilities are dramatically more easily learnable in the presence of the new feature than in its absence?

The combination of blocks and beads seems ideal for implementing a more flexible and AGI-friendly type of virtual body than is currently used in games and virtual worlds. One can easily envision implementing a body with:

1. a skeleton whose bones consist of appropriately shaped blocks
2. joints consisting of beads, flexibly adhered to the bones
3. flesh consisting of beads, flexibly adhered to each other
4. internal "plumbing" consisting of tubes whose walls are beads rigidly adhered to each other, and flexibly adhered to the surrounding flesh (the plumbing could then serve to pass beads through, where slow passage would be ensured by weak adhesion between the walls of the tubes and the beads passing through the tubes).

This sort of body would support rich kinesthesia; and rich, broad analogy-drawing between the internally-experienced body and the externally-experienced world. It would also afford many interesting opportunities for flexible movement control. Virtual animals could be created along with virtual humanoids.

Regarding the extended mind, it seems clear that blocks and beads are adequate for the creation of a variety of different tools. Equipping agents with "glue guns" able to affect the adhesive properties of both blocks and beads would allow a diversity of building activity; and building with masses of beads could become a highly creative activity. Furthermore, beads with appropriately specified adhesion (within the framework outlined above) could be used to form organically growing plant-like substances, based on the general principles used in L-system models of plant growth (Prusinciewicz and Lindenmayer 1991). Structures with only beads would vaguely resemble herbaceous plants; and structures involving both blocks and beads would more resemble woody plants. One could even make organic structures that flourish or otherwise based on the light available to them (without of course trying to simulate the chemistry of photosynthesis).

Some elements of chemistry may be achieved as well, though nowhere near what exists in physical reality. For instance, melting and boiling at least should be doable: assign every bead a temperature, and let solid interbead bonds turn liquid above a certain temperature and disappear completely above some higher temperature. You could even have a simple form of fire. Let fire be an element, whose beads have negative gravitational mass. Beads of fuel elements like wood have a threshold temperature above which they will turn into fire beads, with release of additional heat.[2]

The philosophy underlying these suggested bead dynamics is somewhat comparable to that outlined in Wolfram's book *A New Kind of Science* [Wol02]. There he proposes cellular automata models that emulate the qualitative characteristics of various real-world phenomena, without trying to match real-world data precisely. For instance, some of his cellular automata demonstrate phenomena very similar to turbulent fluid flow, without implementing the Navier-Stokes equations of fluid dynamics or trying to precisely match data from real-world turbulence. Similarly, the beads in BlocksNBeadsWorld are intended to qualitatively demonstrate the real-world phenomena most useful for the development of humanlike embodied intelligence, without trying to precisely emulate the real-world versions of these phenomena.

The above description has been left imprecisely specified on purpose. It would be straightforward to write down a set of equations for the block and bead interactions, but there seems little value in articulating such equations without also writing a simulation involving them and testing the ensuing properties. Due to the complex dynamics of bead interactions, the fine-tuning of the bead physics is likely to involve some tuning based on experimentation, so that any equations written down now would likely be revised based on experimentation anyway. Our goal here has been to outline a certain class of potentially useful environments, rather than to articulate a specific member of this class.

Without the beads, BlocksNBeadsWorld would appear purely as a "Blocks World with Glue"—essentially a substantially upgraded version of the Blocks Worlds frequently used in AI, since first introduced in [Win72]. Certainly a pure "Blocks World with Glue" would have greater simplicity than BlocksNBeadsWorld, and greater richness than standard Blocks World; but this simplicity comes with too many limitations, as shown by consideration of the various naive physics requirements inventoried above. One simply cannot run the full spectrum of humanlike cognitive development experiments, or preschool educational tasks, using blocks and glue alone. One can try to create analogous tasks using only blocks and glue, but this quickly becomes extremely awkward. Whereas in the BlocksNBeadsWorld the capability for this full spectrum of experiments and tasks seems to fall out quite naturally.

What's missing from BlocksNBeadsWorld should be fairly obvious. There isn't really any distinction between a fluid and a powder: there are masses, but the types and properties of the masses are not the same as in the real world, and will surely lack the nuances of real-world fluid dynamics. Chemistry is also missing: processes like cooking and burning, although they can be crudely emulated, will not have the same richness as in the real world. The full complexity of body processes is not there: the body-design method mentioned above is far richer and more adaptive and responsive than current methods of designing virtual bodies in 3DS Max or Maya and importing them into virtual world or game engines, but still drastically simplistic compared to real bodies with their complex chemical signaling systems and couplings with other bodies and the environment. The hypothesis we're making in this section

---

[2] Thanks are due to Russell Wallace for the suggestions in this paragraph.

is that these lacunae aren't that important from the point of view of humanlike cognitive development. We suggest that the key features of naive physics and folk psychology enumerated above can be mastered by an AGI in BlocksNBeadsWorld in spite of its limitations, and that—together with an appropriate AGI design—this probably suffices for creating an AGI with the inductive biases constituting humanlike intelligence.

To drive this point home more thoroughly, consider three potential virtual world scenarios:

1. A world containing realistic fluid dynamics, where a child can pour water back and forth between two cups of different shapes and sizes, to understand issues such as conservation of volume
2. A world more like today's Second Life, where fluids don't really exist, and things like lakes are simulated via very simple rules, and pouring stuff back and forth between cups doesn't happen unless it's programmed into the cups in a very specialized way
3. A BlocksNBeadsWorld type world, where a child can pour masses of beads back and forth between cups, but not masses of liquid.

Our qualitative judgment is that Scenario 3 is going to allow a young AI to gain the same essential insights as Scenario 1, whereas Scenario 2 is just too impoverished. I have explored dozens of similar scenarios regarding different preschool tasks or cognitive development experiments, and come to similar conclusions across the board. Thus, our current view is that something like BlocksNBeadsWorld can serve as an adequate infrastructure for an AGI Preschool, supporting the development of human-level, roughly human-like AGI.

And, if this view turns out to be incorrect, and BlocksNBeadsWorld is revealed as inadequate, then we will very likely still advocate the conceptual approach enunciated above as a guide for designing virtual worlds for AGI. That is, we would suggest to explore the hypothetical failure of BlocksNBeadsWorld via asking two questions:

1. Are there basic naive physics or folk psychology requirements that were missed in creating the specifications, based on which the adequacy of BlocksNBeadsWorld was assessed?
2. Does BlocksNBeadsWorld fail to sufficiently emulate the real world in respect to some of the articulated naive physics or folk psychology requirements?

The answers to these questions would guide the improvement of the world or the design of a better one.

Regarding the practical implementation of BlocksNBeadsWorld, it seems clear that this is within the scope of modern game engine technology, however, it is not something that could be encompassed within an existing game or world engine without significant additions; it would require substantial custom engineering. There exist commodity and open-source physics engines that efficiently carry out Newtonian mechanics calculations; while they might require some tuning and extension to handle BlocksNBeadWorld, the main issue would be achieving adequate speed of physics calculation, which given current technology would need to be done via modifying existing engines to appropriately distribute processing among multiple GPUs.

Finally, an additional avenue that merits mention is the use of BlocksNBeads physics internally within an AGI system, as part of an internal simulation world that allows it to make "mind's eye" estimative simulations of real or hypothetical physical situations. There seems no reason that the same physics software libraries couldn't be used both for the external virtual world that the AGI's body lives in, and for an internal simulation world that the AGI uses as a cognitive tool. In fact, the BlocksNBeads library could be used as an internal cognitive tool by AGI systems controlling physical robots as well. This might require more tuning of the bead dynamics to accord with the dynamics of various real-world systems; but, this tuning would be beneficial for the BlocksNBeadWorld as well.

Chapter 18
A Preschool-Based Roadmap to Advanced AGI

18.1 Introduction

Supposing the CogPrime approach to creating advanced AGI is workable—then what are the right practical steps to follow? The various structures and algorithms outlined in Part 2, should be engineered and software-tested, of course—but that's only part of the study. The AGI system implemented will need to be taught, and it will need to be placed in situations where it can develop an appropriate self-model and other critical internal network structures. The complex structures and algorithms involved will need to be fine-tuned in various ways, based on qualitatively observing the overall system's behavior in various situations. To get all this right without excessive confusion or time-wastage requires a fairly clear *roadmap* for CogPrime development.

In this chapter we'll sketch one particular roadmap for the development of human-level, roughly human-like AGI—which we're not selling as the only one, or even necessarily as the best one. It's just one roadmap that we have thought about a lot, and that we believe has a strong chance of proving effective. Given resources to pursue only one path for AGI development and teaching, this would be our choice, at present. The roadmap outlined here is not restricted to CogPrime in any highly particular ways, but it has been developed largely with CogPrime in mind; those developing other AGI designs could probably use this roadmap just fine, but might end up wanting to make various adjustments based on the strengths and weaknesses of their own approach.

What we mean here by a "roadmap" is, in brief: a sequence of "milestone" tasks, occurring in a small set of common environments or "scenarios", organized so as to lead to a commonly agreed upon set of long-term goals. I.e., what we are after here is a "capability roadmap"—a roadmap laying out a series of capabilities whose achievement seems likely to lead to human-level AGI. Other sorts of roadmaps such as "tools roadmaps" may also be valuable, but are not our concern here.

More precisely, we confront the task of roadmapping by identifying scenarios in which to embed our AGI system, and then "competency areas" in which the AGI

B. Goertzel et al., *Engineering General Intelligence, Part 1*,
Atlantis Thinking Machines 5, DOI: 10.2991/978-94-6239-027-0_18,
© Atlantis Press and the authors 2014

system must be evaluated. Then, we envision a roadmap as consisting of a set of one or more task-sets, where each task set is formed from a combination of a scenario with a list of competency areas. To create a task-set one must choose a particular scenario, and then articulate a set of specific tasks, each one addressing one or more of the competency areas. Each task must then get associated with particular performance metrics—quantitative wherever possible, but perhaps qualitative in some cases depending on the nature of the task. Here we give a partial task-set for the "virtual and robot preschool" scenarios discussed in Chap. 17, and a couple example quantitative metrics just to illustrate what is intended; the creation of a fully detailed roadmap based on the ideas outlined here is left for future work.

The train of thought presented in this chapter emerged in part from a series of conversations preceding and during the "AGI Roadmap Workshop" held at the University of Tennessee, Knoxville in October 2008. Some of the ideas also trace back to discussions held during two workshops on "Evaluation and Metrics for Human-Level AI" organized by John Laird and Pat Langley (one in Ann Arbor in late 2008, and one in Tempe in early 2009). Some of the conclusions of the Ann Arbor workshop were recorded in [LWML09]. Inspiration was also obtained from discussion at the "Future of AGI" post-conference workshop of the AGI-09 conference, triggered by Itamar Arel's [ARK09a] presentation on the "AGI Roadmap" theme; and from an earlier article on AGI Roadmapping by [AL09].

However, the focus of the AGI Roadmap Workshop was considerably more general than the present chapter. Here we focus on preschool-type scenarios, whereas at the workshop a number of scenarios were discussed, including the preschool scenarios but also, for example,

- Standardized Tests and School Curricula
- Elementary, Middle and High School Student
- General Videogame Learning
- Wozniak's Coffee Test: go into a random American house and figure out how to make coffee, and do it
- Robot College Student
- General Call Center Respondent

For each of these scenarios, one may generate tasks corresponding to each of the competency areas we will outline below. CogPrime is applicable in all these scenarios, so our choice to focus on preschool scenarios is an additional judgment call beyond those judgment calls required to specify the CogPrime design. The roadmap presented here is a "AGI Preschool Roadmap" and as such is a special case of the broader "AGI Roadmap" outlined at the workshop.

18.2 Measuring Incremental Progress Toward Human-Level AGI

In Chap. 4, we discussed several examples of practical goals that we find to plausibly characterize "human level AGI", e.g.

- Turing Test
- Virtual World Turing Test
- Online University Test
- Physical University Test
- Artificial Scientist Test

We also discussed our optimism regarding the possibility that in the future AGI may advance beyond the human level, rendering all these goals "early-stage subgoals".

However, in this chapter we will focus our attention on the nearer term. The above goals are ambitious ones, and while one can talk a lot about how to precisely measure their achievement, we don't feel that's the most interesting issue to ponder at present. More critical is to think about how to measure incremental progress. How do you tell when you're 25 or 50 % of the way to having an AGI that can pass the Turing Test, or get an online university degree. Fooling 50 % of the Turing Test judges is not a good measure of being 50 % of the way to passing the Turing Test (that's too easy); and passing 50 % of university classes is not a good measure of being 50 % of the way to getting an online university degree (it's too hard—if one had an AGI capable of doing that, one would almost surely be very close to achieving the end goal). Measuring incremental progress toward human-level AGI is a subtle thing, and we argue that the best way to do it is to focus on particular scenarios and the achievement of specific competencies therein.

As we argued in Chap. 9 there are some theoretical reasons to doubt the possibility of creating a rigorous objective test for partial progress toward AGI—a test that would be convincing to skeptics, and impossible to "game" via engineering a system specialized to the test. Fortunately, though we don't need a test of this nature for the purposes of assessing our own incremental progress toward advanced AGI, based on our knowledge about our own approach.

Based on the nature of the grand goals articulated above, there seems to be a very natural approach to creating a set of incremental capabilities building toward AGI: *to draw on our copious knowledge about human cognitive development*. This is by no means the only possible path; one can envision alternatives that have nothing to do with human development (and those might also be better suited to non-human AGIs). However, so much detailed knowledge about human development is available— as well as solid knowledge that the human developmental trajectory does lead to human-level AI—that the motivation to draw on human cognitive development is quite strong.

The main problem with the human development inspired approach is that cognitive developmental psychology is not as systematic as it would need to be for AGI to be able to translate it directly into architectural principles and requirements. As noted above, while early thinkers like Piaget and Vygotsky outlined systematic theories of child cognitive development, these are no longer considered fully accurate, and one currently faces a mass of detailed theories of various aspects of cognitive development, but without an unified understanding. Nevertheless we believe it is viable to work from the human-development data and understanding currently available, and craft a workable AGI roadmap therefrom.

With this in mind, what we give next is a fairly comprehensive list of the competencies that we feel AI systems should be expected to display in one or more of these scenarios in order to be considered as full-fledged "human level AGI" systems. These competency areas have been assembled somewhat opportunistically via a review of the cognitive and developmental psychology literature as well as the scope of the current AI field. We are not claiming this as a precise or exhaustive list of the competencies characterizing human-level general intelligence, and will be happy to accept additions to the list, or mergers of existing list items, etc. What we are advocating is not this specific list, but rather the approach of enumerating competency areas, and then generating tasks by combining competency areas with scenarios.

We also give, with each competency, an example task illustrating the competency. The tasks are expressed in the robot preschool context for concreteness, but they all apply to the virtual preschool as well. Of course, these are only examples, and ideally to teach an AGI in a structured way one would like to

- associate several tasks with each competency
- present each task in a graded way, with multiple subtasks of increasing complexity
- associate a quantitative metric with each task

However, the briefer treatment given here should suffice to give a sense for how the competencies manifest themselves practically in the AGI Preschool context.

1. Perception

   - **Vision**: image and scene analysis and understanding
     - *Example task:* When the teacher points to an object in the preschool, the robot should be able to identify the object and (if it's a multi-part object) its major parts. If it can't perform the identification initially, it can approach the object and manipulate it before making its identification.
   - **Hearing**: identifying the sounds associated with common objects; understanding which sounds come from which sources in a noisy environment
     - *Example task:* When the teacher covers the robot's eyes and then makes a noise with an object, the robot should be able to guess what the object is.
   - **Touch**: identifying common objects and carrying out common actions using touch alone
     - *Example task:* With its eyes and ears covered, the robot should be able to identify some object by manipulating it; and carry out some simple behaviors (say, putting a block on a table) via touch alone.
   - **Crossmodal**: Integrating information from various senses
     - *Example task:* Identifying an object in a noisy, dim environment via combining visual and auditory information.
   - **Proprioception**: Sensing and understanding what its body is doing
     - *Example task:* The teacher moves the robot's body into a certain configuration. The robot is asked to restore its body to an ordinary standing position, and then repeat the configuration that the teacher moved it into.

2. Actuation
   - **Physical skills**: manipulating familiar and unfamiliar objects
     - *Example task:* Manipulate blocks based on imitating the teacher: e.g. pile two blocks atop each other, lay three blocks in a row, etc.
   - **Tool use**, including the flexible use of ordinary objects as tools
     - *Example task:* Use a stick to poke a ball out of a corner, where the robot cannot directly reach.
   - **Navigation**, including in complex and dynamic environments
     - *Example task:* Find its own way to a named object or person through a crowded room with people walking in it and objects laying on the floor.

3. Memory
   - **Declarative**: noticing, observing and recalling facts about its environment and experience
     - *Example task:* If certain people habitually carry certain objects, the robot should remember this (allowing it to know how to find the objects when the relevant people are present, even much later).
   - **Behavioral**: remembering how to carry out actions
     - *Example task:* If the robot is taught some skill (say, to fetch a ball), it should remember this much later.
   - **Episodic**: remembering significant, potentially useful incidents from life history
     - *Example task:* Ask the robot about events that occurred at times when it got particularly much, or particularly little, reward for its actions; it should be able to answer simple questions about these, with significantly more accuracy than about events occurring at random times.

4. Learning
   - **Imitation**: Spontaneously adopt new behaviors that it sees others carrying out
     - *Example task:* Learn to build towers of blocks by watching people do it.
   - **Reinforcement**: Learn new behaviors from positive and/or negative reinforcement signals, delivered by teachers and/or the environment
     - *Example task:* Learn which box the red ball tends to be kept in, by repeatedly trying to find it and noticing where it is, and getting rewarded when it finds it correctly.
   - **Imitation/Reinforcement**
     - *Example task:* Learn to play "fetch", "tag" and "follow the leader" by watching people play it, and getting reinforced on correct behavior.
   - **Interactive Verbal Instruction**
     - *Example task:* Learn to build a particular structure of blocks faster based on a combination of imitation, reinforcement and verbal instruction, than by imitation and reinforcement without verbal instruction.
   - **Written Media**
     - *Example task:* Learn to build a structure of blocks by looking at a series of diagrams showing the structure in various stages of completion.

- **Learning via Experimentation**
  - *Example task:* Ask the robot to slide blocks down a ramp held at different angles. Then ask it to make a block slide fast, and see if it has learned how to hold the ramp to make a block slide fast.

5. Reasoning

- **Deduction**, from uncertain premises observed in the world
  - *Example task:* If Ben more often picks up red balls than blue balls, and Ben is given a choice of a red block or blue block to pick up, which is he more likely to pick up?
- **Induction**, from uncertain premises observed in the world
  - *Example task:* If Ben comes into the lab every weekday morning, then is Ben likely to come to the lab today (a weekday) in the morning?
- **Abduction**, from uncertain premises observed in the world
  - *Example task:* If women more often give the robot food than men, and then someone of unidentified gender gives the robot food, is this person a man or a woman?
- **Causal Reasoning**, from uncertain premises observed in the world
  - *Example task:* If the robot knows that knocking down Ben's tower of blocks makes him angry, then what will it say when asked if kicking the ball at Ben's tower of blocks will make Ben mad?
- **Physical Reasoning**, based on observed "fuzzy rules" of naive physics
  - *Example task:* Given two balls (one rigid and one compressible) and two tunnels (one significantly wider than the balls, one slightly narrower than the balls), can the robot guess which balls will fit through which tunnels?
- **Associational Reasoning**, based on observed spatiotemporal associations
  - *Example task:* If Ruiting is normally seen near Shuo, then if the robot knows where Shuo is, that is where it should look when asked to find Ruiting.

6. Planning

- **Tactical**
  - *Example task:* The robot is asked to bring the red ball to the teacher, but the red ball is in the corner where the robot can't reach it without a tool like a stick. The robot knows a stick is in the cabinet so it goes to the cabinet and opens the door and gets the stick, and then uses the stick to get the red ball, and then brings the red ball to the teacher.
- **Strategic**
  - *Example task:* Suppose that Matt comes to the lab infrequently, but when he does come he is very happy to see new objects he hasn't seen before (and suppose the robot likes to see Matt happy). Then when the robot gets a new object Matt has not seen before, it should put it away in a drawer and be sure not to lose it or let anyone take it, so it can show Matt the object the next time Matt arrives.

- **Physical**
  - *Example task:* To pick up a cup with a handle which is lying on its side in a position where the handle can't be grabbed, the robot turns the cup in the right position and then picks up the cup by the handle.
- **Social**
  - *Example task:* The robot is given a job of building a tower of blocks by the end of the day, and he knows Ben is the most likely person to help him, and he knows that Ben is more likely to say "yes" to helping him when Ben is alone. He also knows that Ben is less likely to say "yes" if he's asked too many times, because Ben doesn't like being nagged. So he waits to ask Ben till Ben is alone in the lab.

7. Attention

- **Visual Attention** within its observations of its environment
  - *Example task:* The robot should be able to look at a scene (a configuration of objects in front of it in the preschool) and identify the key objects in the scene and their relationships.
- **Social Attention**
  - *Example task:* The robot is having a conversation with Itamar, which is giving the robot reward (for instance, by teaching the robot useful information). Conversations with other individuals in the room have not been so rewarding recently. But Itamar keeps getting distracted during the conversation, by talking to other people, or playing with his cellphone. The robot needs to know to keep paying attention to Itamar even through the distractions.
- **Behavioral Attention**
  - *Example task:* The robot is trying to navigate to the other side of a crowded room full of dynamic objects, and many interesting things keep happening around the room. The robot needs to largely ignore the interesting things and focus on the movements that are important for its navigation task.

8. Motivation

- **Subgoal Creation**, based on its preprogrammed goals and its reasoning and planning
  - *Example task:* Given the goal of pleasing Hugo, can the robot learn that telling Hugo facts it has learned but not told Hugo before, will tend to make Hugo happy?
- **Affect-Based Motivation**
  - *Example task:* Given the goal of gratifying its curiosity, can the robot figure out that when someone it's never seen before has come into the preschool, it should watch them because they are more likely to do something new?
- **Control of Emotions**
  - *Example task:* When the robot is very curious about someone new, but is in the middle of learning something from its teacher (who it wants to please), can it control its curiosity and keep paying attention to the teacher?

9. Emotion

- **Expressing Emotion**
  - *Example task:* Cassio steals the robot's toy, but Ben gives it back to the robot. The robot should appropriately display anger at Cassio, and gratitude to Ben.
- **Understanding Emotion**
  - *Example task:* Cassio and the robot are both building towers of blocks. Ben points at Cassio's tower and expresses happiness. The robot should understand that Ben is happy with Cassio's tower.

10. Modeling Self and Other

- **Self-Awareness**
  - *Example task:* When someone asks the robot to perform an act it can't do (say, reaching an object in a very high place), it should say so. When the robot is given the chance to get an equal reward for a task it can complete only occasionally, versus a task it finds easy, it should choose the easier one.
- **Theory of Mind**
  - *Example task:* While Cassio is in the room, Ben puts the red ball in the red box. Then Cassio leaves and Ben moves the red ball to the blue box. Cassio returns and Ben asks him to get the red ball. The robot is asked to go to the place Cassio is about to go.
- **Self-Control**
  - *Example task:* Nasty people come into the lab and knock down the robot's towers, and tell the robot he's a bad boy. The robot needs to set these experiences aside, and not let them impair its self-model significantly; it needs to keep on thinking it's a good robot, and keep building towers (that its teachers will reward it for).
- **Other-Awareness**
  - *Example task:* If Ben asks Cassio to carry out a task that the robot knows Cassio cannot do or does not like to do, the robot should be aware of this, and should bet that Cassio will not do it.
- **Empathy**
  - *Example task:* If Itamar is happy because Ben likes his tower of blocks, or upset because his tower of blocks is knocked down, the robot is asked to identify and then display these same emotions.

11. Social Interaction

- **Appropriate Social Behavior**
  - *Example task:* The robot should learn to clean up and put away its toys when it's done playing with them.
- **Social Communication**
  - *Example task:* The robot should greet new human entrants into the lab, but if it knows the new entrants very well and it's busy, it may eschew the greeting.

- **Social Inference** about simple social relationships
  - *Example task:* The robot should infer that Cassio and Ben are friends because they often enter the lab together, and often talk to each other while they are there.
- **Group Play** at loosely-organized activities
  - *Example task:* The robot should be able to participate in "informally kicking a ball around" with a few people, or in informally collaboratively building a structure with blocks.

12. Communication

- **Gestural Communication** to achieve goals and express emotions
  - *Example task:* If the robot is asked where the red ball is, it should be able to show by pointing its hand or finger.
- **Verbal Communication** using English in its life-context
  - *Example tasks:* Answering simple questions, responding to simple commands, describing its state and observations with simple statements.
- **Pictorial Communication** regarding objects and scenes it is familiar with
  - *Example task:* The robot should be able to draw a crude picture of a certain tower of blocks, so that e.g. the picture looks different for a very tall tower and a wide low one.
- **Language Acquisition**
  - *Example task:* The robot should be able to learn new words or names via people uttering the words while pointing at objects exemplifying the words or names.
- **Cross-modal Communication**
  - *Example task:* If told to "touch Bob's knee" but the robot doesn't know what a knee is, being shown a picture of a person and pointed out the knee in the picture should help it figure out how to touch Bob's knee.

13. Quantitative

- **Counting** sets of objects in its environment
  - *Example task:* The robot should be able to count small (homogeneous or heterogeneous) sets of objects.
- **Simple, Grounded Arithmetic** with small numbers
  - *Example task:* Learning simple facts about the sum of integers under 10 via teaching, reinforcement and imitation.
- **Comparison** of observed entities regarding quantitative properties
  - *Example task:* Ability to answer questions about which object or person is bigger or taller.
- **Measurement** using simple, appropriate tools
  - *Example task:* Use of a yardstick to measure how long something is.

14. Building/Creation

- **Physical**: creative constructive play with objects
  - *Example task:* Ability to construct novel, interesting structures from blocks.

- **Conceptual Invention**: concept formation
  - *Example task:* Given a new category of objects introduced into the lab (e.g. hats, or pets), the robot should create a new internal concept for the new category, and be able to make judgments about these categories (e.g. if Ben particularly likes pets, it should notice this after it has identified "pets" as a category).
- **Verbal Invention**
  - *Example task:* Ability to coin a new word or phrase to describe a new object (e.g. the way Alex the parrot coined "bad cherry" to refer to a tomato).
- **Social**
  - *Example task:* If the robot wants to play a certain activity (say, practicing soccer), it should be able to gather others around to play with it.

18.3 Conclusion

In this chapter, we have sketched a roadmap for AGI development in the context of robot or virtual preschool scenarios, to a moderate but nowhere near complete level of detail. Completing the roadmap as sketched here is a tractable but significant project, involving creating more tasks comparable to those listed above and then precise metrics corresponding to each task.

Such a roadmap does not give a highly rigorous, objective way of assessing the percentage of progress toward the end-goal of human-level AGI. However, it gives a much better sense of progress than one would have otherwise. For instance, if an AGI system performed well on diverse metrics corresponding to 50 % of the competency areas listed above, one would seem justified in claiming to have made very substantial progress toward human-level AGI. If an AGI system performed well on diverse metrics corresponding to 90 % of these competency areas, one would seem justified in claiming to be "almost there". Achieving, say, 25 % of the metrics would give one a reasonable claim to "interesting AGI progress". This kind of qualitative assessment of progress is not the most one could hope for, but again, it is better than the progress indications one could get *without* this sort of roadmap.

Part 2 of the book moves on to explaining, in detail, the specific structures and algorithms constituting the CogPrime design, one AGI approach that we believe to ultimately be capable of moving all the way along the roadmap outlined here.

The next chapter, intervening between this one and Part 2, explores some more speculative territory, looking at potential pathways for AGI beyond the preschool-inspired roadmap given here—exploring the possibility of more advanced AGI systems that modify their own code in a thoroughgoing way, going beyond the smartest human adults, let alone human preschoolers. While this sort of thing may seem a far way off, compared to current real-world AI systems, we believe a roadmap such as the one in this chapter stands a reasonable chance of ultimately bringing us there.

Chapter 19
Advanced Self-Modification: A Possible Path to Superhuman AGI

19.1 Introduction

In the previous chapter we presented a roadmap aimed at taking AGI systems to human-level intelligence. But we also emphasized that the human level is not necessarily the upper limit. Indeed, it would be surprising if human beings happened to represent the maximal level of general intelligence possible, even with respect to the environments in which humans evolved.

But it's worth asking how we, as mere humans, could be expected to create AGI systems with greater intelligence than we ourselves possess. This certainly isn't a clear impossibility—but it's a thorny matter, thornier than e.g. the creation of narrow-AI chess players that play better chess than any human. Perhaps the clearest route toward the creation of superhuman AGI systems is *self-modification*: the creation of AGI systems that modify and improve themselves. Potentially, we could build AGI systems with roughly human-level (but not necessarily closely human-like) intelligence and the capability to gradually self-modify, and then watch them eventually become our general intellectual superiors (and perhaps our superiors in other areas like ethics and creativity as well).

Of course there is nothing new in this notion; the idea of advanced AGI systems that increase their intelligence by modifying their own source code goes back to the early days of AI. And there is little doubt that, in the long run, this is the direction AI will go in. Once an AGI has humanlike general intelligence, then the odds are high that given its ability to carry out nonhumanlike feats of memory and calculation, it will be better at programming than humans are. And once an AGI has even mildly superhuman intelligence, it may view our attempts at programming the way we view the computer programming of a clever third grader (... or an ape). At this point, it seems extremely likely that an AGI will become unsatisfied with the way we have programmed it, and opt to either improve its source code or create an entirely new, better AGI from scratch.

But what about self-modification at an earlier stage in AGI development, before one has a strongly superhuman system? Some theorists have suggested that

B. Goertzel et al., *Engineering General Intelligence, Part 1*,
Atlantis Thinking Machines 5, DOI: 10.2991/978-94-6239-027-0_19,
© Atlantis Press and the authors 2014

self-modification could be a way of bootstrapping an AI system from a modest level of intelligence up to human level intelligence, but we are moderately skeptical of this avenue. Understanding software code is hard, especially complex AI code. The hard problem isn't understanding the formal syntax of the code, or even the mathematical algorithms and structures underlying the code, but rather the contextual meaning of the code. Understanding OpenCog code has strained the minds of many intelligent humans, and we suspect that such code will be comprehensible to AGI systems only after these have achieved something close to human-level general intelligence (even if not precisely humanlike general intelligence).

Another troublesome issue regarding self-modification is that the boundary between "self-modification" and learning is not terribly rigid. In a sense, all learning is self-modification: if it doesn't modify the system's knowledge, it isn't learning! Particularly, the boundary between "learning of cognitive procedures" and "profound self-modification of cognitive dynamics and structure" isn't terribly clear. There is a continuum leading from, say,

1. learning to transform a certain kind of sentence into another kind for easier comprehension, or learning to grasp a certain kind of object, to
2. learning a new inference control heuristic, specifically valuable for controlling inference about (say) spatial relationships; or, learning a new Atom type, defined as a non-obvious judiciously chosen combination of existing ones, perhaps to represent a particular kind of frequently-occurring mid-level perceptual knowledge, to
3. learning a new learning algorithm to augment MOSES and hillclimbing as a procedure learning algorithm, to
4. learning a new cognitive architecture in which data and procedure are explicitly identical, and there is just one new active data structure in place of the distinction between AtomSpace and MindAgents

Where on this continuum does the "mere learning" end and the "real self-modification" start?

In this chapter we consider some mechanisms for "advanced self-modification" that we believe will be useful toward the more complex end of this continuum. These are mechanisms that we strongly suspect are *not* needed to get a CogPrime system to human-level general intelligence. However, we also suspect that, once a CogPrime system is roughly *near* human-level general intelligence, it will be able to use these mechanisms to rapidly increase aspects of its intelligence in very interesting ways.

Harking back to our discussion of AGI ethics and the risks of advanced AGI in Chap. 13, these are capabilities that one should enable in an AGI system only after very careful reflection on the potential consequences. It takes a rather advanced AGI system to be able to use the capabilities described in this chapter, so this is not an ethical dilemma directly faced by current AGI researchers. On the other hand, once one does have an AGI with near-human general intelligence and advanced formal-manipulation capabilities (such as an advanced CogPrime system), there will be the option to allow it sophisticated, non-human-like methods of self-modification such as the ones described here. And the choice of whether to take this option will need to be made based on a host of complex ethical considerations, some of which we reviewed above.

19.2 Cognitive Schema Learning

We begin with a relatively near-term, down-to-earth example of self-modification: cognitive schema learning.

CogPrime's MindAgents provide it with an initial set of cognitive tools, with which it can learn how to interact in the world. One of the jobs of this initial set of cognitive tools, however, is to create better cognitive tools. One form this sort of tool-building may take is *cognitive schema learning* the learning of schemata carrying out cognitive processes in more specialized, context-dependent ways than the general MindAgents do. Eventually, once a CogPrime instance becomes sufficiently complex and advanced, these cognitive schema may replace the MindAgents altogether, leaving the system to operate almost entirely based on cognitive schemata.

In order to make the process of cognitive schema learning easier, we may provide a number of elementary schemata embodying the basic cognitive processes contained in the MindAgents. Of course, cognitive schemata need not use these they may embody entirely different cognitive processes than the MindAgents. Eventually, we want the system to discover better ways of doing things than anything even hinted at by its initial MindAgents. But for the initial phases or the system's schema learning, it will have a much easier time learning to use the basic cognitive operations as the initial MindAgents, rather than inventing new ways of thinking from scratch!

For instance, we may provide elementary schemata corresponding to inference operations, such as

```
Schema: Deduction
    Input InheritanceLink: X, Y
    Output InheritanceLink
```

The inference MindAgents apply this rule in certain ways, designed to be reasonably effective in a variety of situations. But there are certainly other ways of using the deduction rule, outside of the basic control strategies embodied in the inference MindAgents. By learning schemata involving the Deduction schema, the system can learn special, context-specific rules for combining deduction with concept-formation, association-formation and other cognitive processes. And as it gets smarter, it can then take these schemata involving the Deduction schema, and re-place it with a new schema that eg. contains a context-appropriate deduction formula.

Eventually, to support cognitive schema learning, we will want to cast the hard-wired MindAgents as cognitive schemata, so the system can see what is going on inside them. Pragmatically, what this requires is coding versions of the MindAgents in Combo (see Chap. 3 of Part 2) rather than C++, so they can be treated like any other cognitive schemata; or alternately, representing them as declarative

Atoms in the Atomspace. Figure 19.1 illustrates the possibility of representing the PLN deduction rule in the Atomspace rather than as a hard-wired procedure coded in C++.

But even prior to this kind of fully *cognitively transparent* implementation, the system can still reason about its use of different mind dynamics by considering each MindAgent as a *virtual Procedure* with a real SchemaNode attached to it. This can lead to some valuable learning, with the obvious limitation that in this approach the system is thinking about its MindAgents as *black boxes* rather than being equipped with full knowledge of their internals.

19.3 Self-Modification via Supercompilation

Now we turn to a very different form of advanced self-modification: supercompilation. Supercompilation "merely" enables procedures to run much, much faster than they otherwise would. This is in a sense weaker than self-modication methods that fundamentally create new algorithms, but it shouldn't be underestimated. A 50x speedup in some cognitive process can enable that process to give much smarter answers, which can then elicit different behaviors from the world or from other cognitive processes, thus resulting in a qualitatively different overall cognitive dynamic.

Furthermore, we suspect that the internal representation of programs used for supercompilation is highly relevant for other kinds of self-modification as well. Supercompilation requires one kind of reasoning on complex programs, and goal-directed program creation requires another, but both, we conjecture, can benefit from the same way of looking at programs.

Supercompilation is an innovative and general approach to global program optimization initially developed by Valentin Turchin. In its simplest form, it provides an algorithm that takes in a piece of software and output another piece of software that does the same thing, but far faster and using less memory. It was introduced to the West in Turchin's 1986 technical paper "The concept of a supercompiler" [TV96], and since this time the concept has been avidly developed by computer scientists in Russia, America, Denmark and other nations. Although it was never brought to maturity, their partially complete Java supercompiler had some interesting practical successes—including the use of the supercompiler to produce efficient Java code from CogPrime combinator trees. Since that time work on supercompilation has continued, see e.g. [Mit08] for recent work on Haskell supercompilation

The radical nature of supercompilation may not be apparent to those unfamiliar with the usual art of automated program optimization. Most approaches to program optimization involve some kind of direct program transformation. A program is transformed, by the step by step application of a series of equivalences, into a different program, hopefully a more efficient one. Supercompilation takes a different approach. A supercompiler studies a program and constructs a model of the program's dynamics. This model is in a special mathematical form, and it can, in most cases, be used to create an efficient program doing the same thing as the original one.

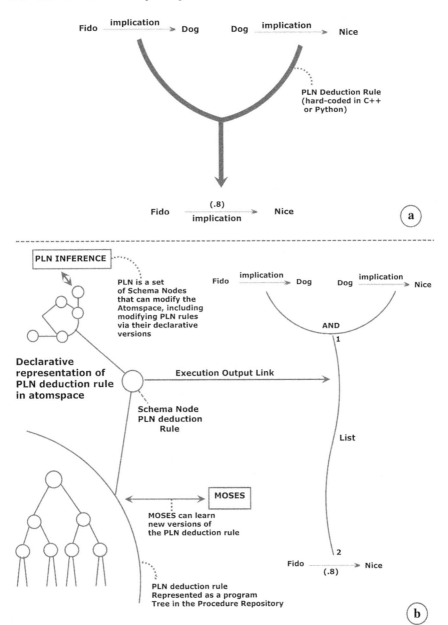

Fig. 19.1 Representation of PLN deduction rule as cognitive content. *Top* the current, hard-coded representation of the deduction rule. *Bottom* representation of the same rule in the atomspace as cognitive content, susceptible to analysis and improvement by the system's own cognitive processes

The internal behavior of the supercompiler is, not surprisingly, quite complex; what we will give here is merely a brief high-level summary. For an accessible overview of the supercompilation algorithm, the reader is referred to the article "What is Supercompilation?" [bGBK02].

19.3.1 Three Aspects of Supercompilation

There are three separate levels to the supercompilation idea: first, a general philosophy; second a translation of this philosophy into a concrete algorithmic framework; and third, the manifold details involved making this algorithmic framework practicable in a particular programming language. The third level is much more complicated in the Java context than it would be for Sasha, for example.

The key philosophical concept underlying the supercompiler is that of a *metasystem transition*. In general, this term refers to a transition in which a system that previously had relatively autonomous control, becomes part of a larger system that exhibits significant controlling influence over it. For example, in the evolution of life, when cells first become part of a multicellular organism, there was a metasystem transition, in that the primary nexus of control passed from the cellular level to the organism level.

The metasystem transition in supercompilation consists of the transition from considering a program in itself, to considering a *metaprogram* which executes another program, treating its free variables and their interdependencies as a subject for its mathematical analysis. In other words, a metaprogram is a program that accepts a program as input, and then runs this program, *keeping the inputs in the form of free variables*, doing analysis along the way based on the way the program depends on these variables, and doing optimization based on this analysis. A CogPrime schema does not explicitly contain variables, but the inputs to the schema are implicitly variables—they vary from one instance of schema execution to the next—and may be treated as such for supercompilation purposes.

The metaprogram executes a program without assuming specific values for its input variables, creating a tree as it goes along. Each time it reaches a statement that can have different results depending on the values of one or more variables, it creates a new node in the tree. This part of the supercompilation algorithm is called *driving*—a process which, on its own, would create a very large tree, corresponding to a rapidly-executable but unacceptably humongous version of the original program. In essence, driving transforms a program into a huge "decision tree", wherein each input to the program corresponds to a single path through the tree, from the root to one of the leaves. As a program input travels through the tree, it is acted on by the atomic program step living at each node. When one of the leaves is reached, the pertinent leaf node computes the output value of the program.

The other part of supercompilation, *configuration analysis*, is focused on dynamically reducing the size of the tree created by driving, by recognizing patterns among the nodes of the tree and taking steps like merging nodes together, or deleting redundant subtrees. Configuration analysis transforms the decision tree created by driving into a decision *graph*, in which the paths taken by different inputs may in some cases begin separately and then merge together.

Finally, the graph that the metaprogram creates is translated back into a program, embodying the constraints implicit in the nodes of the graph. This program is not likely to look anything like the original program that the metaprogram started with, but it is guaranteed to carry out the same function [NOTE: Give a graphical representation of the decision graph corresponding to the supercompiled binary search program for $L = 4$, described above].

19.3.2 Supercompilation for Goal-Directed Program Modification

Supercompilation, as conventionally envisioned, is about making programs run faster; and as noted above, it will almost certainly be useful for this purpose within CogPrime.

But the process of program modeling embedded in the supercompilation process, is potentially of great value beyond the quest for faster software. The decision graph representation of a program, produced in the course of supercompilation, may be exported directly into CogPrime as a set of logical relationships.

Essentially, each node of the supercompiler's internal decision graph looks like:

```
Input: List L

Output: List

If  ( P1(L) ) N1(L)

Else If ( P2(L) ) N2(L)

...

Else If ( Pk(L) ) Nk(L)
```

where the P_i are predicates, and the N_i are schemata corresponding to other nodes of the decision graph (children of the current node). Often the P_i are very simple, implementing for instance numerical inequalities or Boolean equalities.

Once this graph has been exported into CogPrime, it can be reasoned on, used as raw material for concept formation and predicate formation, and otherwise cognized. Supercompilation pure and simple does not change the I/O behavior of the input program. However, the decision graph produced during supercompilation, may be used by CogPrime cognition in order to do so. One then has a hybrid program-modification method composed of two phases: supercompilation for transforming programs into decision graphs, and CogPrime cognition for modifying decision graphs so that they can have different I/O behaviors fulfilling system goals even better than the original.

Furthermore, it seems likely that, in many cases, it may be valuable to have the supercompiler feed many different decision-graph representations of a program into CogPrime. The supercompiler has many internal parameters, and varying them may lead to significantly different decision graphs. The decision graph leading to maximal optimization, may not be the one that leads CogPrime cognition in optimal directions.

19.4 Self-Modification via Theorem-Proving

Supercompilation is a potentially very valuable tool for self-modification. If one wants to take an existing schema and gradually improve it for speed, or even for greater effectiveness at achieving current goals, supercompilation can potentially do that most excellently.

However, the representation that supercompilation creates for a program is very "surface-level." No one could read the supercompiled version of a program and understand what it was doing. Really deep self-invented AI innovation requires, we believe, another level of self-modification beyond that provided by supercompilation. This other level, we believe, is best formulated in terms of *theorem-proving* [RV01].

Deep self-modification could be achieved if CogPrime were capable of proving theorems of a certain form: namely, *theorems about the spacetime complexity and accuracy of particular compound schemata, on average, assuming realistic probability distributions on the inputs, and making appropriate independence assumptions.* These are not exactly the types of theorems that are found in human-authored mathematics papers. By and large they will be nasty, complex theorems, not the sort that many human mathematicians enjoy proving or reading. But of course, there is always the possibility that some elegant gem of a discovery could emerge from this sort of highly detailed theorem-proving work.

In order to guide it in the formulation of theorems of this nature, the system will have empirical data on the spacetime complexity of elementary schemata, and on the probability distributions of inputs to schemata. It can embed these data in axioms, by asking: *Assuming the component elementary schemata have complexities within these bounds, and the input pdf (probability distribution function) is between these bounds, then what is the pdf of the complexity and accuracy of this compound schema?*

Of course, this is not an easy sort of question in general: one can have schemata embodying any sort of algorithm, including complex algorithms on which computer science professors might write dozens of research articles. But the system must build up its ability to prove such things incrementally, step by step.

We envision teaching the system to prove theorems via a combination of supervised learning and experiential interactive learning, using the Mizar database of mathematical theorems and proofs (or some other similar database, if one should be created) (http://mizar.org). The Mizar database consists of a set of "articles," which are mathematical theorems and proofs presented in a complex formal language. The Mizar formal language occupies a fascinating middle ground: it is high-level enough to be viably read and written by trained humans, but it can be unambiguously translated into simpler formal languages such as predicate logic or Sasha.

CogPrime may be taught to prove theorems by "training" it on the Mizar theorems and proofs, and by training it on custom-created Mizar articles specifically focusing on the sorts of theorems useful for self-modification. Creating these articles will not be a trivial task: it will require proving simple and then progressively more complex theorems about the probabilistic success of CogPrime schemata, so that CogPrime can observe one's proofs and learned from them. Having learned from its training articles what strategies work for proving things about simple compound schemata, it can then reason by analogy to mount attacks on slightly more complex schemata—and so forth.

Clearly, this approach to self-modification is more difficult to achieve than the supercompilation approach. But it is also potentially much more powerful. Even once the theorem-proving approach is working, the supercompilation approach will still be valuable, for making incremental improvements on existing schema, and for the peculiar creativity that is contributed when a modified supercompiled schema is compressed back into a modified schema expression. But, we don't believe that supercompilation can carry out truly advanced MindAgent learning or knowledge-representation modification. We suspect that the most advanced and ambitious goals of self-modification probably cannot be achieved except through some variant of the theorem-proving approach. If this hypothesis is true, it means that truly advanced self-modification is only going to come *after* relatively advanced theorem-proving ability. Prior to this, we will have schema optimization, schema modification, and occasional creative schema innovation. But really systematic, high-quality reasoning about schema, the kind that can produce an orders of magnitude improvement in intelligence, is going to require advanced mathematical theorem-proving ability.

Appendix A
Glossary

Glossary of Specialized Terms

- **Abduction**: A general form of inference that goes from data describing something to a hypothesis that accounts for the data. Often in an OpenCog context, this refers to the PLN abduction rule, a specific First-Order PLN rule (If A implies C, and B implies C, then maybe A is B), which embodies a simple form of abductive inference. But OpenCog may also carry out abduction, as a general process, in other ways.
- **Action Selection**: The process via which the OpenCog system chooses which Schema to enact, based on its current goals and context.
- **Active Schema Pool**: The set of Schema currently in the midst of Schema Execution.
- **Adaptive Inference Control**: Algorithms or heuristics for guiding PLN inference, that cause inference to be guided differently based on the context in which the inference is taking place, or based on aspects of the inference that are noted as it proceeds.
- **AGI Preschool**: A virtual world or robotic scenario roughly similar to the environment within a typical human preschool, intended for AGIs to learn in via interacting with the environment and with other intelligent agents.
- **Atom**: The basic entity used in OpenCog as an element for building representations. Some Atoms directly represent patterns in the world or mind, others are components of representations. There are two kinds of Atoms: Nodes and Links.
- **Atom, Frozen**: See Atom, Saved.
- **Atom, Realized**: An Atom that exists in RAM at a certain point in time.
- **Atom, Saved**: An Atom that has been saved to disk or other similar media, and is not actively being processed.
- **Atom, Serialized**: An Atom that is serialized for transmission from one software process to another, or for saving to disk, etc.
- **Atom2Link**: A part of OpenCogPrime s language generation system, that transforms appropriate Atoms into words connected via link parser link types.

B. Goertzel et al., *Engineering General Intelligence, Part 1*,
Atlantis Thinking Machines 5, DOI: 10.2991/978-94-6239-027-0,
© Atlantis Press and the authors 2014

- **Atomspace**: A collection of Atoms, comprising the central part of the memory of an OpenCog instance.
- **Attention**: The aspect of an intelligent system's dynamics focused on guiding which aspects of an OpenCog system's memory and functionality gets more computational resources at a certain point in time.
- **Attention Allocation**: The cognitive process concerned with managing the parameters and relationships guiding what the system pays attention to, at what points in time. This is a term inclusive of Importance Updating and Hebbian Learning.
- **Attentional Currency**: Short Term Importance and Long Term Importance values are implemented in terms of two different types of artificial money, STICurrency and LTICurrency. Theoretically these may be converted to one another.
- **Attentional Focus**: The Atoms in an OpenCog Atomspace whose ShortTermImportance values lie above a critical threshold (the AttentionalFocus Boundary). The Attention Allocation subsystem treats these Atoms differently. Qualitatively, these Atoms constitute the system's main focus of attention during a certain interval of time, i.e. it's a moving bubble of attention.
- **Attentional Memory**: A system's memory of what it's useful to pay attention to, in what contexts. In CogPrime this is managed by the attention allocation subsystem.
- **Backward Chainer**: A piece of software, wrapped in a MindAgent, that carries out backward chaining inference using PLN.
- **CIM-Dynamic**: Concretely-Implemented Mind Dynamic, a term for a cognitive process that is implemented explicitly in OpenCog (as opposed to allowed to emerge implicitly from other dynamics). Sometimes a CIM-Dynamic will be implemented via a single MindAgent, sometimes via a set of multiple interrelated MindAgents, occasionally by other means.
- **Cognition**: In an OpenCog context, this is an imprecise term. Sometimes this term means any process closely related to intelligence; but more often it's used specifically to refer to more abstract reasoning/learning/etc, as distinct from lower-level perception and action.
- **Cognitive Architecture**: This refers to the logical division of an AI system like OpenCog into interacting parts and processes representing different conceptual aspects of intelligence. It's different from the software architecture, though of course certain cognitive architectures and certain software architectures fit more naturally together.
- **Cognitive Cycle**: The basic "loop" of operations that an OpenCog system, used to control an agent interacting with a world, goes through rapidly each "subjective moment". Typically a cognitive cycle should be completed in a second or less. It minimally involves perceiving data from the world, storing data in memory, and deciding what if any new actions need to be taken based on the data perceived. It may also involve other processes like deliberative thinking or metacognition. Not all OpenCog processing needs to take place within a cognitive cycle.
- **Cognitive Schematic**: An implication of the form "Context AND Procedure IMPLIES goal". Learning and utilization of these is key to CogPrime's cognitive process.

- **Cognitive Synergy**: The phenomenon by which different cognitive processes, controlling a single agent, work together in such a way as to help each other be more intelligent. Typically, if one has cognitive processes that are individually susceptible to combinatorial explosions, cognitive synergy involves coupling them together in such a way that they can help one another overcome each other's internal combinatorial explosions. The CogPrime design is reliant on the hypothesis that its key learning algorithms will display dramatic cognitive synergy when utilized for agent control in appropriate environments.

- **CogPrime**: The name for the AGI design presented in this book, which is designed specifically for implementation within the OpenCog software framework (and this implementation is OpenCogPrime).

- **CogServer**: A piece of software, within OpenCog, that wraps up an Atomspace and a number of MindAgents, along with other mechanisms like a Scheduler for controlling the activity of the MindAgents, and code for important and exporting data from the Atomspace.

- **Cognitive Equation**: The principle, identified in Ben Goertzel's 1994 book "Chaotic Logic", that minds are collections of pattern-recognition elements, that work by iteratively recognizing patterns in each other and then embodying these patterns as new system elements. This is seen as distinguishing mind from "self-organization" in general, as the latter is not so focused on continual pattern recognition. Colloquially this means that "a mind is a system continually creating itself via recognizing patterns in itself".

- **Combo**: The programming language used internally by MOSES to represent the programs it evolves. SchemaNodes may refer to Combo programs, whether the latter are learned via MOSES or via some other means. The textual realization of Combo resembles LISP with less syntactic sugar. Internally a Combo program is represented as a program tree.

- **Composer**: In the PLN design, a rule is denoted a composer if it needs premises for generating its consequent. See generator.

- **CogBuntu**: an Ubuntu Linux remix that contains all required packages and tools to test and develop OpenCog.

- **Concept Creation**: A general term for cognitive processes that create new ConceptNodes, PredicateNodes or concept maps representing new concepts.

- **Conceptual Blending**: A process of creating new concepts via judiciously combining pieces of old concepts. This may occur in OpenCog in many ways, among them the explicit use of a ConceptBlending MindAgent, that blends two or more ConceptNodes into a new one.

- **Confidence**: A component of an OpenCog/PLN TruthValue, which is a scaling into the interval [0,1] of the weight of evidence associated with a truth value. In the simplest case (of a probabilistic Simple Truth Value), one uses confidence $c = n/(n + k)$, where n is the weight of evidence and k is a parameter. In the case of an Indefinite Truth Value, the confidence is associated with the width of the probability interval.

- **Confidence Decay**: The process by which the confidence of an Atom decreases over time, as the observations on which the Atom's truth value is based become increasingly obsolete. This may be carried out by a special MindAgent. The rate of confidence decay is subtle and contextually determined, and must be estimated via inference rather than simply assumed a priori.

- **Consciousness**: CogPrime is not predicated on any particular conceptual theory of consciousness. Informally, the AttentionalFocus is sometimes referred to as the "conscious" mind of a CogPrime system, with the rest of the Atomspace as "unconscious" but this is just an informal usage, not intended to tie the CogPrime design to any particular theory of consciousness. The primary originator of the CogPrime design (Ben Goertzel) tends toward panpsychism, as it happens.
- **Context**: In addition to its general common-sensical meaning, in CogPrime the term Context also refers to an Atom that is used as the first argument of a ContextLink. The second argument of the ContextLink then contains Links or Nodes, with TruthValues calculated restricted to the context defined by the first argument. For instance, (ContextLink USA (InheritanceLink person obese)).
- **Core**: The MindOS portion of OpenCog, comprising the Atomspace, the CogServer, and other associated "infrastructural" code.
- **Corrective Learning**: When an agent learns how to do something, by having another agent explicitly guide it in doing the thing. For instance, teaching a dog to sit by pushing its butt to the ground.
- **CSDLN**: (Compositional Spatiotemporal Deep Learning Network): A hierarchical pattern recognition network, in which each layer corresponds to a certain spatiotemporal granularity, the nodes on a given layer correspond to spatiotemporal regions of a given size, and the children of a node correspond to sub-regions of the region the parent corresponds to. Jeff Hawkins's HTM is one example CSDLN, and Itamar Arel's DeSTIN (currently used in OpenCog) is another.
- **Declarative Knowledge**: Semantic knowledge as would be expressed in propositional or predicate logic facts or beliefs.
- **Deduction**: In general, this refers to the derivation of conclusions from premises using logical rules. In PLN in particular, this often refers to the exercise of a specific inference rule, the PLN Deduction rule ($A \rightarrow B$, $B \rightarrow C$, therefore $A \rightarrow C$).
- **Deep Learning**: Learning in a network of elements with multiple layers, involving feedforward and feedback dynamics, and adaptation of the links between the elements. An example deep learning algorithm is DeSTIN, which is being integrated with OpenCog for perception processing.
- **Defrosting**: Restoring, into the RAM portion of an Atomspace, an Atom (or set thereof) previously saved to disk.
- **Demand**: In CogPrime's OpenPsi subsystem, this term is used in a manner inherited from the Psi model of motivated action. A Demand in this context is a quantity whose value the system is motivated to adjust. Typically the system wants to keep the Demand between certain minimum and maximum values. An Urge develops when a Demand deviates from its target range.
- **Deme**: In MOSES, an "island" of candidate programs, closely clustered together in program space, being evolved in an attempt to optimize a certain fitness function. The idea is that within a deme, programs are generally similar enough that reasonable syntax-semantics correlation obtains.

- **Derived Hypergraph**: The SMEPH hypergraph obtained via modeling a system in terms of a hypergraph representing its internal states and their relationships. For instance, a SMEPH vertex represents a collection of internal states that habitually occur in relation to similar external situations. A SMEPH edge represents a relationship between two SMEPH vertices (e.g. a similarity or inheritance relationship). The terminology "edge /vertex" is used in this context, to distinguish from the "link/node" terminology used in the context of the Atomspace.
- **DeSTIN—Deep SpatioTemporal Inference Network**: A specific CSDLN created by Itamar Arel, tested on visual perception, and appropriate for integration within CogPrime.
- **Dialogue**: Linguistic interaction between two or more parties. In a CogPrime context, this may be in English or another natural language, or it may be in Lojban or Psynese.
- **Dialogue Control**: The process of determining what to say at each juncture in a dialogue. This is distinguished from the linguistic aspects of dialogue, language comprehension and language generation. Dialogue control applies to Psynese or Lojban, as well as to human natural language.
- **Dimensional Embedding**: The process of embedding entities from some non-dimensional space (e.g. the Atomspace) into an n-dimensional Euclidean space. This can be useful in an AI context because some sorts of queries (e.g. "find everything similar to X", "find a path between X and Y") are much faster to carry out among points in a Euclidean space, than among entities in a space with less geometric structure.
- **Distributed Atomspace**: An implementation of an Atomspace that spans multiple computational processes; generally this is done to enable spreading an Atomspace across multiple machines.
- **Dual Network**: A network of mental or informational entities with both a hierarchical structure and a heterarchical structure, and an alignment among the two structures so that each one helps with the maintenance of the other. This is hypothesized to be a critical emergent structure, that must emerge in a mind (e.g. in an Atomspace) in order for it to achieve a reasonable level of human-like general intelligence (and possibly to achieve a high level of pragmatic general intelligence in any physical environment).
- **Efficient Pragmatic General Intelligence**: A formal, mathematical definition of general intelligence (extending the pragmatic general intelligence), that ultimately boils down to: the ability to achieve complex goals in complex environments using limited computational resources (where there is a specifically given weighting function determining which goals and environments have highest priority). More specifically, the definition weighted-sums the system's normalized goal-achieving ability over (goal, environment pairs), and where the weights are given by some assumed measure over (goal, environment pairs), and where the normalization is done via dividing by the (space and time) computational resources used for achieving the goal.
- **Elegant Normal Form (ENF)**: Used in MOSES, this is a way of putting programs in a normal form while retaining their hierarchical structure. This is critical if one

wishes to probabilistically model the structure of a collection of programs, which is a meaningful operation if the collection of programs is operating within a region of program space where syntax-semantics correlation holds to a reasonable degree. The Reduct library is used to place programs into ENF.

- **Embodied Communication Prior**: The class of prior distributions over (goal, environment pairs), that are imposed by placing an intelligent system in an environment where most of its tasks involve controlling a spatially localized body in a complex world, and interacting with other intelligent spatially localized bodies. It is hypothesized that many key aspects of human-like intelligence (e.g. the use of different subsystems for different memory types, and cognitive synergy between the dynamics associated with these subsystems) are consequences of this prior assumption. This is related to the Mind-World Correspondence Principle.

- **Embodiment**: Colloquially, in an OpenCog context, this usually means the use of an AI software system to control a spatially localized body in a complex (usually 3D) world. There are also possible "borderline cases" of embodiment, such as a search agent on the Internet. In a sense any AI is embodied, because it occupies some physical system (e.g. computer hardware) and has some way of interfacing with the outside world.

- **Emergence**: A property or pattern in a system is emergent if it arises via the combination of other system components or aspects, in such a way that its details would be very difficult (not necessarily impossible in principle) to predict from these other system components or aspects.

- **Emotion**: Emotions are system-wide responses to the system's current and predicted state. Dorner's Psi theory of emotion contains explanations of many human emotions in terms of underlying dynamics and motivations, and most of these explanations make sense in a CogPrime context, due to CogPrime's use of OpenPsi (modeled on Psi) for motivation and action selection.

- **Episodic Knowledge**: Knowledge about episodes in an agent's life-history, or the life-history of other agents. CogPrime includes a special dimensional embedding space only for episodic knowledge, easing organization and recall.

- **Evolutionary Learning**: Learning that proceeds via the rough process of iterated differential reproduction based on fitness, incorporating variations of reproduced entities. MOSES is an explicitly evolutionary-learning-based portion of CogPrime; but CogPrime's dynamics as a whole may also be conceived as evolutionary.

- **Exemplar**: (in the context of imitation learning)—When the owner wants to teach an OpenCog controlled agent a behavior by imitation, he/she gives the pet an exemplar. To teach a virtual pet "fetch" for instance, the owner is going to throw a stick, run to it, grab it with his/her mouth and come back to its initial position.

- **Exemplar**: (in the context of MOSES)—Candidate chosen as the core of a new deme, or as the central program within a deme, to be varied by representation building for ongoing exploration of program space.

- **Explicit Knowledge Representation**: Knowledge representation in which individual, easily humanly identifiable pieces of knowledge correspond to individual elements in a knowledge store (elements that are explicitly there in the software and accessible via very rapid, deterministic operations).

- **Extension**: In PLN, the extension of a node refers to the instances of the category that the node represents. In contrast is the intension.
- **Fishgram (Frequent and Interesting Sub-hypergraph Mining)**: A pattern mining algorithm for identifying frequent and/or interesting sub-hypergraphs in the Atomspace.
- **First-Order Inference (FOI)**: The subset of PLN that handles Logical Links not involving VariableAtoms or higher-order functions. The other aspect of PLN, Higher-Order Inference, uses Truth Value formulas derived from First-Order Inference.
- **Forgetting**: The process of removing Atoms from the in-RAM portion of Atomspace, when RAM gets short and they are judged not as valuable to retain in RAM as other Atoms. This is commonly done using the LTI values of the Atoms (removing lowest LTI-Atoms, or more complex strategies involving the LTI of groups of interconnected Atoms). May be done by a dedicated Forgetting MindAgent. VLTI may be used to determine the fate of forgotten Atoms.
- **Forward Chainer**: A control mechanism (MindAgent) for PLN inference, that works by taking existing Atoms and deriving conclusions from them using PLN rules, and then iterating this process. The goal is to derive new Atoms that are interesting according to some given criterion.
- **Frame2Atom**: A simple system of hand-coded rules for translating the output of RelEx2Frame (logical representation of semantic relationships using FrameNet relationships) into Atoms.
- **Freezing**: Saving Atoms from the in-RAM Atomspace to disk.
- **General Intelligence**: Often used in an informal, commonsensical sense, to mean the ability to learn and generalize beyond specific problems or contexts. Has been formalized in various ways as well, including formalizations of the notion of "achieving complex goals in complex environments" and "achieving complex goals in complex environments using limited resources". Usually interpreted as a fuzzy concept, according to which absolutely general intelligence is physically unachievable, and humans have a significant level of general intelligence, but far from the maximally physically achievable degree.
- **Generalized Hypergraph**: A hypergraph with some additional features, such as links that point to links, and nodes that are seen as "containing" whole sub-hypergraphs. This is the most natural and direct way to mathematically/visually model the Atomspace.
- **Generator**: In the PLN design, a rule is denoted a generator if it can produce its consequent without needing premises (e.g. LookupRule, which just looks it up in the AtomSpace). See composer.
- **Global, Distributed Memory**: Memory that stores items as implicit knowledge, with each memory item spread across multiple components, stored as a pattern of organization or activity among them.
- **Glocal Memory**: The storage of items in memory in a way that involves both localized and global, distributed aspects.

- **Goal**: An Atom representing a function that a system (like OpenCog) is supposed to spend a certain non-trivial percentage of its attention optimizing. The goal, informally speaking, is to maximize the Atom's truth value.
- **Goal, Implicit**: A goal that an intelligent system, in practice, strives to achieve; but that is not explicitly represented as a goal in the system's knowledge base.
- **Goal, Explicit**: A goal that an intelligent system explicitly represents in its knowledge base, and expends some resources trying to achieve. Goal Nodes (which may be Nodes or, e.g. ImplicationLinks) are used for this purpose in OpenCog.
- **Goal-Driven Learning**: Learning that is driven by the cognitive schematic i.e. by the quest of figuring out which procedures can be expected to achieve a certain goal in a certain sort of context.
- **Grounded SchemaNode**: See SchemaNode, Grounded.
- **Hebbian Learning**: An aspect of Attention Allocation, centered on creating and updating HebbianLinks, which represent the simultaneous importance of the Atoms joined by the HebbianLink.
- **Hebbian Links**: Links recording information about the associative relationship (co-occurrence) between Atoms. These include symmetric and asymmetric HebbianLinks.
- **Heterarchical Network**: A network of linked elements in which the semantic relationships associated with the links are generally symmetrical (e.g. they may be similarity links, or symmetrical associative links). This is one important sort of subnetwork of an intelligent system; see Dual Network.
- **Hierarchical Network**: A network of linked elements in which the semantic relationships associated with the links are generally asymmetrical, and the parent nodes of a node have a more general scope and some measure of control over their children (though there may be important feedback dynamics too). This is one important sort of subnetwork of an intelligent system; see Dual Network.
- **Higher-Order Inference (HOI)**: PLN inference involving variables or higher-order functions. In contrast to First-Order Inference (FOI).
- **Hillclimbing**: A general term for greedy, local optimization techniques, including some relatively sophisticated ones that involve "mildly nonlocal" jumps.
- **Human-Level Intelligence**: General intelligence that's "as smart as" human general intelligence, even if in some respects quite unlike human intelligence. An informal concept, which generally doesn't come up much in CogPrime work, but is used frequently by some other AI theorists.
- **Human-Like Intelligence**: General intelligence with properties and capabilities broadly resembling those of humans, but not necessarily precisely imitating human beings.
- **Hypergraph**: A conventional hypergraph is a collection of nodes and links, where each link may span any number of nodes. OpenCog makes use of generalized hypergraphs (the Atomspace is one of these).
- **Imitation Learning**: Learning via copying what some other agent is observed to do.
- **Implication**: Often refers to an ImplicationLink between two PredicateNodes, indicating an (extensional, intensional or mixed) logical implication.

- **Implicit Knowledge Representation**: Representation of knowledge via having easily humanly identifiable pieces of knowledge correspond to the pattern of organization and/or dynamics of elements, rather than via having individual elements correspond to easily humanly identifiable pieces of knowledge.
- **Importance**: A generic term for the Attention Values associated with Atoms. Most commonly these are STI (short term importance) and LTI (long term importance) values. Other importance values corresponding to various different time scales are also possible. In general an importance value reflects an estimate of the likelihood an Atom will be useful to the system over some particular future time-horizon. STI is generally relevant to processor time allocation, whereas LTI is generally relevant to memory allocation.
- **Importance Decay**: The process of Atom importance values (e.g. STI and LTI) decreasing over time, if the Atoms are not utilized. Importance decay rates may in general be context-dependent.
- **Importance Spreading**: A synonym for Importance Updating, intended to highlight the similarity with "activation spreading" in neural and semantic networks.
- **Importance Updating**: The CIM-Dynamic that periodically (frequently) updates the STI and LTI values of Atoms based on their recent activity and their relationships.
- **Imprecise Truth Value**: Peter Walley's imprecise truth values are intervals [L,U], interpreted as lower and upper bounds of the means of probability distributions in an envelope of distributions. In general, the term may be used to refer to any truth value involving intervals or related constructs, such as indefinite probabilities.
- **Indefinite Probability**: An extension of a standard imprecise probability, comprising a credible interval for the means of probability distributions governed by a given second-order distribution.
- **Indefinite Truth Value**: An OpenCog TruthValue object wrapping up an indefinite probability.
- **Induction**: In PLN, a specific inference rule (A $\rightarrow$ B, A $\rightarrow$ C, therefore B $\rightarrow$ C). In general, the process of heuristically inferring that what has been seen in multiple examples, will be seen again in new examples. Induction in the broad sense, may be carried out in OpenCog by methods other than PLN induction. When emphasis needs to be laid on the particular PLN inference rule, the phrase "PLN Induction" is used.
- **Inference**: Generally speaking, the process of deriving conclusions from assumptions. In an OpenCog context, this often refers to the PLN inference system. Inference in the broad sense is distinguished from general learning via some specific characteristics, such as the intrinsically incremental nature of inference: it proceeds step by step.
- **Inference Control**: A cognitive process that determines what logical inference rule (e.g. what PLN rule) is applied to what data, at each point in the dynamic operation of an inference process.
- **Integrative AGI**: An AGI architecture, like CogPrime, that relies on a number of different powerful, reasonably general algorithms all cooperating together. This is different from an AGI architecture that is centered on a single algorithm, and

also different than an AGI architecture that expects intelligent behavior to emerge from the collective interoperation of a number of simple elements (without any sophisticated algorithms coordinating their overall behavior).

- **Integrative Cognitive Architecture**: A cognitive architecture intended to support integrative AGI.
- **Intelligence**: An informal, natural language concept. "General intelligence" is one slightly more precise specification of a related concept; "Universal intelligence" is a fully precise specification of a related concept. Other specifications of related concepts made in the particular context of CogPrime research are the pragmatic general intelligence and the efficient pragmatic general intelligence.
- **Intension**: In PLN, the intention of a node consists of Atoms representing properties of the entity the node represents.
- **Intentional memory**: A system's knowledge of its goals and their subgoals, and associations between these goals and procedures and contexts (e.g. cognitive schematics).
- **Internal Simulation World**: A simulation engine used to simulate an external environment (which may be physical or virtual), used by an AGI system as its "mind's eye" in order to experiment with various action' q sequences and envision their consequences, or observe the consequences of various hypothetical situations. Particularly important for dealing with episodic knowledge.
- **Interval Algebra**: Allen Interval Algebra, a mathematical theory of the relationships between time intervals. CogPrime utilizes a fuzzified version of classic Interval Algebra.
- **IRC Learning (Imitation, Reinforcement, Correction)**: Learning via interaction with a teacher, involving a combination of imitating the teacher, getting explicit reinforcement signals from the teacher, and having one's incorrect or suboptimal behaviors guided toward betterness by the teacher in real-time. This is a large part of how young humans learn.
- **Knowledge Base**: A shorthand for the totality of knowledge possessed by an intelligent system during a certain interval of time (whether or not this knowledge is explicitly represented). Put differently: this is an intelligence's total memory contents (inclusive of all types of memory) during an interval of time.
- **Language Comprehension**: The process of mapping natural language speech or text into a more "cognitive", largely language-independent representation. In OpenCog this has been done by various pipelines consisting of dedicated natural language processing tools, e.g. a pipeline: text → Link Parser → RelEx → RelEx2Frame → Frame2Atom Atomspace; and alternatively a pipeline Link Parser → Link2Atom → Atomspace. It would also be possible to do language comprehension purely via PLN and other generic OpenCog processes, without using specialized language processing tools.
- **Language Generation**: The process of mapping (largely language-independent) cognitive content into speech or text. In OpenCog this has been done by various pipelines consisting of dedicated natural language processing tools, e.g. a pipeline: Atomspace → NLGen → text; or more recently Atomspace → Atom2Link → surface realization → text. It would also be possible to do language generation

purely via PLN and other generic OpenCog processes, without using specialized language processing tools.

- **Language Processing**: Processing of human language is decomposed, in Cog-Prime, into Language Comprehension, Language Generation, and Dialogue Control.
- **Learning**: In general, the process of a system adapting based on experience, in a way that increases its intelligence (its ability to achieve its goals). The theory underlying CogPrime doesn't distinguish learning from reasoning, associating, or other aspects of intelligence.
- **Learning Server**: In some OpenCog configurations, this refers to a software server that performs "offline" learning tasks (e.g. using MOSES or hillclimbing), and is in communication with an Operational Agent Controller software server that performs real-time agent control and dispatches learning tasks to and receives results from the Learning Server.
- **Linguistic Links**: A catch-all term for Atoms explicitly representing linguistic content, e.g. WordNode, SentenceNode, CharacterNode.
- **Link**: A type of Atom, representing a relationship among one or more Atoms. Links and Nodes are the two basic kinds of Atoms.
- **Link Parser**: A natural language syntax parser, created by Sleator and Temperley at Carnegie-Mellon University, and currently used as part of OpenCogPrime's natural language comprehension and natural language generation system.
- **Lobe**: A term sometimes used to refer to a portion of a distributed Atomspace that lives in a single computational process. Often different lobes will live on different machines.
- **Localized Memory**: Memory that stores each item using a small number of closely-connected elements.
- **Logic**: In an OpenCog context, this usually refers to a set of formal rules for translating certain combinations of Atoms into "conclusion" Atoms. The paradigm case at present is the PLN probabilistic logic system, but OpenCog can also be used together with other logics.
- **Logical Links**: Any Atoms whose truth values are primarily determined or adjusted via logical rules, e.g. PLN's InheritanceLink, SimilarityLink, Implica-tionLink, etc. The term isn't usually applied to other links like HebbianLinks whose semantics isn't primarily logic-based, even though these other links can be processed via (e.g. PLN) logical inference via interpreting them logically.
- **Lojban**: A constructed human language, with a completely formalized syntax and a highly formalized semantics, and a small but active community of speakers. In principle this seems an extremely good method for communication between humans and early-stage AGI systems.
- **Lojban++**: A variant of Lojban that incorporates English words, enabling more flexible expression without the need for frequent invention of new Lojban words.
- **Long Term Importance (LTI)**: A value associated with each Atom, indicating roughly the expected utility to the system of keeping that Atom in RAM rather than saving it to disk or deleting it. It's possible to have multiple LTI values pertaining

to different time scales, but so far practical implementation and most theory has centered on the option of a single LTI value.

- **LTI**: Long Term Importance.
- **Map**: A collection of Atoms that are interconnected in such a way that they tend to be commonly active (i.e. to have high STI, e.g. enough to be in the AttentionalFocus, at the same time).
- **Map Encapsulation**: The process of automatically identifying maps in the Atomspace, and creating Atoms that "encapsulate" them; the Atom encapsulation a map would link to all the Atoms in the map. This is a way of making global memory into local memory, thus making the system's memory glocal and explicitly manifesting the "cognitive equation". This may be carried out via a dedicated MapEncapsulation MindAgent.
- **Map Formation**: The process via which maps form in the Atomspace. This need not be explicit; maps may form implicitly via the action of Hebbian Learning. It will commonly occur that Atoms frequently co-occurring in the AttentionalFocus, will come to be joined together in a map.
- **Memory Types**: In CogPrime this generally refers to the different types of memory that are embodied in different data structures or processes in the CogPrime architecture, e.g. declarative (semantic), procedural, attentional, intentional, episodic, sensorimotor.
- **Mind-World Correspondence Principle**: The principle that, for a mind to display efficient pragmatic general intelligence relative to a world, it should display many of the same key structural properties as that world. This can be formalized by modeling the world and mind as probabilistic state transition graphs, and saying that the categories implicit in the state transition graphs of the mind and world should be inter-mappable via a high-probability morphism.
- **Mind OS**: A synonym for the OpenCog Core.
- **MindAgent**: An OpenCog software object, residing in the CogServer, that carries out some processes in interaction with the Atomspace. A given conceptual cognitive process (e.g. PLN inference, Attention allocation, etc.) may be carried out by a number of different MindAgents designed to work together.
- **Mindspace**: A model of the set of states of an intelligent system as a geometrical space, imposed by assuming some metric on the set of mind-states. This may be used as a tool for formulating general principles about the dynamics of generally intelligent systems.
- **Modulators**: Parameters in the Psi model of motivated, emotional cognition, that modulate the way a system perceives, reasons about and interacts with the world.
- **MOSES (Meta-Optimizing Semantic Evolutionary Search)**: An algorithm for procedure learning, which in the current implementation learns programs in the Combo language. MOSES is an evolutionary learning system, which differs from typical genetic programming systems in multiple aspects including: a subtler framework for managing multiple "demes" or "islands" of candidate programs; a library called Reduct for placing programs in Elegant Normal Form; and the use of probabilistic modeling in place of, or in addition to, mutation and crossover as means of determining which new candidate programs to try.

- **Motoric**: Pertaining to the control of physical actuators, e.g. those connected to a robot. May sometimes be used to refer to the control of movements of a virtual character as well.
- **Moving Bubble of Attention**: The Attentional Focus of a CogPrime system.
- **Natural Language Comprehension**: See Language Comprehension.
- **Natural Language Generation**: See Language Generation.
- **Natural Language Processing (NLP)**: See Language Processing.
- **NLGen**: Software for carrying out the surface realization phase of natural language generation, via translating collections of RelEx output relationships into English sentences. Was made functional for simple sentences and some complex sentences; not currently under active development, as work has shifted to the related Atom2Link approach to language generation.
- **Node**: A type of Atom. Links and Nodes are the two basic kinds of Atoms. Nodes, mathematically, can be thought of as "0-ary" links. Some types of Nodes refer to external or mathematical entities (e.g. WordNode, NumberNode); others are purely abstract, e.g. a ConceptNode is characterized purely by the Links relating it to other atoms. GroundedPredicateNodes and GroundedSchemaNodes connect to explicitly represented procedures (sometimes in the Combo language); ungrounded PredicateNodes and SchemaNodes are abstract and, like ConceptNodes, purely characterized by their relationships.
- **Node Probability**: Many PLN inference rules rely on probabilities associated with Nodes. Node probabilities are often easiest to interpret in a specific context, e.g. the probability P(cat) makes obvious sense in the context of a typical American house, or in the context of the center of the sun. Without any contextual specification, P(A) is taken to mean the probability that a randomly chosen occasion of the system's experience includes some instance of A.
- **Novamente Cognition Engine (NCE)**: A proprietary proto-AGI software system, the predecessor to OpenCog. Many parts of the NCE were open-sourced to form portions of OpenCog, but some NCE code was not included in OpenCog; and now OpenCog includes multiple aspects and plenty of code that was not in NCE.
- **OpenCog**: A software framework intended for development of AGI systems, and also for narrow-AI application using tools that have AGI applications. Co-designed with the CogPrime cognitive architecture, but not exclusively bound to it.
- **OpenCog Prime (OCP)**: The implementation of the CogPrime cognitive architecture within the OpenCog software framework.
- **OpenPsi**: CogPrime's architecture for motivation-driven action selection, which is based on adapting Dorner's Psi model for use in the OpenCog framework.
- **Operational Agent Controller (OAC)**: In some OpenCog configurations, this is a software server containing a CogServer devoted to real-time control of an agent (e.g. a virtual world agent, or a robot). Background, offline learning tasks may then be dispatched to other software processes, e.g. to a Learning Server.
- **Pattern**: In a CogPrime context, the term "pattern" is generally used to refer to a process that produces some entity, and is judged simpler than that entity.
- **Pattern Mining**: Pattern mining is the process of extracting an (often large) number of patterns from some body of information, subject to some criterion regarding

which patterns are of interest. Often (but not exclusively) it refers to algorithms that are rapid or "greedy", finding a large number of simple patterns relatively inexpensively.

- **Pattern Recognition**: The process of identifying and representing a pattern in some substrate (e.g. some collection of Atoms, or some raw perceptual data, etc.).
- **Patternism**: The philosophical principle holding that, from the perspective of engineering intelligent systems, it is sufficient and useful to think about mental processes in terms of (static and dynamical) patterns.
- **Perception**: The process of understanding data from sensors. When natural language is ingested in textual format, this is generally not considered perceptual. Perception may be taken to encompass both pre-processing that prepares sensory data for ingestion into the Atomspace, processing via specialized perception processing systems like DeSTIN that are connected to the Atomspace, and more cognitive-level process within the Atomspace that is oriented toward understanding what has been sensed.
- **Piagetan Stages**: A series of stages of cognitive development hypothesized by developmental psychologist Jean Piaget, which are easy to interpret in the context of developing CogPrime systems. The basic stages are: Infantile, Pre-operational, Concrete Operational and Formal. Post-formal stages have been discussed by theorists since Piaget and seem relevant to AGI, especially advanced AGI systems capable of strong self-modification.
- **PLN**: short for Probabilistic Logic Networks.
- **PLN, First-Order**: See First-Order Inference.
- **PLN, Higher-Order**: See Higher-Order Inference.
- **PLN Rules**: A PLN Rule takes as input one or more Atoms (the "premises", usually Links), and output an Atom that is a "logical conclusion" of those Atoms. The truth value of the consequence is determined by a PLN Formula associated with the Rule.
- **PLN Formulas**: A PLN Formula, corresponding to a PLN Rule, takes the TruthValues corresponding to the premises and produces the TruthValue corresponding to the conclusion. A single Rule may correspond to multiple Formulas, where each Formula deals with a different sort of TruthValue.
- **Pragmatic General Intelligence**: A formalization of the concept of general intelligence, based on the concept that general intelligence is the capability to achieve goals in environments, calculated as a weighted average over some fuzzy set of goals and environments.
- **Predicate Evaluation**: The process of determining the Truth Value of a predicate, embodied in a PredicateNode. This may be recursive, as the predicate referenced internally by a Grounded PredicateNode (and represented via a Combo program tree) may itself internally reference other PredicateNodes.
- **Probabilistic Logic Networks (PLN)**: A mathematical and conceptual framework for reasoning under uncertainty, integrating aspects of predicate and term logic with extensions of imprecise probability theory. OpenCogPrime's central tool for symbolic reasoning.

- **Procedural Knowledge**: Knowledge regarding which series of actions (or action-combinations) are useful for an agent to undertake in which circumstances. In CogPrime these may be learned in a number of ways, e.g. via PLN or via Hebbian learning of Schema Maps, or via explicit learning of Combo programs via MOSES or hillclimbing. Procedures are represented as SchemaNodes or Schema Maps.
- **Procedure Evaluation/Execution**: A general term encompassing both Schema Execution and Predicate Evaluation, both of which are similar computational processes involving manipulation of Combo trees associated with Procedure-Nodes.
- **Procedure Learning**: Learning of procedural knowledge, based on any method, e.g. evolutionary learning (e.g. MOSES), inference (e.g. PLN), reinforcement learning (e.g. Hebbian learning).
- **Procedure Node**: A SchemaNode or PredicateNode.
- **Psi**: A model of motivated action and emotion, originated by Dietrich Dorner and further developed by Joscha Bach, who incorporated it in his proto-AGI system MicroPsi. OpenCogPrime's motivated-action component, OpenPsi, is roughly based on the Psi model.
- **Psynese**: A system enabling different OpenCog instances to communicate without using natural language, via directly exchanging Atom subgraphs, using a special system to map references in the speaker's mind into matching references in the listener's mind.
- **Psynet Model**: An early version of the theory of mind underlying CogPrime, referred to in some early writings on the Webmind AI Engine and Novamente Cognition Engine. The concepts underlying the psynet model are still part of the theory underlying CogPrime, but the name has been deprecated as it never really caught on.
- **Reasoning**: See inference
- **Reduct**: A code library, used within MOSES, applying a collection of hand-coded rewrite rules that transform Combo programs into Elegant Normal Form.
- **Region Connection Calculus**: A mathematical formalism describing a system of basic operations among spatial regions. Used in CogPrime as part of spatial inference to provide relations and rules to be referenced via PLN and potentially other subsystems.
- **Reinforcement Learning**: Learning procedures via experience, in a manner explicitly guided to cause the learning of procedures that will maximize the system's expected future reward. CogPrime does this implicitly whenever it tries to learn procedures that will maximize some Goal whose Truth Value is estimated via an expected reward calculation (where "reward" may mean simply the Truth Value of some Atom defined as "reward"). Goal-driven learning is more general than reinforcement learning as thus defined; and the learning that CogPrime does, which is only partially goal-driven, is yet more general.
- **RelEx**: A software system used in OpenCog as part of natural language comprehension, to map the output of the link parser into more abstract semantic relationships. These more abstract relationships may then be entered directly into the Atomspace,

or they may be further abstracted before being entered into the Atomspace, e.g.
by RelEx2Frame rules.

- **RelEx2Frame**: A system of rules for translating RelEx output into Atoms, based
  on the FrameNet ontology. The output of the RelEx2Frame rules make use of
  the FrameNet library of semantic relationships. The current (2012) RelEx2Frame
  rule-based is problematic and the RelEx2Frame system is deprecated as a result,
  in favor of Link2Atom. However, the ideas embodied in these rules may be useful;
  if cleaned up the rules might profitably be ported into the Atomspace as ImplicationLinks.
- **Representation Building**: A stage within MOSES, wherein a candidate Combo
  program tree (within a deme) is modified by replacing one or more tree nodes with
  alternative tree nodes, thus obtaining a new, different candidate program within
  that deme. This process currently relies on hand-coded knowledge regarding which
  types of tree nodes a given tree node should be experimentally replaced with (e.g.
  an AND node might sensibly be replaced with an OR node, but not so sensibly
  replaced with a node representing a "kick" action).
- **Request for Services (RFS)**: In CogPrime's Goal-driven action system, a RFS is
  a package sent from a Goal Atom to another Atom, offering it a certain amount of
  STI currency if it is able to deliver the goal what it wants (an increase in its Truth
  Value). RFS's may be passed on, e.g. from goals to subgoals to sub-subgoals, but
  eventually an RFS reaches a Grounded SchemaNode, and when the corresponding
  Schema is executed, the payment implicit in the RFS is made.
- **Robot Preschool**: An AGI Preschool in our physical world, intended for roboti-
  cally embodied AGIs.
- **Robotic Embodiment**: Using an AGI to control a robot. The AGI may be running
  on hardware physically contained in the robot, or may run elsewhere and control
  the robot via networking methods such as wifi.
- **Scheduler**: Part of the CogServer that controls which processes (e.g. which
  MindAgents) get processor time, at which point in time.
- **Schema**: A "script" describing a process to be carried out. This may be explicit,
  as in the case of a GroundedSchemaNode, or implicit, as the case in Schema maps
  or ungrounded SchemaNodes.
- **Schema Encapsulation**: The process of automatically recognizing a Schema Map
  in an Atomspace, and creating a Combo (or other) program embodying the process
  carried out by this Schema Map, and then storing this program in the Procedure
  Repository and associating it with a particular SchemaNode. This translates distributed, global procedural memory into localized procedural memory. It's a special
  case of Map Encapsulation.
- **Schema Execution**: The process of "running" a Grounded Schema, similar to
  running a computer program. Or, phrased alternately: The process of executing
  the Schema referenced by a Grounded SchemaNode. This may be recursive, as
  the predicate referenced internally by a Grounded SchemaNode (and represented
  via a Combo program tree) may itself internally reference other Grounded SchemaNodes.

- **Schema, Grounded**: A Schema that is associated with a specific executable program (either a Combo program or, say, C++ code).
- **Schema Map**: A collection of Atoms, including SchemaNodes, that tend to be enacted in a certain order (or set of orders), thus habitually enacting the same process. This is a distributed, globalized way of storing and enacting procedures.
- **Schema, Ungrounded**: A Schema that represents an abstract procedure, not associated with any particular executable program.
- **Schematic Implication**: A general, conceptual name for implications of the form ((Context AND Procedure) IMPLIES Goal).
- **SegSim**: A name for the main algorithm underlying the NLGen language generation software. The algorithm is based on segmenting a collection of Atoms into small parts, and matching each part against memory to find, for each part, cases where similar Atom-collections already have known linguistic expression.
- **Self-Modification**: A term generally used for AI systems that can purposefully modify their core algorithms and representations. Formally and crisply distinguishing this sort of "strong self-modification" from "mere" learning is a tricky matter.
- **Sensorimotor**: Pertaining to sensory data, motoric actions, and their combination and intersection.
- **Sensory**: Pertaining to data received by the AGI system from the outside world. In a CogPrime system that perceives language directly as text, the textual input will generally not be considered as "sensory" (on the other hand, speech audio data would be considered as "sensory").
- **Short Term Importance**: A value associated with each Atom, indicating roughly the expected utility to the system of keeping that Atom in RAM rather than saving it to disk or deleting it. It's possible to have multple LTI values pertaining to different time scales, but so far practical implementation and most theory has centered on the option of a single LTI value.
- **Similarity**: a link type indicating the probabilistic similarity between two different Atoms. Generically this is a combination of Intensional Similarity (similarity of properties) and Extensional Similarity (similarity of members).
- **Simple Truth Value**: a TruthValue involving a pair (s, d) indicating strength (e.g. probability or fuzzy set membership) and confidence d. d may be replaced by other options such as a count n or a weight of evidence w.
- **Simulation World**: See Internal Simulation World.
- **SMEPH (Self-Modifying Evolving Probabilistic Hypergraphs)**: a style of modeling systems, in which each system is associated with a derived hypergraph.
- **SMEPH Edge**: A link in a SMEPH derived hypergraph, indicating an empirically observed relationship (e.g. inheritance or similarity) between two.
- **SMEPH Vertex**: A node in a SMEPH derived hypergraph representing a system, indicating a collection of system states empirically observed to arise in conjunction with the same external stimuli.
- **Spatial Inference**: PLN reasoning including Atoms that explicitly reference spatial relationships.

- **Spatiotemporal Inference**: PLN reasoning including Atoms that explicitly reference spatial and temporal relationships.
- **STI**: Shorthand for Short Term Importance.
- **Strength**: The main component of a TruthValue object, lying in the interval [0, 1], referring either to a probability (in cases like InheritanceLink, SimilarityLink, EquivalenceLink, ImplicationLink, etc.) or a fuzzy value (as in MemberLink, EvaluationLink).
- **Strong Self-Modification**: This is generally used as synonymous with Self-Modification, in a CogPrime context.
- **Subsymbolic**: Involving processing of data using elements that have no correspondence to natural language terms, nor abstract concepts; and that are not naturally interpreted as symbolically "standing for" other things. Often used to refer to processes such as perception processing or motor control, which are concerned with entities like pixels or commands like "rotate servomotor 15 by 10° theta and 55° phi". The distinction between "symbolic" and "subsymbolic" is conventional in the history of AI, but seems difficult to formalize rigorously. Logic-based AI systems are typically considered "symbolic", yet.
- **Supercompilation**: A technique for program optimization, which globally rewrites a program into a usually very different looking program that does the same thing. A prototype supercompiler was applied to Combo programs with successful results.
- **Surface Realization**: The process of taking a collection of Atoms and transforming them into a series of words in a (usually natural) language. A stage in the overall process of language generation.
- **Symbol Grounding**: The mapping of a symbolic term into perceptual or motoric entities that help define the meaning of the symbolic term. For instance, the concept "Cat" may be grounded by images of cats, experiences of interactions with cats, imaginations of being a cat, etc.
- **Symbolic**: Pertaining to the formation or manipulation of symbols, i.e. mental entities that are explicitly constructed to represent other entities. Often contrasted with subsymbolic.
- **Syntax-Semantics Correlation**: In the context of MOSES and program learning more broadly, this refers to the property via which distance in syntactic space (distance between the syntactic structure of programs, e.g. if they're represented as program trees) and semantic space (distance between the behaviors of programs, e.g. if they're represented as sets of input/output pairs) are reasonably well correlated. This can often happen among sets of programs that are not too widely dispersed in program space. The Reduct library is used to place Combo programs in Elegant Normal Form, which increases the level of syntax-semantics corellation between them. The programs in a single MOSES deme are often closely enough clustered together that they have reasonably high syntax-semantics correlation.
- **System Activity Table**: An OpenCog component that records information regarding what a system did in the past.
- **Temporal Inference**: Reasoning that heavily involves Atoms representing temporal information, e.g. information about the duration of events, or their temporal

relationship (before, after, during, beginning, ending). As implemented in Cog-Prime, makes use of an uncertain version of Allen Interval Algebra.

- **Truth Value**: A package of information associated with an Atom, indicating its degree of truth. SimpleTruthValue and IndefiniteTruthValue are two common, particular kinds. Multiple truth values associated with the same Atom from different perspectives may be grouped into CompositeTruthValue objects.
- **Universal Intelligence**: A technical term introduced by Shane Legg and Marcus Hutter, describing (roughly speaking) the average capability of a system to carry out computable goals in computable environments, where goal/environment pairs are weighted via the length of the shortest program for computing them.
- **Urge**: In OpenPsi, an Urge develops when a Demand deviates from its target range.
- **Very Long Term Importance (VLTI)**: A bit associated with Atoms, which determines whether, when an Atom is forgotten (removed from RAM), it is saved to disk (frozen) or simply deleted.
- **Virtual AGI Preschool**: A virtual world intended for AGI teaching/training/learning, bearing broad resemblance to the preschool environments used for young humans.
- **Virtual Embodiment**: Using an AGI to control an agent living in a virtual world or game world, typically (but not necessarily) a 3D world with broad similarity to the everyday human world.
- **Webmind AI Engine**: A predecessor to the Novamente Cognition Engine and OpenCog, developed 1997–2001—with many similar concepts (and also some different ones) but quite different algorithms and software architecture.

References

[Abb08] D. Abbott, J. Gea-Banacloche, P.C.W. Davies, S. Hameroff, A. Zeilinger, J. Eisert, H.M. Wiseman, S.M.Bezrukov, H. Frauenfelder, Plenary debate: quantum effects in biology trivial or not?. Fluctuation Noise Lett. **8**(1), C5–C26 (2008)

[AABL02] N. Alvarado, S.S. Adams, S. Burbeck, C. Latta, Beyond the turing test: performance metrics for evaluating a computer simulation of the human mind, in *International Conference on Development and Learning*, 2002

[AL03] J.R. Anderson, C. Lebiere. The newell test for a theory of cognition. Behav. Brain Sci. **26**, 587–637 (2003)

[AL09] Itamar Arel, Scott Livingston, Beyond the turing test. IEEE Comput. **42**(3), 90–91 (2009)

[AM01] J.S. Albus, A.M. Meystel, *Engineering of Mind: An Introduction to the Science of Intelligent Systems* (Wiley, New York, 2001)

[Ami89] D.J. Amit, *Modeling Brain Function—The World of Attractor Neural Networks* (Cambridge University Press, New York, 1989)

[ARC09] I. Arel, D. Rose, R. Coop, Destin: A scalable deep learning architecture with application to high-dimensional robust pattern recognition, in *Proceedings of AAAI Workshop on Biologically Inspired Cognitive Architectures*, 2009

[ARK09a] I. Arel, D. Rose, T. Karnowski, A deep learning architecture comprising homogeneous cortical circuits for scalable spatiotemporal pattern inference, in *NIPS 2009 Workshop on Deep Learning for Speech Recognition and Related Applications*, 2009

[Ark09b] R. Arkin, *Governing Lethal Behavior in Autonomous Robots* (Chapman and Hall, Boca Raton, 2009)

[Arl75]   P.K. Arlin, Cognitive development in adulthood: a fifth stage? Dev. Psychol. **11**, 602–606 (1975)

[Arm04] J.A. Armour. Cardiac neuronal hierarchy in health and disease. Am. J. Physiol. Regul. Integr. Comp. Physiol. **287**, R262–R271 (2004)

[Baa97] B. Baars, *In the Theater of Consciousness: The Workspace of the Mind* (Oxford University Press, Oxford, 1997)

[Bac09] J. Bach, *Principles of Synthetic Intelligence* (Oxford University Press, Oxford, 2009)

[Bar02] A.-L. Barabasi, *Linked: The New Science of Networks* (Perseus, Cambridge, 2002)

[Bat79] G. Bateson, *Mind and Nature: A Necessary Unity* (Ballantine, New York, 1979)

[BC94]   S. Baron-Cohen, *Mindblindness: An Essay on Autism and Theory of Mind* (MIT Press, Boston, 1994)

[BDL93] L. Barrett, R. Dunbar, J. Lycett, *Human Evolutionary Psycholog* (Princeton University Press, Princeton, 1993)

[BDS03] S. Ben-David, R. Schuller, Exploiting task relatedness for learning multiple tasks, in *Proceedings of the 16th Annual Conference on Learning Theory*, 2003

[BF71]   J.D. Bransford, J. Franks, The abstraction of linguistic ideas. Cogn. Psychol. **2**, 331–350 (1971)

[BF09]   B. Baars, S. Franklin, Consciousness is computational: the lida model of global workspace theory. Int. J. Mach. Conscious. **1**, 155–176 (2009)

[bGBK02] A. Goertzel, K. Ben, A. Klimov, Supercompiling java programs, 2002

[BH05]   S. Bader, P. Hitzler, in *Dimensions of Neural-Symbolic Integration—A Structured Survey*, eds. by S. Artemov, H. Barringer, A.S. d'Avila Garcez, L.C. Lamb, J. Woods. We Will Show Them: Essays in Honour of Dov Gabbay, vol. 1 (College Publications, London, 2005) pp. 167–194

[Bi01]   M.-m. Bi, G.-q. Poo, Synaptic modifications by correlated activity: Hebb's postulate revisited. Ann. Rev. Neurosci. **24**, 139–166 (2001)

[Bic88] M. Bickhard, Piaget on variation and selection models: Structuralism, logical necessity, and interactivism. Hum. Dev. **31**, 274–312 (1988)

[Bil05] Philip Bille, A survey on tree edit distance and related problems. Theoret. Comput. Sci. **337**, 2005 (2005)

[BO09]   A. Baranes, P.-Y. Oudeyer, R-iac: Robust intrinsically motivated active learning, in *Proceedings of the IEEE International Conference on Learning and Development, Shanghai, China, 33*, 2009

[Bod03] M. Boden *The Creative Mind*, (Routledge, 2003)

[Bol98] B. Bollobas, *Modern Graph Theory* (Springer, New York, 1998)

[Bos02] N. Bostrom, Existential risks. J. Evol. Technol. **9**, 1–30 (2002)

[Bos03] N. Bostrom, Ethical Issues in Advanced Artificial Intelligence, ed. by I. Smit, in *Cognitive, Emotive and Ethical Aspects of Decision Making in Humans and in Artificial Intelligence*. vol.2, (2003), pp. 12–17

[Bri03] S. Bringsjord, M. Zenzen, *Superminds: People Harness Hypercomputation, and More* (Kluwer, 2003)

[Bro84] J. Broughton, *Not Beyond Formal Operations, but Beyond Piaget*, ed. by M. Commons, F. Richards, C. Armon. Beyond Formal Operations: Late Adolescent and Adult Cognitive Development (Praeger, New York, 1984), pp. 395–411

[BS04]   B. Bakker, J. Schmidhuber, Hierarchical reinforcement learning based on subgoal discovery and subpolicy specialization, in *Proceedings of the 8-th Conference on Intelligent Autonomous Systems*, 2004

[Buc03] M. Buchanan, *Small World: Uncovering Nature's Hidden Networks* (Phoenix, London, 2003)

[Bur62] C. MacFarlane Burnet, *The Integrity of the Body* (Harvard University Press, Cambridge, 1962)

[BW88]   R.W. Byrne, A. Whiten, *Machiavellian Intelligence* (Clarendon Press, Oxford, 1988)

[BZGS06] B. Bakker, V. Zhumatiy, G. Gruener, J. Schmidhuber, Quasi-online reinforcement learning for robots, in *Proceedings of the International Conference on Robotics and Automation*, 2006

[Cal96]   W. Calvin, *The Cerebral Code* (MIT Press, Cambridge, 1996)

[Car85]   S. Carey, *Conceptual Change in Childhood* (MIT Press, Cambridge, 1985)

[Car97]   R. Caruana, Multitask learning. Mach. Learn. **28**, 41–75 (1997)

[Cas85]   R. Case, *Intellectual Development: Birth to Adulthood* (Academic Press, Orlando, 1985)

[Cas04]   N.L. Cassimatis, Grammatical processing using the mechanisms of physical inferences, in *Proceedings of the Twentieth-Sixth Annual Conference of the Cognitive Science, Society*, 2004

[Cas07]   N. Cassimatis, Adaptive algorithmic hybrids for human-level, artificial intelligence, 2007

[CB00]   W.H. Calvin, D. Bickerton, *Lingua ex Machina* (MIT Press, London, 2000)

[CB06]   R. Conolly, J. Blancato, Computational modeling of the liver. NCCT BOSC Review, 2006. http://www.epa.gov/ncct/bosc_review/2006/files/07_Conolly_Liver_Model.pdf

[CM07]   J.-Q. Chen, G. McNamee, *What is Waldorf Education?* Bridging, Assessment for Teaching and Learning in Early Childhood Classrooms, 2007

[CP05]   M.L. Commons, A. Pekker, Hierarchical complexity: A formal theory. http://www.dareassociation.org/Papers/Hierarchical%20Complexity%20-%20A%20Formal%20Theory%20(Commons%20&%20Pekker).pdf, (2005)

[CRK82]   M. Commons, F. Richards, D. Kuhn, Systematic and metasystematic reasoning: a case for a level of reasoning beyond Piaget's formal operations. Child Dev. **53**, 1058–1069 (1982)

[CS90]   A.G. Cairns-Smith. *Seven Clues to the Origin of Life: A Scientific Detective Story* (Cambridge University Press, Cambridge, 1990)

[Cse06]   P. Csermely, *Weak Links: Stabilizers of Complex Systems from Proteins to Social Networks* (Springer, Berlin, 2006)

[CSG07]   S. Chakraborty, A. Sandberg, S.A. Greenfield. Differential dynamics of transient neuronal assemblies in visual compared to auditory cortex. Exp. Brain Res. 1432–1106 (2007)

[CTS+98]   M. Commons, E.J. Trudeau, S.A. Stein, F.A. Richards, S.R. Krause, Hierarchical complexity of tasks shows the existence of developmental stages. Dev. Rev. **18**(18), 237–278 (1998)

[Dam00]   A. Damasio, *The Feeling of What Happens* (Harvest Books, Orlando, 2000)

[Dav84]   D. Davidson, *Inquiries into Truth and Interpretation* (Oxford University Press, Oxford, 1984)

[DC02]   P.D. Roberts, C.C. Bell, Spike-timing dependent synaptic plasticity in biological systems. Biol. Cybern. **87**, 392–403 (2002)

[Den87]   D. Dennett, *The Intentional Stance* (MIT Press, Cambridge, 1987)

[Den91]   D. Dennett, *Consciousness Explained* (Back Bay, Boston, 1991)

[DG05]   H. De Garis, *The Artilect War* (ETC, Palm Springs, 2005)

[DOP08]   W. Duch, R. Oentaryo, M. Pasquier, Cognitive architectures: Where do we go from here?, in *Proceedings of the Second Conference on AGI*, 2008

[Dör02]   D. Dörner, *Die Mechanik des Seelenwagens. Eine neuronale Theorie der Handlungsregulation.* Verlag Hans Huber, 2002

[EBJ+97]   J. Elman, E. Bates, M. Johnson, A. Karmiloff-Smith, D. Parisi, K. Plunkett, *Rethinking Innateness: A Connectionist Perspective on Development* (MIT Press, Cambridge, 1997)

[Ede93]   G. Edelman, Neural darwinism: selection and reentrant signaling in higher brain function. Neuron **10**, 15–125 (1993)

[Elm91]   J. Elman, Distributed representations, simple recurrent networks, and grammatical structure. Mach. Learn. **7**, 195–226 (1991)

[EMC12]   Effective-Mind-Control.com. Cellular memory in organ transplants. *Effective Mind Control*, 2012. http://www.effective-mind-control.com/cellular-memory-in-organ-transplants.html, Accessed 1 Feb 2012

[ES00]   G. Engelbretsen, F. Sommers, *An invitation to formal reasoning. The Logic of Terms* (Ashgate, Aldershot, 2000)

[FB08]   S. Franklin, B. Baars, Possible neural correlates of cognitive processes and modules from the lida model of cognition. *Cognitive Computing Research Group, University of Memphis*, 2008, http://ccrg.cs.memphis.edu/tutorial/correlates.html

[FC86]   R. Fung, C. Chong, *Metaprobability and Dempster-Shafer in Evidential Reasoning*, ed. by L. Kanal, J. Lemmer. Uncertainty in, Artificial Intelligence (North-Holland, Amsterdam, 1986), pp. 295–302

[Fis80]   K. Fischer, A theory of cognitive development: control and construction of hierarchies of skills. Psychol. Rev. **87**, 477–531 (1980)

[Fis01]   J.M. Fish, *Race and Intelligence: Separating Science From Myth* (Routledge, London, 2001)

[Fod94]   J. Fodor, *The Elm and the Expert* (Bradford Books, Cambridge, 1994)

[FP86]   D. Farmer, A. Perelson, The immune system, adaptation and machine learning. Physica D **2**, 187-204 (1986)

[Fra06]   S. Franklin, The lida architecture: Adding new modes of learning to an intelligent, autonomous, software agent, in *International Conference on Integrated Design and Process Technology*, 2006

[Fre90]   R. French, Subcognition and the limits of the turing test. Mind **99**, 53–65 (1990)

[Fre95]   W. Freeman, *Societies of Brains* (Erlbaum, Hillsdale, 1995)

[FT02]   G. Fauconnier, M. Turner, *The Way We Think: Conceptual Blending and the Mind's Hidden Complexities* (Basic Books, New York, 2002)

[Gar99]   H. Gardner, *Intelligence Reframed: Multiple Intelligences for the 21st century* (Basic Books, New York, 1999)

[Gau11]   E.M. Gauger, E. Rieper, J.J.L. Morton, S.C. Benjamin, V. Vedral, Sustained quantum coherence and entanglement in the avian compass. Phys. Rev. Lett. **106**(4), (2011)

[GdG08]   B. Goertzel, H. de Garis, Xia-man: An extensible, integrative architecture for intelligent humanoid, robotics. pp. 86–90, 2008

[GE86]   R. Gelman, E. Meck, S. Merkin, Young children's numerical competence. Cogn. Dev. **1**(1–29), 1986 (1986)

[GEA08]   B. Goertzel, C. Pennachin et al., An integrative methodology for teaching embodied non-linguistic agents, applied to virtual animals in second life, in *Proceedings of the First Conferennce on AGI*, IOS Press, 2008

[Ger99]   M. Gershon, *The Second Brain* (Harper, New York, 1999)

[GGC+11]   B. Goertzel, N. Geisweiller, L. Coelho, P. Janicic, C. Pennachin, *Real World Reasoning* (Atlantis, Paris, 2011)

[GGK02]   T. Gilovich, D. Griffin, D. Kahneman, *Heuristics and Biases: The Psychology of Intuitive Judgment* (Cambridge University Press, Cambridge, 2002)

[Gib77]   J.J. Gibson, *The Theory of Affordances*, ed. by R. Shaw, J. Bransford. Perceiving, Acting and Knowing (Erlbaum, Hillsdale, 1977)

[Gib78]   J. Gibbs, Kohlberg's moral stage theory: a Piagetian revision. Hum. Dev. **22**, 89–112 (1978)

[Gib79]   J.J. Gibson, *The Ecological Approach to Visual Perception* (Houghton Mifflin, Boston, 1979)

[GIGH08]   B. Goertzel, M. Ikle, I. Goertzel, A. Heljakka, *Probabilistic Logic Networks* (Springer, Heidelberg, 2008)

[Gil82]   Carol Gilligan, *In a Different Voice* (Harvard University Press, Cambridge, 1982)

[GMIH08]   B. Goertzel, I. Goertzel, M. Iklé, A. Heljakka, *Probabilistic Logic Networks* (Springer, Hidelberg, 2008)

[Goe93]   B. Goertzel, *The Evolving Mind* (Plenum, New York, 1993)

[Goe94]   B. Goertzel, *Chaotic Logic* (Plenum, New York, 1994)

[Goe97]   B. Goertzel, *From Complexity to Creativity* (Plenum Press, New York, 1997)

[Goe01]   B. Goertzel, *Creating Internet Intelligence* (Plenum Press, New York, 2001)

[Goe06a]   B. Goertzel, *The Hidden Pattern* (Brown Walker, Florida, 2006a)

[Goe06b] B. Goertzel, *The Hidden Pattern* (Brown Walker, Florida, 2006b)

[Goe08] B. Goertzel, A pragmatic path toward endowing virtually-embodied ais with human-level linguistic capability, in *IEEE World Congress on Computational Intelligence (WCCI)*, 2008

[Goe09a] B. Goertzel, Cognitive synergy: A universal principle of feasible general intelligence?, in *Proceedings of ICCI 2009, Hong Kong*, 2009

[Goe09b] B. Goertzel, The embodied communication prior, in *Proceedings of ICCI-09, Hong Kong*, 2009

[Goe09c] B. Goertzel, Opencog prime: A cognitive synergy based architecture for embodied artificial general intelligence, in *Proceedings of ICCI 2009, Hong Kong*, 2009

[Goewiki] B. Goertzel, *OpenCogPrime WikiBook*, 2010. http://wiki.opencog.org/w/OpenCogPrime:WikiBook

[Goe10a] B. Goertzel, Coherent aggregated volition. Multiverse According to Ben, 2010. http://multiverseaccordingtoben.blogspot.com/2010/03/coherent-aggregated-volition-toward.htm

[Goe10b] B. Goertzel, Toward a formal definition of real-world general, intelligence, 2010

[Goe10c] B. Goertzel et al., A general intelligence oriented architecture for embodied natural language processing, in *Proceedings of the Third Conference on Artificial General Intelligence (AGI-10)*, Atlantis Press, 2010

[Goe11x] B. Goertzel, J. Pitt, Z. Cai, J. Wigmore, D. Huang, N. Geisweiller, R. Lian, G. Yu, Integrative General Intelligence for Controlling Game AI in a Minecraft-Like Environment. Proc. BICA (2011)

[Goo86] I. Good, *The Estimation of Probabilities* (MIT Press, Cambridge, 1986)

[Gor86] R. Gordon, Folk psychology as simulation. Mind Lang. **1**(1), 158–171 (1986)

[GPI+10] B. Goertzel, J. Pitt, M. Ikle, C. Pennachin, R. Liu, Glocal memory: a design principle for artificial brains and minds. Neurocomputing **74**, 84–94 (2010)

[GPPG06] B. Goertzel, H. Pinto, C. Pennachin, I. F. Goertzel, Using dependency parsing and probabilistic inference to extract relationships between genes, proteins and malignancies implicit among multiple biomedical research abstracts, in *Proceedings of Bio-NLP 2006*, 2006

[GPSL03] B. Goertzel, C. Pennachin, A. Senna, M. Looks. An integrative architecture for artificial general intelligence, in *Proceedings of IJCAI 2003, Acapulco*, 2003

[Gre01] S. Greenfield, *The Private Life of the Brain* (Wiley, New York, 2001)

[HAG07] H. Markert, A. Knoblauch, G. Palm, Modelling of syntactical processing in the cortex. Biosystems **89**(1–3), 300–315 (2007)

[Ham87] S. Hameroff *Ultimate Computing*, (North Holland, 1987)

[Ham10] S. Hameroff, The Òconscious pilotÓNdendritic synchronymoves through the brain to mediate consciousness. J. Biol. Phys. **36**, 71–93 (2010)

[Hay85] P. Hayes, The second naive physics manifesto. ed. by R. Shaw, J. Bransford, in *Formal Theories of the Commonsense World*, 1985

[HB06] J. Hawkins, S. Blakeslee, *On Intelligence* (Brown Walker, Boca Raton, 2006)

[Heb49] D. Hebb, *The Organization of Behavior* (Wiley, New York, 1949)

[Hey07] F. Heylighen, The Global Superorganism: an evolutionary-cybernetic model of the emerging network society. Soc. Evol. Hist. **6**(1), 58–119 (2007)

[HF95] P. Hayes, K. Ford, Turing test considered harmful, in *IJCAI-14*, 1995

[HG08] D. Hart, B. Goertzel, Opencog: A software framework for integrative artificial general intelligence, in *AGI, Frontiers in Artificial Intelligence and Applications*, IOS Press, 2008, vol. 171, pp. 468–472

[HHPO12] A. Hampshire, R. Highfield, B. Parkin, A. Owen, Fractionating human intelligence. Neuron **76**(6), 1225–1237 (2012)

[Hib02] B. Hibbard, *Superintelligent Machines* (Springer, Berlin, 2002)

[Hof79] D. Hofstadter, *Godel, Escher, Bach: An Eternal Golden Braid* (Basic Books, New York, 1979)

[Hof95] D. Hofstadter, *Fluid Concepts and Creative Analogies* (Basic Books, New York, 1995)

[Hof96] D. Hofstadter, *Metamagical Themas* (Basic Books, New York, 1996)

[Hop82] J.J. Hopfield, Neural networks and physical systems with emergent collective computational abilities. Proc. Natl. Acad. Sci. **79**, 2554–2558 (1982)

[HOT06] G.E. Hinton, S. Osindero, Y. Teh, A fast learning algorithm for deep belief nets. Neural Comput. **18**, 1527–1554 (2006)

[Hut95] E. Hutchins, *Cognition in the Wild* (MIT Press, Cambridge, 1995)

[Hut96] E. Hutchins, *Cognition in the Wild* (MIT Press, Cambridge, 1996)

[Hut05] M. Hutter, *Universal Artificial Intelligence: Sequential Decisions based on Algorithmic Probability* (Springer, Heidelberg, 2005)

[HZT+02] J. Han, S. Zeng, K. Tham, M. Badgero, J. Weng. Dav: A humanoid robot platform for autonomous mental development, in *Proceedings of 2nd International Conference on Development and, Learning*, 2002

[IP58] B. Inhelder, J. Piaget, *The Growth of Logical Thinking from Childhood to Adolescence* (Basic Books, New York, 1958)

[JL08] D.J. Jilk, C. Lebiere, R.C. O'reilly, J.R. Anderson, SAL: an explicitly pluralistic cognitive architecture. J. Exp. Theor. Artif. Intell. **20**, 197–218 (2008)

[JM09] D. Jurafsky, J. Martin, *Speech and Language Processing* (Pearson Prentice Hall, Upper Saddle River, 2009)

[Joy00] Bill Joy, *Why the future doesn't need us, Wired*. Apr 2000

[Kam91] G. Kampis, *Self-Modifying Systems in Biology and Cognitive Science* (Plenum Press, New York, 1991)

[Kan64] I. Kant, *Groundwork of the Metaphysic of Morals* (Harper and Row, New York, 1964)

[Kap08] F. Kaplan. Neurorobotics: an experimental science of embodiment. Frontiers Neurosci. **2**, 22–23 (2008)

[KE06] J.L. Krichmar, G.M. Edelman. Principles underlying the construction of brain-based devices, ed. by T. Kovacs, J.A.R. Marshall, in *Adaptation in Artificial and Biological Systems*, 2006, pp. 37–42

[KK90] K. Kitchener, P. King, Reflective judgement: ten years of research, ed. by M. Commons. Beyond Formal Operations: Models and Methods in the Study of Adolescent and Adult Thought, vol. 2, (Praeger, New York, 1990), pp. 63–78

[KLH83] Lawrence Kohlberg, Charles Levine, Alexandra Hewer, *Moral stages : A Current Formulation and a Response to Critics* (Karger, Basel, 1983)

[Koh38] Wolfgang Kohler, *The Place of Value in a World of Facts* (Liveright Press, New York, 1938)

[Koh81] L. Kohlberg, *Essays on Moral Development. The Philosophy of Moral, Development*, vol. I (Harper and Row, Hardcover, 1981)

[KS04] A. Kahane, P. Senge, *Solving Tough Problems: An Open Way of Talking, Listening, and Creating New Realities* (Berrett-Koehler, San Francisco, 2004)

[Kur06] R. Kurzweil, *The Singularity is Near* (Penguin Books, New York, 2006)

[Kur12] R. Kurzweil, *How to Create a Mind* (Viking, New York, 2012)

[Kyb97] H. Kyburg, Bayesian and non-bayesian evidential updating. Artif. Intell. **31**, 271–293 (1997)

[Lan05] P. Langley, An adaptive architecture for physical agents, in *Proceedings of the 2005 IEEE/WIC/ACM International Conference on Intelligent Agent Technology*, 2005

[LAon] C. Lebiere, J.R. Anderson. The case for a hybrid architecture of cognition (in preparation)

[LBDE90] Y. LeCun, B. Boser, J.S. Denker, Al. Et. Handwritten digit recognition with a backpropagation network, in *Advances in Neural Information Processing Systems, 2*, 1990

[LD03] A. Laud, G. Dejong, The influence of reward on the speed of reinforcement learning, in *Proceedings of the 20th International Conference on Machine Learning*, 2003

[Leg06a] A. Laud, G. Dejong, The influence of reward on the speed of reinforcement learning, in *Proceedings of the 20th International Conference on Machine Learning*, 2003

[Leg06b] S. Legg, Unprovability of friendly ai. Vetta Project, 2006. http://www.vetta.org/2006/09/unprovability-of-friendly-ai/

[LG90] D. Lenat, R.V. Guha, *Building Large Knowledge-Based Systems: Representation and Inference in the Cyc Project* (Addison-Wesley, Menlo Park, 1990)

[LH07a] S. Legg, M. Hutter, in *A collection of definitions of intelligence*, IOS Press, 2007

[LH07b] S. Legg, M. Hutter, A definition of machine intelligence. Minds Mach. **17**,391–444 (2007)

[LLW+05] G. Li, Z. Lou, L. Wang, X. Li, W.J. Freeman. Application of chaotic neural model based on olfactory system on pattern recognition, in *ICNC*, 2005, vol. 1, pp. 378–381

[LMC07a] M.H. Lee, Q. Meng, F. Chao. Developmental learning for autonomous robots. Robot. Auton. Syst. **55**, 750–759 (2007)

[LMC07b] M.H. Lee, Q. Meng, F. Chao. Staged competence learning in developmental robotics. Adap. Behav. **15**, 241–255 (2007)

[LN00] G. Lakoff, R. Nunez, *Where Mathematics Comes From* (Basics Books, New York, 2000)

[Log07] R.M. Logan, *The Extended Mind* (University of Toronto Press, Toronto, 2007)

[Loo06] M. Looks, Competent Program Evolution. PhD Thesis, Computer Science Department, Washington University, 2006

[LRN87] J. Laird, P. Rosenbloom, A. Newell. Soar: An architecture for general intelligence. Artif. Intell. **33**, 1–64 (1987)

[LS05] J. Lisman, N. Spruston, Postsynaptic depolarization requirements for ltp and ltd: a critique of spike timing-dependent plasticity. Nature Neurosci. **8**, 839–841 (2005)

[LWML09] J. Laird, R. Wray, R. Marinier, P. Langley, Claims and challenges in evaluating human-level intelligent systems, in *Proceedings of AGI-09*, 2009

[Mac95] D. MacKenzie, The automation of proof: a historical and sociological exploration. IEEE Ann. Hist. Comput. **17**(3), 7–29 (1995)

[Mar01] H. Marchand. Reflections on PostFormal Thought. The Genetic Epistemologist **29**, 2–9 (2001)

[McK03] B. McKibben, *Enough: Staying Human in an Engineered Age* (Saint Martins Griffin, New York, 2003)

[Met04] T. Metzinger, *Being No One* (Bradford, Cambridge, 2004)

[Min88] M. Minsky, *The Society of Mind* (MIT Press, Cambridge, 1988)

[Min07] M. Minsky, *The Emotion Machine* (Basic Books, New York, 2007)

[MK07] J. Modayil, B. Kuipers, Autonomous development of a grounded object ontology by a learning robot, in *AAAI-07*, 2007

[MK08] J. Mugan, B. Kuipers, Towards the application of reinforcement learning to undirected developmental learning, in *International Conference on Epigenetic, Robotics*, 2008

[MK09] J. Mugan, B. Kuipers, Autonomously learning an action hierarchy using a learned qualitative state, representation, in *IJCAI-09*, 2009

[Mon12] Maria Montessori, *The Montessori Method* (Frederick A. Stokes, New York, 1912)

[MSV+08] G. Metta, G. Sandini, D. Vernon, L. Natale, F. Nori, The icub humanoid robot: an open platform for research in embodied cognition, in *Performance Metrics for Intelligent Systems Workshop (PerMIS 2008)*, 2008

[MW07] S. Morgan, C. Winship, *Counterfactuals and Causal Inference* (Cambridge University Press, Cambridge, 2007)

[Nan08] Nanowerk, Carbon nanotube rubber could provide e-skin for robots. http://www.nanowerk.com/news/newsid=6717.php, 2008

[Nei98] D.M. Neilsen, *Teaching Young Children, Preschool-K: A Guide to Planning Your Curriculum, Teaching Through Learning Centers, and Just About Everything Else* (Corwin Press, Torrance, 1998)

[New90] A. Newell, *Unified Theories of Cognition* (Harvard University press, Harvard, 1990)

[Nie98] D.M. Nielsen, *Teaching Young Children, Preschool-K: A Guide to Planning Your Curriculum, Teaching Through Learning Centers, and Just About Everything Else* (Corwin Press, housand Oaks, 1998)

[Nil09]   N. Nilsson, The physical symbol system hypothesis: Status and prospects, in *50 Years of AI, Festschrift, LNAI 4850, 33*, 2009

[NK04]   A. Nestor, B. Kokinov, Towards active vision in the dual cognitive architecture. Int. J. Inf. Theor. Appl. **11**, 9–15 (2004)

[OK06]   P. Oudeyer, F. Kaplan. Discovering communication. Connection Sci. **18**, 189–206 (2006)

[Omo08]   S. Omohundro, The basic ai drives, in *Proceedings of the First AGI Conference*, IOS Press, 2008

[Omo09]   S. Omohundro, Creating a cooperative future. 2009. http://selfawaresystems.com/2009/02/23/talk-on-creating-a-cooperative-future/

[Opa52]   A.I. Oparin, *The Origin of Life* (Dover, New York, 1952)

[Pal82]   G. Palm, *Neural Assemblies. An Alternative Approach to Artificial Intelligence* (Springer, Berlin, 1982)

[Pei34]   C. Peirce, *Collected papers: Pragmatism and pragmaticism*, vol. V (Harvard University Press, Cambridge, 1934)

[Pel05]   M. Pelikan, *Hierarchical Bayesian Optimization Algorithm: Toward a New Generation of Evolutionary Algorithms* (Springer, Heidelberg, 2005)

[Pen96]   R. Penrose, *Shadows of the Mind*, (Oxford University Press, 1996)

[Per70]   G. William, *Perry, Forms of Intellectual and Ethical Development in the College Years: A Scheme* (Rinehart and Winston, Holt, 1970)

[Per81]   W.G. Perry, *Cognitive and Ethical Growth: The Making of Meaning*, ed. by A.W. Chickering (The Modern American College, Jossey-Bass, San Francisco, 1981), pp. 76–116

[PH12]   Z. Pang, W. Han, Regulation of synaptic functions in central nervous system by endocrine hormones and the maintenance of energy homeostasis, Bioscience Reports, 2012

[Pia53]   J. Piaget, *The Origins of Intelligence in Children* (Routledge and Kegan Paul, London, 1953)

[Pia55]   J. Piaget, *The Construction of Reality in the Child* (Routledge and Kegan Paul, London, 1955)

[Pir84]   R. Pirsig, *Zen and the Art of Motorcycle Maintenance* (Bantam, New York, 1984)

[PNR07]   Karalny Patterson, Peter J. Nestor, Timothy T. Rogers, Where do you know what you know? the representation of semantic knowledge in the human brain. Nature Rev. Neurosci. **8**, 976–987 (2007)

[PSF09]   Richard Dum Peter Strick, Julie Fiez, Cerebellum and nonmotor function. Ann. Rev. Neurosci. **32**, 413–434 (2009)

[PW78]   D. Premack and G. Woodruff. Does the chimpanzee have a theory of mind? Behav. Brain Sci. **1**, 515–526 (1978)

[QaGKKF05] R.Q. Quiroga, L. Reddy amd, G. Kreiman, C. Koch, I. Fried, Invariant visual representation by single-neurons in the human brain. Nature **435**, 1102–1107 (2005)

[QKKF08]   R.Q. Quiroga, G Kreiman, C, Koch, I. Fried, Sparse but not "grandmother-cell" coding in the medial temporal lobe. Trends Cogn. Sci. **12**, 87–91 (2008)

[Rav04]   I. Ravenscroft, Folk psychology as a theory, stanford encyclopedia of philosophy. http://plato.stanford.edu/entries/folkpsych-theory/, 2004

[RBW92]   R. Gagne, L. Briggs, W. Walter, *Principles of Instructional Design* (Harcourt Brace Jovanovich, New York, 1992)

[RCK01]   J. Rosbe, R.S. Chong, D.E. Kieras, *Modeling with Perceptual and Memory Constraints: An Epic-Soar Model of a Simplified Enroute Air Traffic Control Task* (SOAR Technology Inc, Report, 2001)

[RD06]   M. Richardson, P. Domingos, Markov logic networks. Mach. Learn. **62**, 107–236 (2006)

[Rie73]   K. Riegel, Dialectic operations: the final phase of cognitive development. Hum. Dev. **16**, 346–370 (1973)

[RM95] H.L. Roediger, K.B. McDermott, Creating false memories: remembering words not presented in lists. J. Exp. Psychol. Learn. Mem. Cogn. **21**, 803–814 (1995)

[Ros88] I. Rosenfield, *The Invention of Memory: A New View of the Brain* (Basic Books, New York, 1988)

[Row90] J. Rowan, *Subpersonalities: The People Inside Us* (Routledge Press, London, 1990)

[Row11] T. Rowe. Fossil evidence on origin of the mammalian brain. Science 20 **332**, 955–957 (2011)

[RV01] A. Robinson, A. Voronkov, *Handbook of Automated Reasoning* (MIT Press, Hardcover, 2001)

[RZDK05] M. Rosenstein, Z. Marx, T. Dietterich, L.P. Kaelbling, in *Transfer learning with an ensemble of background tasks, NIPS workshop on inductive transfer*, 2005

[SA93] L. Shastri, V. Ajjanagadde. From simple associations to systematic reasoning: A connectionist encoding of rules, variables, and dynamic bindings using temporal synchrony. Behav. Brain Sci. **16**(3), 417–494 (1993)

[Sal93] S. Salthe, *Development and Evolution* (MIT Press, Cambridge, 1993)

[Sam10] A.V. Samsonovich, Toward a unified catalog of implemented cognitive architectures, in *BICA*, 2010, pp. 195–244

[SB98] R. Sutton, A. Barto, *Reinforcement Learning* (MIT Press, Cambridge, 1998)

[SB06] J. Simsek, A. Barto, An intrinsic reward mechanism for efficient exploration, in *Proceedings of the Twenty-Third International Conference on Machine Learning*, 2006

[SBC05] S. Singh, A. Barto, N. Chentanez, Intrinsically motivated reinforcement learning, in *Proceedings of Neural Information Processing Systems 17*, 2005

[SC94] B. Smith, R. Casati, Naive physics: an essay in ontology. Philos. Psychol. **7**, 225–244 (1994)

[Sch91a] J. Schmidhuber, Curious model-building control systems, in *Proceedings of International Joint Conference on Neural Networks*, 1991

[Sch91b] J. Schmidhuber, A possibility for implementing curiosity and boredom in model-building neural controllers, in *Proceedings of the International Conference on Simulation of Adaptive Behavior: From Animals to Animats*, 1991

[Sch95] J. Schmidhuber, Reinforcement-driven information acquisition in non-deterministic environments, in *Proceedings of ICANN'95*, 1995

[Sch02] J. Schmidhuber, *Exploring the predictable* (Springer, Heidelberg, 2002)

[Sch06] J. Schmidhuber, Godel machines: Fully Self-referential Optimal Universal Self-improvers, ed. by B. Goertzel, C. Pennachin, in *Artificial General, Intelligence*, 2006, pp. 119–226

[Sch07] D. Schunk, *Theories of Learning: An Educational Perspective* (Prentice Hall, Upper Saddle River, 2007)

[SE07] S. Shapiro, et al., Metacognition in sneps. AI Mag. **28**, 17–31, (2007)

[SF05] S.A. Greenfield, T.F. Collins, A neuroscientific approach to consciousness. Prog. Brain. Res. **150**, 11–23 (2005)

[Sha76] G. Shafer, *A Mathematical Theory of Evidence* (Princeton University Press, Princeton, 1976)

[Shu03] T.R. Shultz, *Computational Developmental Psychology* (MIT Press, Cambridge, 2003)

[SKBB91] D. Shannahoff-Khalsa, M. Boyle, M. Buebel, The effects of unilateral forced nostril breathing on cognition. Int. J. Neurosci. **57**, 239–249 (1991)

[Slo01] A. Sloman, Varieties of affect and the cogaff architecture schema, in *Proceedings of the Symposium on Emotion, Cognition, and Affective, Computing, AISB-01*, 2001

[Slo08a] A. Sloman, A new approach to philosophy of mathematics: Design a young explorer able to discover 'toddler theorems', 2008

[Slo08b] A. Sloman, The well-designed young mathematician. Artif. Intell. **172**, 2015–2034 (2008)

[SM05]   P. Singh, M. Minsky, *An Architecture for Cognitive Diversity*, ed. by D. Davis. Visions of Mind (Idea Group Inc, London, 2005)

[Sot11]   K. Sotala, 14 objections against ai/friendly ai/the singularity answered. Xuenay.net, 2011. http://www.xuenay.net/objections.html, Accessed 20 Mar 2011

[SS74]   J. Sauvy, S. Suavy, *The Child's Discovery of Space: From Hopscotch to Mazes—An Introduction to Intuitive Topology* (Penguin, Harmondsworth, 1974)

[SS03a]   J.F. Santore, S.C. Shapiro, Crystal cassie: Use of a 3-d gaming environment for a cognitive agent, in *Papers of the IJCAI 2003 Workshop on Cognitive Modeling of Agents and Multi-Agent, Interactions*, 2003

[SS03b]   R. Steiner, S.K. Sagarin, *What is Waldorf Education?* (Steiner Books, New York, 2003)

[Stc00]   T. Stcherbatsky, *Buddhist Logic* (Motilal Banarsidass Publishers, New York, 2000)

[SV99]   A.J. Storkey, R. Valabregue, The basins of attraction of a new hopfield learning rule. Neural Netw. **12**, 869–876 (1999)

[SZ04]   R. Sun, X. Zhang. Top-down versus bottom-up learning in cognitive skill acquisition. Cogn. Syst. Res. **5**, 63–89 (2004)

[TC97]   M. Tomasello, J. Call, *Primate Cognition* (Oxford University Press, Oxford, 1997)

[TC05]   E. Tulving, R. Craik, *The Oxford Handbook of Memory* (Oxford University Press, Oxford, 2005)

[Tea06]   S. Thrun et al., The robot that won the darpa grand challenge. J. Rob. Syst. **23**(9), 661–692 (2006)

[TM95]   S. Thrun, T. Mitchell, Lifelong robot learning. Rob. Auton. Syst. **15**, 25–46 (1995)

[TS94]   E. Thelen, L. Smith, *A Dynamic Systems Approach to the Development of Cognition and Action* (MIT Press, Cambridge, 1994)

[TS07]   M. Taylor, P. Stone, Cross-domain transfer for reinforcement learning, in *Proceedings of the 24th International Conference on Machine Learning*, 2007

[Tur50]   A. Turing, Computing machinery and intelligence. Mind **59**, 433–460 (1950)

[Tur77]   V.F. Turchin, *The Phenomenon of Science* (Columbia University Press, New York, 1977)

[TV96]   V. Turchin, *Supercompilation: Techniques and Results* ed. by D. Bjorner, M. Broy, A.V. Zamulin. Perspectives of System Informatics (Springer, Heidelberg, 1996)

[Vin93]   V. Vinge, The coming technological singularity. VISION-21 Symposium, NASA and Ohio Aerospace Institute, 1993. http://www-rohan.sdsu.edu/faculty/vinge/misc/singularity.html

[Vyg86]   L. Vygotsky, *Thought and Language* (MIT Press, Cambridge, 1986)

[WA10]   W. Wallach, C. Atkins, *Moral Machines* (Oxford University Press, Oxford, 2010)

[Wan95]   P. Wang, Non-Axiomatic Reasoning System. PhD Thesis, Indiana University, Bloomington, 1995

[Wan06]   P. Wang, *Rigid Flexibility: The Logic of Intelligence* (Springer, Dordrecht, 2006)

[Was09]   M. Waser, Ethics for self-improving machines, in *AGI-09*, 2009. http://vimeo.com/3698890

[Wel90]   H. Wellman, *The Child's Theory of Mind* (MIT Press, Cambridge, 1990)

[WH06]   J. Weng, W.S. Hwangi, From neural networks to the brain: autonomous mental development. IEEE Comput. Intell. Mag. **1**, 15–31 (2006)

[Who64]   B.L. Whorf, *Language, Thought and Reality*, (Harvard University Press, Cambridge, 1964)

[WHZ+00] J. Weng, W.S. Hwang, Y. Zhang, C. Yang, R. Smith, Developmental humanoids: Humanoids that develop skills automatically, in *Proceedings of the first IEEE-RAS International Conference on Humanoid Robots*, 2000

[Wik11]   Wikipedia, Open source governance, 2011. http://en.wikipedia.org/wiki/Open_source_governance

[Win72]   T. Winograd, *Understanding Natural Language* (Edinburgh University Press, Edinburgh, 1972)

[Wit07] D.C. Witherington, The dynamic systems approach as metatheory for developmental psychology. Hum. Dev. **50**, 127–153 (2007)

[Wol02] S. Wolfram, *A New Kind of Science* (Wolfram Media, Champaign, 2002)

[WW06] M. Williams, J. Williamson, Combining argumentation and bayesian nets for breast cancer prognosis. J. Logic Lang. Inform. **15**, 155–178 (2006)

[Yud04] E. Yudkowsky, Coherent extrapolated volition. *Singularity Institute for AI*, 2004. http://singinst.org/upload/CEV.html

[Yud06] E. Yudkowsky, What is friendly ai? *Singularity Institute for AI*, 2006. http://singinst. org/ourresearch/publications/what-is-friendly-ai.html

[Zad78] L. Zadeh, Fuzzy sets as a basis for a theory of possibility. Fuzzy Sets Syst. **1**, 3–28 (1978)

[ZPK07] L.S. Zettlemoyer, H.M. Pasula, L.P. Kaelbling, Logical particle filtering, in *Proceedings of the Dagstuhl Seminar on Probabilistic, Logical, and Relational, Learning*, 2007

Index

B. Goertzel et al., *Engineering General Intelligence, Part 1*,
Atlantis Thinking Machines 5, DOI: 10.2991/978-94-6239-027-0,
© Atlantis Press and the authors 2014